ADVANCES IN

GEOPHYSICS

VOLUME 23

Contributors to This Volume

SUBIR K. BANERJEE
G. T. CSANADY
ROBERT T. LANGAN
RICHARD E. MORITZ
ROGER A. PIELKE
NORMAN H. SLEEP
ALFONSO SUTERA

Advances in
GEOPHYSICS

Edited by

BARRY SALTZMAN

Department of Geology and Geophysics
Yale University
New Haven, Connecticut

VOLUME 23

1981

Academic Press
A Subsidiary of Harcourt Brace Jovanovich, Publishers

New York   London   Toronto   Sydney   San Francisco

Copyright © 1981, by Academic Press, Inc.
ALL RIGHTS RESERVED.
NO PART OF THIS PUBLICATION MAY BE REPRODUCED OR
TRANSMITTED IN ANY FORM OR BY ANY MEANS, ELECTRONIC
OR MECHANICAL, INCLUDING PHOTOCOPY, RECORDING, OR ANY
INFORMATION STORAGE AND RETRIEVAL SYSTEM, WITHOUT
PERMISSION IN WRITING FROM THE PUBLISHER.

ACADEMIC PRESS, INC.
111 Fifth Avenue, New York, New York 10003

United Kingdom Edition published by
ACADEMIC PRESS, INC. (LONDON) LTD.
24/28 Oval Road, London NW1 7DX

Library of Congress Catalog Card Number: 52-12266

ISBN 0-12-018823-6

PRINTED IN THE UNITED STATES OF AMERICA

81 82 83 84   9 8 7 6 5 4 3 2 1

CONTENTS

LIST OF CONTRIBUTORS vii

Thermal Evolution of the Earth: Some Recent Developments
NORMAN H. SLEEP AND ROBERT T. LANGAN

1. Introduction	1
2. Thermal History Models	3
3. Core Formation	14
4. Conclusions and Future Work	19
References	20

Experimental Methods of Rock Magnetism and Paleomagnetism
SUBIR K. BANERJEE

1. Introduction	25
2. Natural Remanent Magnetization	29
3. Artificial or Laboratory-Imparted Remanent Magnetizations ...	56
4. Parameters Measured in the Presence of a Field	76
5. Future Trends	92
References	95

Circulation in the Coastal Ocean
G. T. CSANADY

1. Introduction	101
2. Wind-Driven Transient Currents	103
3. Upwelling, Downwelling, and Coastal Jets	119
4. Trapped Waves and Propagating Fronts	126
5. Steady Currents	139
6. Conclusion	176
References	177

Mesoscale Numerical Modeling
ROGER A. PIELKE

1. Introduction	186
2. Basic Set of Equations	187
3. Simplification of the Basic Equations	189
4. Averaging the Conservation Relations	198

5. Types of Models 204
6. Coordinate Representation 221
7. Planetary Boundary-Layer Parameterization 226
8. Radiation Parameterization 239
9. Moist Thermodynamics 248
10. Methods of Solution 253
11. Boundary and Initial Conditions 269
12. Model Evaluation 285
13. Examples of Mesoscale Models 290
14. Conclusions 317
References 321

The Predictability Problem: Effects of Stochastic Perturbations in Multiequilibrium Systems

RICHARD E. MORITZ AND ALFONSO SUTERA

1. Introduction 345
2. The Predictability Problem 347
3. Prototype Model and Mathematical Concepts 349
4. Computational Techniques 359
5. Numerical Results 360
6. Concluding Remarks 374
Appendix 376
References 381

INDEX 385

LIST OF CONTRIBUTORS

Numbers in parentheses indicate the pages on which the authors' contributions begin.

SUBIR K. BANERJEE, *Department of Geology and Geophysics, University of Minnesota, Minneapolis, Minnesota 55455* (25)

G. T. CSANADY, *Woods Hole Oceanographic Institution, Woods Hole, Massachusetts 02543* (101)

ROBERT T. LANGAN, *Department of Civil Engineering, Northwestern University, Evanston, Illinois 60201* (1)

RICHARD E. MORITZ,* *Department of Geology and Geophysics, Yale University, New Haven, Connecticut 06511* (345)

ROGER A. PIELKE, *Department of Environmental Sciences, University of Virginia, Charlottesville, Virginia 22903* (185)

NORMAN H. SLEEP, *Department of Geophysics, Stanford University, Stanford, California 94305* (1)

ALFONSO SUTERA, *The Center for the Environment and Man, Inc., Hartford, Connecticut 06120* (345)

---

* Present address: Polar Science Center, University of Washington, Seattle, Washington 98105.

THERMAL EVOLUTION OF THE EARTH: SOME RECENT DEVELOPMENTS

NORMAN H. SLEEP

Department of Geophysics
Stanford University
Stanford, California

AND

ROBERT T. LANGAN

Department of Civil Engineering
Northwestern University
Evanston, Illinois

1.	Introduction	1
2.	Thermal History Models	3
	2.1. Kinematic Convective Models	5
	2.2. Parameterized Convection	6
	2.3. Stratified Convection	11
	2.4. Discussion of Thermal Modeling	13
3.	Core Formation	14
	3.1. Timing of Core Formation	15
	3.2. Mechanism of Core Formation	15
	3.3. Ferric Iron and Water in Mantle	18
4.	Conclusions and Future Work	19
	References	20

1. INTRODUCTION

In this contribution we review recently developed methods for thermal modeling of the Earth and show that the Earth has been cooling over the last 3 B.Y. The earlier history is less clear. We examine the mechanism of core formation and find that total melting and eruption of mantle material may have cooled the Earth after core formation and that the temperature increased again during the Archean from radioactive heating. We then discuss the future of thermal modeling. The history of thermal modeling of the earth prior to 1970 is well known and not repeated here.

Thermal models constructed during the last three years have indicated that the earth is cooling and that radioactivity is not in balance with surface heat flow. The discovery of komatiitic lavas in the Archean has provided a direct, not yet completely understood, constraint on the rate the Earth has cooled. The ease with which global thermal history models can be computed along with better direct information on radioactive heating in the mantle and the current means of heat loss have led to vigorous investigation.

The heat loss from the Earth is dominated by the formation of new oceanic lithosphere at midoceanic ridges. The global rate of heat loss is thus controlled by the global rate of accretion of oceanic crust or the "global spreading rate." This rate, including crustal formation in intraarc basins, is about 3.5 km^2/yr (Sclater *et al.*, 1980). Estimates of the global rate of heat loss are given in Table I.

Simplifications in thermal modeling result because a single upper mantle temperature can be used to represent the material emplaced at ridges and the temperature below the lithosphere. If the entire mantle is convecting, this upper mantle temperature can be used to represent the heat content in the Earth. This representation is best if the whole mantle is well stirred by convection. It may be adequate, however, if upper and lower mantle convections are separated by a thin boundary layer (see Jeanloz and Richter, 1979).

The controversial assumption of mantle-wide convection is made so that upper-mantle-derived compositions as well as temperatures can be

TABLE I. GLOBAL HEAT FLOW ESTIMATES

Author	Global spreading rate (km^2/yr)	Global average heat flow from mantle[a]	Heat flow of 1 M.Y. crust[b]	Comment
Sleep (1979)	3.0	1.42	10.4	Square root of time model for whole Earth
	3.7	1.57	10.4	Subduction probability is independent of crustal age
Davies (1980b)	2.94	1.68	12.2	Square root of time for oceans basal heat flow in continents. Subduction occurs at same age. This increases heat flow by a factor of 1.13 from Sleep's formulation
Sclater *et al.* (1980)	3.55	1.67	11.3	Actual distribution of seafloor and continents is used
This article	3.55	1.80	12.2	Sleep's method, most reliable values from Davies and Sclater

[a] HFU. Includes basal heat flow in continents, and all heat lost in ocean basins. Heat generated by continental radioactivity is excluded.

[b] HFU. The heat flow depends on the square root of age. Differences for old oceans and continents do not affect global averages significantly.

extrapolated to the rest of the Earth. Note that this restriction is unnecessary if "hot spot" lavas actually sample the lower mantle.

The geophysical arguments on whether convection is mantle-wide are reviewed by Davies (1977) and Peltier (1972, 1980, 1981) and are not repeated here. Chemical arguments for stratification have been based on models for the accretion, core formation, and stability of mineral phases in the Earth (Smith, 1981), models for the postsubduction distribution of oceanic lithosphere (Anderson, 1979), and models of radiogenic ^{143}Nd abundances in mantle-derived and crustal rocks (O'nions *et al.*, 1979a,b; DePaolo, 1980). It is not yet clear whether a thermal history with mantle-wide convection can be made internally consistent with geochemical knowledge on the abundances of radioactive elements in the Earth. Continued use of mantle-wide thermal models is warranted at present, because information on the interrelationship of flow rates, the thermal history, and the radioactive abundances are obtained. These calculations may conceivably lead to the rejection of mantle-wide flow for all or part of the Earth's history. More complex thermal history models will then be necessary.

It is tacitly assumed in the calculations that the flow in the past was not drastically different from current plate tectonics. Current geological evidence is compatible with, but does not unequivocally establish plate tectonics in the Archean (e.g., Windley, 1977). Nonplate tectonic schemes are thus preferred by some investigators (e.g., Hargraves, 1978).

2. Thermal History Models

Conservation of energy implies that the change in the heat content in the Earth is the difference between heat produced by radioactive decay and the heat that escapes through the surface. Mathematically this relationship is

$$P(\partial T_e/\partial t) = -q_e + q_h, \tag{1}$$

where P is the product of heat capacity and the mass of the Earth divided by the surface area of the Earth; T_e is the average temperature of the Earth's interior; t is time; q_e is the average surface heat flow from the mantle, including heat lost at midoceanic ridges; and q_h is the total radioactivity heat generation in the Earth (excluding the continental crust) divided by surface area.

If radioactive heat production is balanced with surface heat flux, the temperature of the Earth's interior will remain constant. Several recent studies (Sharp and Peltier, 1979; Sleep, 1979; Daly, 1980; Davies, 1980a;

McKenzie and Weiss, 1980; Schubert *et al.*, 1980; Stacey, 1980; Turcotte, 1980) have concluded that this balance is unlikely. The reasons may be summarized as follows:

(1) The parameter P^{-1} is about 100°C/B.Y./HFU; i.e., cooling of the earth by 100°C in the last billion years would provide a 1 HFU imbalance on the right-hand side of Eq. (1). The major uncertainty in P is the extent to which the core contributes heat to convection. The specific heat in the core is about $\frac{1}{5}$ of the total in the Earth. Energy for formation of the inner core may be significant (see Gubbins *et al.*, 1979). The existence of a magnetic field throughout geological time does not currently provide a hard limit on the cooling rate of the Earth. The specific heat in the Earth is tabulated by Stacey (1977).

(2) The geologically inferred rate of cooling of the Earth is about 200–300°C in the last 3 B.Y. (Sleep, 1979; Davies, 1980a; McKenzie and Weiss, 1980). This estimate is obtained from differences between Archean komatiites and modern ridge basalts. The actual value depends on how these rocks formed in the mantle.

(3) The assumption of a balance between radioactivity and heat flow throughout the Earth's evolution leads to a contradiction because the radioactive heat generation has decreased markedly during the Earth's history. Because the heat lost from the Earth correlates with mantle temperature, higher temperatures would be expected in the past than in the present if heat flow and radioactivity were in balance. The heat released to reduce the temperature from the past to the present would contribute to heat flow and thus imply an imbalance between radioactivity and heat flow.

This general cooling of the Earth is in conflict with the hypothesis of Tozer (see, e.g., 1972) that the Earth's temperature is controlled by the temperature dependence of viscosity, so that the flow rate adjusts to maintain the temperature at the viscosity necessary for flow. Tozer's reasoning is inapplicable because the time needed to circulate the mantle through the lithosphere (and thus cool the Earth) is 2–3 B.Y., while the turnover time of a cell is less than 200 M.Y. The feedback to flow rates provided by the Earth's viscosity is thus exceedingly sluggish.

Equation (1) can be used to construct a thermal history of the earth. In "parameterized convection" calculations the radioactive heat generation is assumed and the dependence of surface heat flow on temperature is obtained from the physics of convection as outlined in the latter part of this article. Alternatively, the rate of convection can be obtained from geochemical considerations, such as ^{40}Ar degassing, and then used to obtain surface heat flux (Sleep, 1979).

Modeling of convection in the Earth is simplified because most of the heat loss is associated with the formation of oceanic lithosphere.

The global average of heat flow from the mantle is given adequately by the cooling of a half-space:

$$(2) \qquad q_e = kT_\lambda/(Kt_e)^{1/2} \equiv kT_\lambda/L$$

where k is the thermal conductivity, T_λ is the temperature at the base of the lithosphere as well as the intrusion temperature at ridges, and K is the thermal diffusivity. The time t_e, which is the area of the Earth divided by the global rate of seafloor spreading (Sleep, 1979), is small compared with the time scale on which the Earth cools or heats up. The parameter L is a typical thickness of oceanic lithosphere. This relationship is explicit or implicit in parameterized convection models of the earth's thermal evolution.

2.1. Kinematic Convective Models

It is conceivable that the global rate of seafloor spreading in the Earth's past can be constrained more reliably from observational evidence than consideration of the mechanics of convection. It is also useful to have thermal models that assume a constant global spreading rate for comparison with more complex models. The situation is favorable if the objective is to compute heat flow and, by means of Eq. (1), mantle radioactivity, because global heat flow depends only on the square root of global spreading rate. Estimates of global spreading rate in the Archean and later times can be obtained by considering the degassing of ^{40}Ar from the mantle, paleomagnetic estimates of continental drift rates, and variations in sea level and continental freeboard.

A minimal rate of continental drift can be obtained from polar-wandering curves (Gordon et al., 1979). The oldest polar-wandering path from Australia would imply rates somewhat more than 1 cm/yr in the Archean (McElhinny and Senanayake, 1980). This is somewhat lower than the current spreading rate.

At times of rapid global seafloor spreading the volume of midoceanic ridges is large and seawater is displaced onto the continents (Pitman, 1978). Turcotte and Burke (1978) have ruled out large changes in seafloor spreading rate during the Paleozoic using this relationship. Application of this relationship to the Archean is highly dependent on assumptions about the rates of continent and ocean formation and the temperature of continental lithosphere. The current evidence is at least compatible with spreading rates similar to the current rate throughout the Archean.

The ^{40}Ar degassing calculations are advantageous because radiogenic heating by ^{40}K decay and the rate of seafloor spreading are involved (Sleep, 1979). If an excessive amount of K is assumed to be in the mantle, large amounts of ^{40}Ar are generated. The rapid spreading rate required to cool the Earth leads to an excessive venting of ^{40}Ar. If too little ^{40}K is assumed to be present in the mantle, little ^{40}Ar is produced. The low spreading rate needed to balance heat allows little of the ^{40}Ar to escape. At an intermediate amount of K (about 120 ppm) in the mantle yields acceptable Ar values and an average spreading rate over the last 3 B.Y. close to the current rate. The spreading rate early in the Archean is less well constrained by this argument because the early thermal history is less constrained and because much ^{40}Ar had not been generated by then and thus could not have escaped. This method of constraining global spreading rates is dependent on knowledge of the ^{40}Ar in the mantle at present and the assumption that little Ar is subducted. Both these assumptions are controversial.

A thermal history with constant spreading at the rate of 4.4 km^2/yr is shown by the dashed line in Fig. 1. For the parameters we used, a nearly flat temperature curve occurs between 4.5 B.Y. and 3.5 B.Y. A gradual decrease to the present mantle temperature occurs thereafter. To obtain an increasing temperature in the early Archean, a lower spreading rate would have to be assumed.

The sensitivity of Sleep's (1979) results to input parameters can be seen in Table II. Increasing the heat flow of 1 M.Y. crust (or equivalently conductivity) results in a slightly higher global heat flow at a lower spreading rate. If the temperature drop is assumed to be 200°C rather than 300°C, the radioactivity is somewhat higher and the heat flow and spreading rate are lower. If ^{40}Ar degassing is assumed to be less efficient, a higher spreading rate and heat flow result.

2.2. Parameterized Convection

In parameterized convection calculations, the global spreading rate and the residence time t_e are obtained from the physics of convective flow. We present a derivation based on the force balance of plate tectonics. The relevance of physical parameters is more patent than in more rigorous (cellular convection) derivations used by Turcotte and Oxburgh (1967), Daly (1980), Davies (1980a), and Schubert et al. (1980). Stacey (1980) derives his formulation by determining the amount of thermodynamic work and balancing it against shear heating. Sharp and Peltier (1978) present models with stratified viscosity in addition to global parameterized convection models. Most treatments are extensions of the work of McKenzie and Weiss (1975).

First, we consider only the forces on a horizontal lithospheric plate. The elevation of the ridge crest is dimensionally

$$E \sim \alpha T_\lambda L \tag{3}$$

where α is the volume coefficient of thermal expansion. The driving force exerted by the ridge is dimensionally

$$F_D \sim (\rho g L E)L = \rho g \alpha T_\lambda L^2 \tag{4}$$

where the term in parentheses is the vertical excess pressure from the topography of the ridge, and ρg is the specific weight of the lithosphere. Rigorous derivations that yield expressions of this form are given by Artyushkov (1973, 1974) and Hager (1978). Note that this driving force is proportional to lithospheric age because $L^2 \equiv K t_e$.

The force resisting plate motion is caused by the viscosity of the mantle. The stress on the base of the lithosphere is

$$\sigma_R \sim \eta V/h \sim \eta V/R \tag{5}$$

where h is the scale of the depth of the flow. For mantle-wide flow h is of the order of the radius of the Earth, R. This stress acts along the full length of the plate, a distance also of the order of R. Thus the resisting force is

$$F_R \sim \sigma_R R \sim \eta V \tag{6}$$

The velocity is obtained by noting that the plate covers the radius of the earth in a time proportional to t_e

$$V \sim R/t_e = RK/L^2 \tag{7}$$

Using this expression and balancing the resistive and driving forces yields

$$\rho g \alpha T_\lambda L^2 \sim RK\eta/L^2 \tag{8}$$

This expression may be rewritten as

$$\frac{R^4}{L^4} \sim \frac{\rho g \alpha T_\lambda R^3}{K\eta} \equiv \text{Ra} \tag{9}$$

where the right-hand term is the Rayleigh number. Combining this expression with the expression for global heat flow yields

$$q_e \sim (kT/R)(\text{Ra})^{1/4} \tag{10}$$

Noting that convection begins at the critical Rayleigh number Ra_c, the expression can be written as

TABLE II. PARAMETERIZED CONVECTION MODELS

Author	Model	Viscosity formulation	Present heat flow (HFU)	Radiogenic heat[a]	Mantle temperature change in last 3 B.Y., °C	Global spreading rate at 3 B.Y.[b]	Comment
Daly (1980)		Constant	n.g.[f]	<50	n.g.	n.g.	Summary of nondimensional results. Cellular model checked with parameterized model
Davies (1980a)	Fig. 7	$(\nu/\nu_0) = (T/T_0)^{-n}$ $n = 29$	1.68	56	200[c]	25[c]	Various models computed
McKenzie and Weiss (1980)	Fig. 6	$\nu = DT_a \exp(E/RT_a)$ $E = 4.25$ eV $\nu = 2 \times 10^{21}$ st at 1350°C	1.88	78	340[c]	6.6	Heated from within. Model is also given for heating from below
Schubert et al. (1980)	Figs. 2 and 3	$\nu = \bar{\nu} \exp(A/T_a)$ $\bar{\nu} = 1.65 \times 10^6$ cm^2/sec $A = 7 \times 10^{4}$°K	1.33	75	170	9	

Stacey (1980)		$\dot{\epsilon}$ = constant × σ^N exp($-gT_{ma}/T_a$) $g = 20 T_{ma} = 2750°K$ $N = 1$ (linear viscosity)	1.12	71	165	n.g.	
Stacey (1980)		Same except $N = 5$ (power law viscosity)	1.12	58	185	n.g.	
Turcotte (1980)		$(\eta/\eta_0) = \exp(-E \Delta T/RT_{0a}^2)$ $E/RT_{0a} = 42$ $T_{0a} = 2250°K$	1.75	83	146	4.6	
Sleep (1979)	Fig. 2	Kinematic model 4.4 km²/yr	1.75	38	300[c]	1	Constant spreading rate model constrained with Ar degassing
This article	Fast	Kinematic model 3.5 km²/yr	1.80	37	300[c]	1	
	Slow	Kinematic model 3.1 km²/yr	1.68	45	200[c]	1	
	Slow	Kinematic model 6.2 km²/yr	2.38	37	300[c]	1	
	Degassing[g]						Sleep's method with Davies (1980b) heat flow relationship

[a] Present radiogenic heat as percent of present heat flow.
[b] Relative to current rate. Result is presented as heat flow in some articles.
[c] Obtained from differences between komatiitic and recent lavas.
[d] Geologically obtained upper limit. Lower limit is about 4.
[e] A 15-km layer is degassed of ^{40}Ar at ridges; a 30-km layer is used in other models.
[f] n.g., Not given in results.

Mathematical Symbols (other than constants defined in table): ν, kinematic viscosity, viscosity divided by density; T, temperature relative to surface temperature; $\dot{\epsilon}$, strain rate; σ, stress; η, viscosity; ΔT, difference between past and present mantle temperature; *Subscripts*: 0, present value; a, absolute temperature.

$$(11) \qquad q_e = \frac{kT_c}{(R - R_c)} \left(\frac{\text{Ra}}{\text{Ra}_c}\right)^{1/4}$$

where T_c and R_c are the temperature and radius of the core.

A second similar expression can be obtained by balancing forces on a cross section through the downgoing slab. The excess weight per area in the section is

$$(12) \qquad \sigma_s \sim \rho g \alpha T_\lambda L$$

This force is balanced by viscous forces at the edges of the slab. This resistive stress is again

$$(13) \qquad \sigma_R \sim \eta V/R$$

Using the previous expression for V one obtains

$$(14) \qquad \rho g \alpha T_\lambda L \sim \eta K/L^2$$

or

$$(15) \qquad \frac{R^3}{L^3} \sim \frac{\rho g \alpha T_\lambda R^3}{K\eta} = \text{Ra}$$

This gives

$$(16) \qquad q_e \sim (kT_\lambda/R)(\text{Ra})^{1/3}$$

As reviewed by Davies (1980a) and Turcotte (1980), the surface heat flux varies between $\frac{1}{4}$Ra and $\frac{1}{3}$Ra in physical and numerical experiments, as expected from Eqs. (11) and (16). It can be shown that the thermodynamic treatment of Stacey (1980) would yield Eq. (16) for a linear viscosity.

An alternate form of the Rayleigh number for a body such as the earth, which is heated from within, can be obtained by noting that

$$T_H \equiv \rho H R^2/k$$

(where H is the heat generation per unit mass) has dimensions of temperature. It is probably not advisable to use this for the Earth, because the flow rate at a given time depends on the actual temperature contrasts of ridges and slabs and not on the ultimate temperature that would be reached in a steady-state static body. The heat generated during the history of the earth by radioactive decay is a small fraction of that needed to heat the Earth to T_H. Use of this formulation, however, gives results similar (see Turcotte, 1980) to those of other treatments, because the viscosity of the Earth is the most variable parameter in the Rayleigh number.

Computation of a parameterized convection model is simplified because the current heat flow from the mantle is well constrained. Equation (11) or (16) thus can be written as

(17)
$$q_e = q_0 \frac{T_\lambda}{T_0} \left(\frac{\text{Ra}}{\text{Ra}_0}\right)^p$$

where the subscript 0 denotes current values and p is a constant between $\frac{1}{4}$ and $\frac{1}{3}$. The only parameters in the Rayleigh number that are likely to vary significantly are viscosity η and temperature. Thus Eq. (17) becomes

(18)
$$q_e = q_0 \left(\frac{T}{T_0}\right)^{1+p} \left(\frac{\eta_0}{\eta}\right)^p$$

A differential equation in temperature as a function of time is obtained by combining (1) and (18) and expressing viscosity as a function of temperature. The mantle is assumed to be sufficiently well mixed such that the upper-mantle temperature T is indicative of the temperature in the rest of the Earth. Use of a simple expression for the viscosity–temperature function yields an analytic solution of (1) that can be computed backwards in time (Davies, 1980a). The global spreading rate can be obtained from (2):

(19)
$$U \equiv \frac{4\pi R^2}{t_e} = U_0 \left(\frac{q_e}{q_0}\right)^2$$

since the variation in mantle temperature is fairly small. This expression is useful because global spreading rate is not explicitly computed in many articles on parameterized convection.

The thermal history obtained from a parameterized convection calculation depends on the amount of radioactive heating in the mantle (Davies, 1980a). If radioactive heating is small, the model Earth is currently cooling, and the higher previous temperatures implied by the cooling imply higher heat flow and more rapid temperature decrease in the past. Unreasonably high temperatures are thus obtained for early in the Earth's history. Excessively high radioactivity implies that the model Earth is heating up from unreasonably low temperatures. Acceptable models either cool monotonically or heat up because of the higher radioactive heat generation early in the Earth's history and then cool down. Heat flows 2–5 times greater than current heat flow are obtained for Archean times. Thus a global spreading rate in the Archean 4–25 times greater than the current rate is thus implied (Table II).

2.3. Stratified Convection

The kinematic and parameterized convection models discussed above assume that the mantle is sufficiently well mixed such that a temperature in the upper mantle can be used to represent the temperature variation of the earth as a whole. More complex possibilities are that the upper mantle convects separately from the lower mantle or that cold material in slabs sinks to great depth and returns to the surface slowly. We present here an idealized scheme that represents the second possibility. Although a kinematic model is presented here, the upper-mantle temperatures computed in the model could be used to determine the flow rate through parameterized convection.

We assume that the material in slabs (as well as some entrained asthenosphere) sinks all the way to the core mantle boundary and then thermally equilibrates with the core. Subsequent slab material would displace the previously subducted material upwards. Therefore, the ambient temperature in the mantle is computed from a model with slow vertical advection. In the uppermost regions, where cold material is being subducted to the core–mantle boundary, the average vertical velocity decreases gradually to zero at the surface

There are three free parameters in this type of model: (1) the initial adiabatic temperature after core formation, (2) the average spreading rate through time, and (3) the amount of material entrained with the slab and hence the mantle turnover rate. A feature of these models is that a long period of time occurs between when material is subducted and when it is returned to the surface. The mantle thus heats up from radioactivity early in the Earth's history, while the core and lower mantle are cooled from material that was subducted from the surface. A rapid drop in the uppermost-mantle temperature occurs when material that was part of slabs reaches the surface for the second time.

An example of a calculation is shown by the solid line in Fig. 1. The temperature of magma eruption has been computed from the upper-mantle temperature. The curve has been fit to three data points at 3.5 B.Y., 2.7 B.Y., and the present. The point at 3.5 B.Y. fixes the initial temperature in the model because the material at the surface at that time has risen from some depth and heated up radioactively. Therefore, the magma temperature at 3.5 B.Y. is independent of spreading rate. The average spreading rate between the start and 2.7 B.Y. is fixed by the temperature at 2.7 B.Y. The spreading rate affects the amount of cooling between 3.5 B.Y. and 2.7 B.Y. when cold material first returns to the surface. The temperature at the present fixes the spreading rate between 2.7 B.Y. and the present. By adjusting the thickness of material entrained with the slab, we obtained a model that had a constant global spreading rate through time.

An intended feature of this kinematic scheme is the rapid decrease in upper-mantle temperatures at the end of the Archean. This rapid decrease does not occur in a well-mixed model unless the convection rate is quite high compared to the present (Fig. 1). A rapid change in mantle chemistry as inferred from lava might occur when the material at the surface during

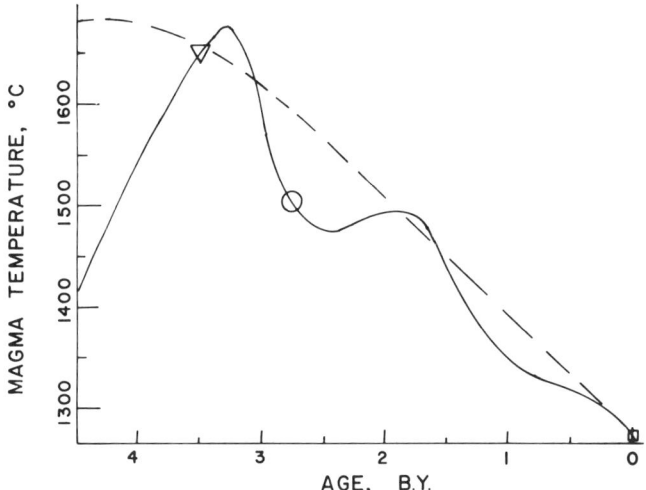

FIG. 1. The temperature of primitive basalt magmas was computed for a constant global spreading rate of 4.4 km²/yr. A well-mixed kinematic model (dashed line) gives a steady upper-mantle temperature before 3.8 B.Y. and a steady decrease thereafter. The stratified model (solid line) gives (1) a gradual increase before 3.1 B.Y. as material that was never part of a slab ascends to the surface, (2) a sharp drop between 3.1 and 2.6 B.Y. when material that was subducted first returned to the surface, and (3) a somewhat irregular temperature history that approaches the well-mixed model. The stratified model is forced to fit data on komatiites by Green et al. (1975) (△), by Arndt (1976, 1977) (○), and modern ridge basalts (Bender et al., 1978) (□). The well-mixed model has the same parameters as the stratified model and is forced to fit only the modern ridge basalt datum. It is uncertain whether the data on the lavas is representative of the average upper mantle. The inference of magma temperature from lava temperature is dependent on the mechanism of lava generation. Both curves are acceptable at the current level of understanding.

start-up returned a second time. This feature is intended to be compatible with the observation that early Archean lavas have chondritic Nd isotopic ratios (O'nions et al., 1979a,b; DePaolo, 1980).

2.4. Discussion of Thermal Modeling

The kinematic constraints as well as the successful stratified model indicate that the global spreading rate in the past is similar to the rate today. The parameterized convection models indicate that the global spreading rate in the Archean was much faster than in the present. This difficulty probably occurs because the chemical stratification of oceanic lithosphere was not included in the parameterized convection.

Although the method of generation of modern ridge basalts and Archean komatiites is still controversial, any kinematically viable mechanism should include a significantly greater thickness of Archean oceanic crust than present oceanic crust. At present the oceanic crust is about 5 km thick and the layer of depleted residuum from which the basalt is generated is 25 km thick (Kay *et al.*, 1970). In the Archean, the higher temperatures of ascending mantle would intersect the melting curve at greater depths. In a schematic diagram by Sleep (1979), extensive melting begins at a depth of 85 km to yield a 20-km-thick crust with an average melting of 20%. This example uses a 230°C temperature difference between Archean and Recent below the zone of magma generation.

The oceanic crust and also the depleted residuum are less dense than primary mantle (McGetchin and Smyth, 1978). This contrast would tend to retard subduction and thus seafloor spreading. Where subduction was established, the oceanic crust would revert to ecologite at some depth and thus provide a driving force. Significant cooling of the lithosphere, and therefore much time, would be needed to get the thicker crust cold enough for the phase change to occur. The lower viscosity of the Archean upper mantle would cause established slabs to detach rather than drive primarily the surface plates.

The variation of at least 50% in global spreading rate in the last 100 M.Y. inferred by Pitman (1978) implies that the geometry of surface plates rather than mantle temperature controls the flow rate at least on an intermediate time scale. No cause-and-effect mechanism has been established for these variations. Once one is formulated and the present difficulties in subduction of thick oceanic crust beneath aseismic ridges are quantified, more sophisticated parameterized convection models can be constructed for the Archean.

3. Core Formation

An estimate of the conditions after core formation is obtained by determining which starting conditions in a thermal history model lead to the present state of the Earth. Direct constraints on the core forming process are not evident, but the general range of plausible conditions can be deduced from the mechanics of the process. We develop here a scheme of core formation that has been modified from Smith (1981) to yield a well-stirred mantle at the end of core formation. We do this to be compatible with our assumption of mantle-wide flow. Except for our modifications, the conclusions are equally applicable to the stratified mantle developed by Smith (1981). We do not consider schemes where the core accretes separately from the mantle (see, e.g., Matsui, 1979).

3.1. Timing of Core Formation

Hard evidence on the timing of core formation comes largely from isotopic ratios of elements produced by relatively short-lived parents. An early time of core formation is indicated, but the event probably was contemporaneous with and somewhat after accretion. Both accretion and core formation probably ended by around 4.4 B.Y. ago.

The systematics of Pb isotopes provide the most direct chemical constraint on the time of core formation. The stable isotope ^{204}Pb is strongly depleted in the mantle relative to the isotopic ratios expected for a chondrite mixture of U and Pb. Much of the Earth's Pb thus entered the core before large amounts of radiogenic lead had time to form. The data are compatible with core formation occurring about 100 M.Y. after accretion (Oversby and Ringwood, 1971). Continuous formation of the core over a longer period is also possible (Vollmer, 1977).

Volatile elements provide less direct constraints, since degassing need not have been necessarily associated with core formation. The stable isotope ^{36}Ar is much more strongly concentrated in the air than ^{40}Ar, which is produced by potassium decay (Dymond and Hogan, 1978). The degassing must have been concentrated in a period much shorter than the half-life of ^{40}K for this to be true. The radiogenic isotope ^{129}Xe is also preferentially enriched in the air relative to the mantle. The initial degassing, therefore, occurred in at most a few half-lives of ^{129}I—17 M.Y. (Thomsen, 1980).

An efficient cycling of material through the surficial regions is needed to degas the Earth. A loss of material into space is not indicated since the atmospheric abundance is approximately that of ordinary chondrites (Fanale, 1971). Water, carbon dioxide, and chloride may have also been degassed in the Earth of that time.

3.2. Mechanism of Core Formation

The mechanics of core formation can be constrained by considering that the Earth was likely to be heated heterogeneously by accretion. The large impacts were the most important heat source. The available heat would increase as the Earth grew in size (Wetherill, 1976, 1980; Kaula, 1979, 1980).

Segregation of iron from silicate would begin near the surface. Molten iron could efficiently propagate downward by the instabilities of crack formation, as water propagates downward through ice (Weertman, 1971). The dikes of iron thus formed would go down until they reached the core. The tendency to form dikes would preclude the buildup of large bodies of molten iron except in the core.

As stated by Smith (1981), most of the accretional heat is liberated in the upper 100 km. Volcanism would remove areas that were molten and hence would preclude temperatures in the deeper regions that exceeded the adiabat within the source zone of the volcanics. Cold regions, some of which might become mixed with iron metal, would be produced as lava cooled on the surface. Additional cool regions would be produced early by accretion and by areas that escaped the heterogeneous heating by large impacts.

These initial cool regions would be slow to heat up. Shear–strain heat was most important in hot regions that could flow. Tens of million years would be required for conduction to warm 100-km-size regions. Contact with hot descending iron and hot ascending silicate would warm some areas but not others.

Once a sizable core formed, the highest mantle temperatures would be at the core boundary (see also Shaw, 1978). A stable stratification of hot silicate near the surface, cool undifferentiated material, and the hot core would develop (Fig. 2). New differentiated silicate would be produced at the core–mantle boundary and flow upward toward the surface. The chemistry of this material would differ somewhat from that of the silicate differentiated near the surface.

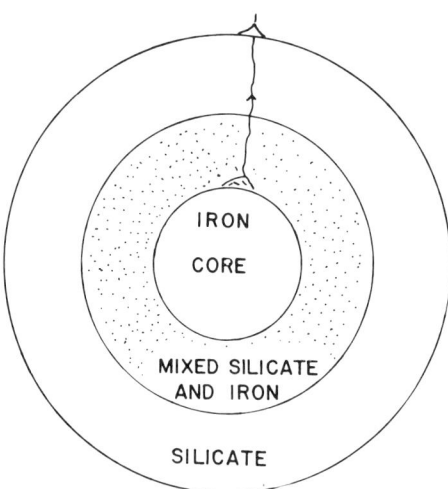

FIG. 2. A schematic drawing of the Earth at a time when accretion was essentially complete and the core was partly formed. New differentiated silicate formed at the core–mantle boundary from mixed silicate and iron, ascended to the surface, and lost its heat by melting and volcanism. The freshly erupted material is buried by subsequent material. The differentiated silicate layer did not come into contact with the core until core formation was complete.

Our scheme differs from Smith's (1981) by the presence of the layer of undifferentiated silicate above the core. The ascending silicate is intended to stir the mantle and thus permit mantle-wide convection later in Earth history. The presence of mantle material that equilibrated with iron at great depths is a geochemical effect that is potentially testable by high-pressure experimental petrology. In Smith's scheme the lower mantle remains stagnant and does not mix with later accreting and more oxidized material.

The energy release by core formation is sufficient to heat the entire earth somewhere between 1200°C (Murthy and Hall, 1972) and 2100°C (Flaser and Birch, 1973) depending on the composition of the core. The amount of heat retained for some time during accretion is comparable (see, e.g., Kaula, 1979). Unless the initial temperature of the Earth was quite cold, a heat loss causing about 100 times the present rate of cooling (10°C/M.Y. versus 0.1°C/M.Y.) is necessary for the initial heat to escape in the first 100 M.Y. of Earth history. A plate tectonic or bulk convective rate of 10^4 times the present rate would be needed for this heat to escape.

Total melting and bulk eruption of hot ascending material (and excessively hot material buried at the surface during impacts) is a more probable mechanism for venting the early heat of the Earth (Smith, 1981). Volcanism is now an inefficient heat-loss mechanism of the Earth because only about 30% melting is possible at present temperatures and the rate of volcanism is limited by the rate of plate tectonics in removing the depleted residuum. Totally melted mantle would have an extremely low viscosity and erupt as thin flows on the surface. The rate of return to depth would be controlled by burial by subsequent flows and not by the removal of minor amounts of residuum at depth.

Hot material erupted in flows would totally cool before being returned to depth. If a significant portion of the mantle was cycled through the surface in this manner, the mantle would be cooler it was than later in the Earth's history. An overcooling of this type does not occur in plate tectonics because most of the circulating lithosphere is not near the surface and can cool only a limited amount. The rapid motions implied by core formation would leave the mantle adiabatic except at the surface and near the core at the end of core formation.

As noted by Walker (1977) and Jakosky and Ahrens (1979), the heat loss from the Earth during core formation and accretion is less than current solar flux. The surface and atmosphere were therefore never particularly hot during early Earth history.

The cycling of much of the mantle through lava flows is an effective mechanism for degassing of the Earth. Some H_2O, HCl, and CO_2 would be expected to react with cooled lava and return into the Earth. The escape of rare gases would be more nearly complete. An early formation of carbonates and oceans is implied by this scheme (Smith, 1981).

The temperatures of mantle silicates necessary for total melting can be constrained by using forsterite (90)–fayalite (10) as a model for the mantle. The liquidus of forsterite (90) is about 1810°C at the surface (Bradley, 1962). The ascending adiabat must exceed this temperature by the latent heat (120.9 kJ/mole, Bradley, 1962) divided by the specific heat (195 J/mole deg, Robie *et al.*, 1978), 620°C, to give an extrapolated temperature of 2430°C. This temperature would be reduced somewhat by the presence of pyroxene and aluminous components in the mantle. Thermal models of accretion (Kaula, 1979) and core formation (Shaw, 1978) give mantle temperatures far above this adiabat. Total melting of significant parts of the mantle is thus feasible, both during accretion and from material ascending from the core–mantle boundary. As noted above, the occurrence of volcanism during accretion would keep temperatures below that estimated by Kaula and Shaw.

3.3. Ferric Iron and Water in Mantle

The H_2O and Fe_2O_3 in the upper mantle are strongly out of equilibrium with metallic iron of meteorites or the core at surface conditions. These oxidized materials must have formed (or accreted) late in the Earth's accretion or else mixing and back reaction with incoming iron would have occurred. The formation of the oxidized material may have occurred in space, near the surface, or at great depths in the Earth.

Smith (1981) prefers heterogeneous accretion of oxidized material late in the earth's accretion after the core and lower mantle had formed. The compositional differences and the low temperatures of the already accreted material (due to cooling by volcanism) prevented mantle-wide convection. The oceans and CO_2 in the Earth accreted with the more oxidized material.

Oxidation of ferrous iron by water followed by loss of H_2 to space is conceivable. Since this is about 1% Fe_2O_3 in the mantle, a volume of water similar to the present ocean would need to react. Unless material erupted at extreme temperature, the yield of H_2 would be about 1% of the H_2O (see summary of data by Ringwood, 1977). This process is likely to be too slow to oxidize a significant part of the mantle and therefore can be rejected.

At great depths ferrous iron in silicates (or oxides) may be partly soluble in the core and partly disproportionate into ferric iron in a silicate phase (Ringwood, 1977). The iron-depleted silicate thus created would be less dense than more primitive silicate and tend to rise in the mantle. The mantle would be efficiently stirred and oxidized by this process. In any case whole mantle flow at some stage in the Earth's history is necessary to get the ferric iron back to the surface where it is now sampled.

This scheme would produce oceans if it occurred when accretion was nearly complete. The H_2O and CO_2 released then would not be reduced by additional infall of metallic iron or buried back into the mantle.

It should be noted that accretion is quite ineffective at forming oceans and carbonate deposits (Jakosky and Ahrens, 1979). As the temperatures during accretion were similar to those of the present, any water present was in liquid or hydrated silicates. Any CO_2 was in carbonates. Any transient water bodies were filled with material if not destroyed outright by impact mixing. Silicate and carbonate phases were efficiently buried. Volcanism during accretion returned some H_2O and CO_2 to the surface but this material was probably buried again.

The heterogeneous accretion and the disproportionation of ferric iron hypotheses should imply different abundances of trace elements in the upper mantle. Future work in experimental petrology should be able to appraise the disproportionation hypothesis.

4. Conclusions and Future Work

The conclusions of this article are as follows:

(1) The Earth has been cooling since Archean time and a substantial amount of the Earth's heat flow comes from cooling rather than radioactivity.

(2) The rate of plate motions in Archean and Early Proterozoic times is poorly constrained. Kinematic models using Ar degassing give rates similar to present rates. Parameterized convection models indicate rates many times the present rate because of the low viscosity of the hot Archean mantle. Thicker oceanic crust, which probably existed in the Archean, may have retarded plate motions to rates similar to those of today.

(3) Extensive heat was lost from the Earth by volcanism during core formation and accretion. Cool or warm starting conditions are thus plausible, but hot conditions, with much of the mantle near the melting point, can be excluded.

(4) It is unclear whether convection is mantle-wide or not. The thermal evolution would not be extensively modified if heat is efficiently transformed between the lower and the upper mantle. This issue, therefore,

will be determined largely from chemical data and its consistency with thermal models.

The nature of further studies of the thermal history of the Earth will depend largely on the objective of these studies. The variation in upper-mantle temperatures for the last 3.5 B.Y. is probably best constrained by the variation of mantle-derived igneous rocks through time. The average rate of plate motions for the last 3.5 B.Y. may similarly be constrained using paleomagnetics. A thermal model for the last 3.5 B.Y. thus will be used with isotopic systematics to obtain an optimal thermal history compatible with various kinematic assumptions. It is likely that the assumption of mantle-wide convection can be tested from the internal consistency of various lines of evidence.

For the period extending from the formation of the core to the time corresponding to the oldest reliable samples, the radioactivity (and to some extent the convection rate) will be deduced from the subsequent behavior. The models will be run backward to obtain temperature shortly after core formation.

There is some hope that high-pressure experimental petrology may give information on the physics of core formation. Until then, either cool or warm starting conditions (after core formation) are justifiable in a thermal model. Very hot conditions that would lead to total melting can be excluded.

Increasing sophistication of models will include more complex kinematics and variable mantle radioactivity as radioactive elements become concentrated in the crust with time. Parameterized convection models will include variations in the thickness of the oceanic crust. Progress will be more difficult if mantle-wide models are found to be unacceptable. The abundance of radioactive elements in the deeper layers, which never see the surface, will be subject to little constraint. The upper-mantle temperature will also be less representative of the thermal state of the Earth.

Acknowledgments

This work was supported by National Science Foundation Grant #EAR80-01076.

References

Anderson, D. L. (1979). Chemical stratification of the mantle. *J. Geophys. Res.* **84,** 6297–6298.

Arndt, N. T. (1976). Melting relations of ultramafic lavas (komatittes) at one atmosphere and high pressure. *Year Book—Carnegie Inst. Washington* **75,** 555–561.

Arndt, N. T. (1977). Ultrabasic magmas and high-degree melting of the mantle. *Contrib. Mineral. Petrol.* **64,** 205–221.

Artyushkov, E. (1973). Stresses in the lithosphere caused by crustal thickness inhomogeneities. *J. Geophys. Res.* **78**, 7675–7708.

Artyushkov, E. (1974). Can the earth's crust be in a state of isostasy? *J. Geophys. Res.* **79**, 741–752.

Bender, J. F., Hodges, F. N., and Bence, A. E. (1978). Petrogenesis of basalts from the project FAMOUS area: Experimental study from 0 to 15 kbar. *Earth Planet. Sci. Lett.* **41**, 277–302.

Bradley, R. S. (1962). Thermodynamic calculations on phase equilibria involving fused salts. Part II. Solid solutions and application the olivines. *Am. J. Sci.* **260**, 550–554.

Daly, S. F. (1980). Convection with decaying heat sources: Constant viscosity. *Geophys. J. R. Astron. Soc.* **61**, 519–548.

Davies, G. F. (1977). Whole mantle convection and plate tectonics. *Geophys. J. R. Astron. Soc.* **49**, 459–486.

Davies, G. F. (1980a). Thermal histories of convective earth models and constraints on radiogenic heat production in the earth. *J. Geophys. Res.* **85**, 2517–2530.

Davies, G. F. (1980b). Review of oceanic and global heat flow estimates, *Rev. Geophys. Space Phys.* **18**, 718–722.

DePaolo, D. J. (1980). Crustal growth and mantle evolution: Inferences from models of element transport and Nd and Sr isotopes. *Geochim. Cosmochim. Acta* **44**, 1185–1196.

Dymond, J., and Hogan, L. (1978). Factors controlling the Noble gas abundance patterns of deep-sea basalts. *Earth Planet. Sci. Lett.* **38**, 117–128.

Fanale, F. P. (1971). A case for catastrophic early degassing of the earth. *Chem. Geol.* **8**, 79–105.

Flaser, F. M., and Birch, F. (1973). Energetics of core formation: A correction. *J. Geophys. Res.* **78**, 6101–6103.

Gordon, R. G., McWilliams, M. O., and Cox, A. (1979). Pre-tertiary velocities of the continents: A lower bound from paleomagnetic data. *J. Geophys. Res.* **84**, 5480–5486.

Green, D. H., Nicholls, I. A., Viljoen, M., and Viljoen, R. (1975). Experimental demonstration of the existence of peridotite liquids in earliest Archean magmatism. *Geology* **3**, 11–14.

Gubbins, D., Masters, T. G., and Jacobs, J. A. (1979). Thermal evolution of the earth's core. *Geophys. J. R. Astron. Soc.* **59**, 57–100.

Hager, B. H. (1978). Oceanic plate motions driven by lithospheric thickening and subducted slabs. *Nature (London)* **276**, 156–159.

Hargraves, R. G. (1978). Punctuated evolution of tectonic style. *Nature (London)* **276**, 459–461.

Jakosky, B. M., and Ahrens, T. J. (1979). The history of an atmosphere of impact origin. *Geochim. Cosmochim. Acta, Suppl.*, **10**, 2727–2739.

Jeanloz, R., and Richter, F. M. (1979). Convection, composition, and the thermal state of the lower mantle. *J. Geophys. Res.* **84**, 5497–5503.

Kaula, W. M. (1979). Thermal evolution of earth and moon growing by planetesimal impacts. *J. Geophys. Res.* **84**, 999–1008.

Kaula, W. M. (1980). The beginning of the earth's thermal evolution. *Spec. Pap.—Geol. Assoc. Can.* **20**, 25–34.

Kay, R., Hubbard, N., and Gast, P. (1970). Chemical characteristics and the origin of oceanic ridge volcanic rocks. *J. Geophys. Res.* **75**, 1585–1613.

McElhinny, M. W., and Senanayake, W. E. (1980). Paleomagnetic evidence for the existence of the geomagnetic field 3.5 Ga ago. *J. Geophys. Res.* **85**, 3523–3528.

McGetchin, T., and Smyth, J. R. (1978). The mantle of Mars: Some possible geological implications of high density. *Icarus* **34**, 512–536.

McKenzie, D., and Weiss, N. (1975). Speculations on the thermal and tectonic history of the earth. *Geophys. J. R. Astron. Soc.* **42,** 131–174.

McKenzie, D., and Weiss, N. (1980). Thermal history of the earth. *Spec. Pap.—Geol. Assoc. Can.* **20,** 555–590.

Matsui, T. (1979). Collisional evolution of the mass-distribution spectrum of planetesimals. II. *Geochim. Cosmochim. Acta, Suppl.* **10,** 1881–1895.

Murthy, V. R., and Hall, H. T. (1972). The origin and chemical composition of the earth's core. *Phys. Earth Planet. Int.* **6,** 123–130.

O'nions, R. K., Carter, S. R., Evensen, N. M., and Hamilton, P. J. (1979a). Geochemical and cosmochemical applications of Nd isotope analysis, *Annu. Rev. Earth Planet. Sci.* **7,** 11–38.

O'nions, R. K., Evensen, J. N., and Hamilton, P. J. (1979b). Geochemical modelling of mantle differentiation and crustal growth. *J. Geophys. Res.* **84,** 6091–6101.

Oversby, V., and Ringwood, A. E. (1971). Time of formation of the earth's core. *Nature (London)* **234,** 463–465.

Peltier, W. R. (1972). Penetrative convection in the planetary mantle. *Geophys. Fluid Dyn.* **5,** 47–88.

Peltier, W. R. (1980). Mantle convection and viscosity. *In* "Physics of the Earth's Interior" (A. M. Dziewonski and E. Boschi, eds.), pp. 362–427. Elsevier/North-Holland Publ., New York.

Peltier, W. R. (1981). Ice-age dynamics. *Annu. Rev. Earth Planet. Sci.* **9,** 199–226.

Pitman, W. C., III (1978). The relationship between eustacy and stratigraphic sequences of passive margins. *Geol. Soc. Am. Bull.* **89,** 1389–1403.

Ringwood, A. E. (1977). Composition of the core and implications for origin of the earth. *Geochem. J.* **11,** 111–135.

Robie, R. A., Hemingway, B. S., and Fisher, J. R. (1978). Thermodynamic properties of minerals and related substances at 298.11 K and 1 bar (10^5 pascals) pressure and at higher temperatures. *Geol. Surv. Bull. (U.S.)* **1452,** 1–456.

Schubert, G., Stevenson, D., and Cassen, P. (1980). Whole planet cooling and the radiogenic heat source contents of the earth and moon. *J. Geophys. Res.* **85,** 2531–2538.

Sclater, J. G., Jaupart, C., and Galson, D. (1980). The heat flow through oceanic and continental crust and the heat loss of the earth. *Rev. Geophys. Space. Phys.* **18,** 269–312.

Sharp, H. N., and Peltier, W. R. (1978). Parameterized mantle convection and the earth's thermal history. *Geophys. Res. Lett.* **5,** 737–744.

Sharp, H. N., and Peltier, W. R. (1979). A thermal history model of the earth with parameterized convection. *Geophys. J. R. Astron. Soc.* **59,** 171–204.

Shaw, G. H. (1978). Effects of core formation. *Phys. Earth Planet Int.* **16,** 361–369.

Sleep, N. H. (1979). The thermal history and degassing of the earth: Some simple calculations. *J. Geol.* **87,** 671–686.

Smith, J. V. (1981). The first 800 million years of earth's history. *Philos. Trans. R. Soc. London* (in press).

Stacey, F. D. (1977). A thermal model of the earth. *Phys. Earth Planet. Int.* **15,** 341–348.

Stacey, F. D. (1980). The cooling earth: A reappraisal. *Phys. Earth Planet. Int.* **22,** 89–96.

Thomsen, L. (1980). ^{129}Xe on the outgassing of the atmosphere. *J. Geophys. Res.* **85,** 4374–4378.

Tozer, D. C. (1972). The present thermal state of the terrestrial planets. *Phys. Earth Planet. Int.* **5,** 187–197.

Turcotte, D. L. (1980). On the thermal evolution of the earth. *Earth Planet. Sci. Lett.* **48,** 53–58.

Turcotte, D. L., and Burke, K. (1978). Global sea-level changes and thermal structure of the earth. *Earth Planet. Sci. Lett.* **41,** 341–346.

Turcotte, D. L., and Oxburgh, E. R. (1967). Finite amplitude convection cells and continental drift. *J. Fluid Mech.* **28,** 29–42.

Vollmer, R. (1977). Terrestrial lead isotopic evolution and formation time of the earth's core. *Nature (London)* **270,** 144–147.

Walker, J. C. G. (1977). "Evolution of the Atmosphere." Macmillan, New York.

Weertman, J. (1971). Theory of water-filled crevasses in glaciers applied to vertical magma transport beneath oceanic ridges. *J. Geophys. Res.* **76,** 1171–1183.

Wetherill, G. W. (1976). The role of large bodies in the formation of the earth and moon. *Geochim. Cosmochim. Acta, Suppl.* **7,** 3245–3257.

Wetherill, G. W. (1980). Numerical calculations relevant to the accumulation of the terrestrial planets. *Spec. Pap.—Geol. Assoc. Can.* **20,** 3–24.

Windley, B. F. (1977). "The Evolving Continents." Wiley, New York.

EXPERIMENTAL METHODS OF ROCK MAGNETISM AND PALEOMAGNETISM*

Subir K. Banerjee

Department of Geology and Geophysics
University of Minnesota
Minneapolis, Minnesota

1. Introduction 25
2. Natural Remanent Magnetization 29
   2.1. Static Samples 31
   2.2. Rotating Samples 36
   2.3. Ballistic and Resonance Magnetometers 41
   2.4. Selective Demagnetization or "Cleaning" Methods 43
3. Artificial or Laboratory-Imparted Remanent Magnetizations 56
   3.1. Viscous Remanent Magnetization 56
   3.2. Isothermal Remanent Magnetization 60
   3.3. Anhysteretic Remanent Magnetization 63
   3.4. Thermoremanent Magnetization 68
4. Parameters Measured in the Presence of a Field 76
   4.1. Magnetic Susceptibility 76
   4.2. Hysteresis Parameters 81
   4.3. Thermomagnetic Properties 83
   4.4. Magnetocrystalline Anisotropy and Magnetostriction 87
   4.5. Magnetic Domain Structure 91
5. Future Trends 92
   5.1. Studies on Single Grains 93
   5.2. Comprehensive Instruments 94
   5.3. Continuous Measurement of Long Cores 94
   5.4. On-Line Signal Processing 95
   References 95

1. Introduction

Experimental methods of rock magnetism and paleomagnetism have become fairly well known, by reputation at least, to geologists and geophysicists. Also the basic instruments, e.g., a spinner magnetometer and an alternating-field (AF) demagnetizer, although not very cheap, are still within the reaches of many an earth scientist. Especially when compared to isotope geology—another experimental technique that has acquired recent general popularity—the instrumentation for rock magnetism and paleomagnetism is definitely cheaper and the interpretation of the results may be generally less complex. One goal of this brief article, therefore, is

---

* This is Publication No. 1036 from The School of Earth Sciences, University of Minnesota.

to reach as many geologists and geophysicists as possible and to acquaint them with the principles and potential of this tool. The hope is that a larger number of them than at present will begin to utilize this relatively simple but powerful tool in diverse areas of research ranging from earthquake prediction to Precambrian plate tectonics to the construction of a polarity time scale for rocks of Eocene and older ages. In order not to duplicate the contents of two fairly recent reviews (Collinson, 1975; Goree and Fuller, 1976), which are totally oriented towards instruments per se, this article will emphasize the physical principles behind the experimental methods and their broad problem-solving capabilities. The aforementioned article by Collinson (1975) deals with a wide range of commercial instruments, while Goree and Fuller (1976) have discussed in detail both the physical principles and the technical parameters of an attractive newcomer to the instrument field—the superconducting magnetometer, which utilizes SQUIDs (superconducting quantum interference devices). Thus, I expect that the emphasis on basic physical principles in the present article will make it useful for a wide audience and, hopefully, for some time to come.

It is appropriate to begin with a general description of the goals and terminologies of rock magnetism and paleomagnetism; indeed first I ought to explain the difference between the two terms, even if they have been mistakenly regarded as the same by some fairly astute earth scientists in the past. This is also an appropriate time to define the various abbreviations and acronyms commonly used in rock magnetic and paleomagnetic literature.

Rock magnetism is a literal translation of the German word "Gesteinsmagnetismus," which was also the title of Haalck's pioneer book on the subject published in 1942. Rock magnetism has come to mean the magnetic properties of both minerals and rocks, measured either in the presence or absence of a magnetic field. It also includes the magnetic properties of compounds synthesized to simulate minerals. The goal of rock magnetist is to understand the basic physiochemical mechanisms that lead to the observed magnetic parameters. Thus, the physical principles, tools, and approaches for interpreting the data depend heavily on the past and present studies by the physicists of the various magnetic elements and compounds. However, in the last decade or so a reverse trend has become obvious in which pure physicists in their study of complex synthetic compounds have found it convenient to dip into the accumulated rock magnetic theories and experimental observations in order to explain their own data. The most recent example of this is the explanation of the magnetic properties of the so-called "spin glasses" whose magnetic behavior is very similar to that of fine-grained magnetic minerals.

Saturation magnetization can be determined for a pure mineral, e.g., magnetite, by measuring the induced magnetization of a sample in the presence of a large (10,000–20,000 Oe) magnetic field. When saturation magnetization is expressed per unit volume, the symbol is M_s, while the symbol σ_s is used for saturation magnetization per unit mass. Saturation magnetization is a macroscopic laboratory determination of the hypothetical spontaneous magnetization that is postulated to exist inside a uniformly magnetized microscopic volume because of the magnetic ordering forces between neighboring atoms or ions. Saturation magnetization of a rock or a synthesized sample that has a given volume fraction of a magnetic material is usually expressed by the symbol J_s. Upon the complete withdrawal of a large saturating field, the magnetization of a sample does not decrease to zero, even if the applied field is zero, because of the phenomenon of magnetic hysteresis (see Fig. 1). The observed magnetization in such a case is saturation remanence (expressed by the alternative symbols M_{rs}, σ_{rs}, or J_{rs}). It is also called isothermal remanent magnetization (IRM$_s$) by some workers. [The words "remanence" and "remanent" have gained predominance over the more common English words "remnance" and "remnant" simply because of the prior entry into the rock magnetic literature of the German word "remanenz," thanks to early important contributions of the German-speaking scientists.] The reversed field required to reduced saturation remanence to zero when the measurement is made in the presence of the field is called saturation coercivity

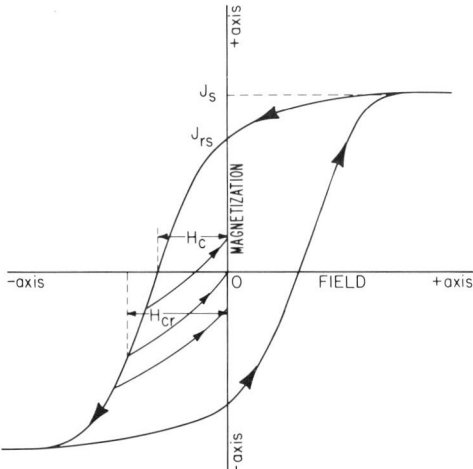

Fig. 1. Magnetic hysteresis loop showing saturation magnetization (J_s), saturation remanent magnetization (J_{rs}), coercive force (H_c), and coercivity of remanence (H_{cr}).

(H_c). If, however, all these measurements are made with a field that is not large enough to saturate the sample, the subscript "s" is dropped from the symbols M_{rs}, σ_{rs}, J_{rs} to indicate that fact and, although H_c is still used, it now is termed "coercive force" and not coercivity. The related parameter remanence coercivity (H_{cr}) describes the reverse magnetic field (dc, i.e., steady field), which when applied and later withdrawn leaves a single grain or a sample with $J_r = 0$. The term susceptibility and symbol χ are used normally to denote the observed magnetization in the presence of a low field on the order of the Earth's field (0.5–1.0 Oe).

When saturation magnetization is measured in an applied large field and as a function of temperature, a critical temperature is reached above and below which the saturation magnetization (J_s) and susceptibility (χ) fall to near-zero values, while coercivity (H_c) and saturation remanence (J_{rs}) are indeed equal to zero. This critical point is called the Curie temperature (T_c) (see Fig. 2). In general, coercivity and saturation remanence may attain zero values *before* T_c is reached and that (those) temperature(s) is (are) called blocking temperature(s).

Paleomagnetism deals with the record of past directions and magnitudes of the earth's field left in a rock as natural remanent magnetizations (NRMs). I shall now discuss the terms and definitions related to NRM and its laboratory analogs. As is obvious from its name, NRM is the magnetization of a rock or mineral measured in the absence of an applied field. This NRM is usually due to a combination of primary or characteristic remanences and some others, which are called secondary, meaning the latter originated at a time well after the origin of the primary remanence(s). A primary remanence could be a thermoremanent magnetization (TRM), a chemical remanent magnetization (CRM), or a depositional re-

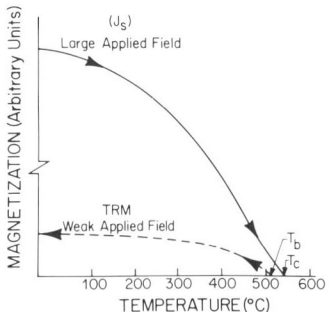

FIG. 2. Temperature dependence of saturation magnetization (J_s) and thermoremanent magnetization (TRM). Note the difference between Curie temperature (T_c) and blocking temperature (T_b) where J_s and TRM, respectively, go to zero.

manent magnetization (DRM). A TRM is created when the sample cools through its Curie temperature in the presence of a field such as the Earth's field. A CRM is created when either the preexisting magnetic phase undergoes a volume change or a new magnetic phase is created, both in the presence of a finite magnetic field, e.g., the Earth's field. A DRM is created when a net depositional alignment of NRM-bearing magnetic grains in a wet sediment takes place around a small ambient field direction, such as the Earth's field. The alignment is preserved when the water content decreases to moderately small values, e.g., 40% or less.

Secondary components of NRM may be many. A viscous remanent magnetization (VRM) is acquired when a sample carrying a primary remanence is exposed to a new field such as that due to a new direction and/or magnitude of the Earth's field or due to the removal of a natural sample in the laboratory after collection in the field. On occasion VRM can account for as much as 90-99% of the NRM and it must be preferentially removed to isolate the small primary components. A secondary IRM (isothermal remanent magnetization) can be acquired by a sample when it is exposed, for example, to lightning-induced magnetic fields. Secondary partial thermoremanent magnetizations (PTRMs) can be acquired if a sample such as a lava flow carrying a primary TRM is later reheated to moderate temperature *below* the Curie temperature in the presence of an altered Earth's field. In order to evaluate the ability of a sample to acquire secondary remanences, it is exposed in the laboratory to fields and conditions to generate a VRM, IRM, or PTRM. In order to simulate a primary remanence such as a TRM, a sample can be given an anhysteretic remanent magnetization (ARM) in the laboratory. In this process a large (≥ 1000 Oe) alternating field is smoothly decreased to zero in the presence of a small steady field. The alternating field simulates the role of temperature in inducing magnetization fluctuations, while the steady field simulates the Earth's field.

The physical theory of rock magnetism has been discussed by Nagata (1961) and by Stacey and Banerjee (1974). The methods of statistical analysis applied to the directions of the primary components of NRM have been discussed by Irving (1964) and McElhinny (1973).

2. NATURAL REMANENT MAGNETIZATION

Natural remanent magnetizations (NRM, J_n) of a wide variety of rocks, sediments, and even wood have been measured. [It is only fair to add that the NRM of wood has proven to be both extremely weak and unstable, and because of the question regarding the age of the magnetization vis-à-vis that of the wood proper, it has proved difficult to interpret the observed NRMs.] The scalar magnitude of the total magnetic moment (**M**) is related to J_n as follows:

$$|\mathbf{M}| = J_n V$$

where V is the volume of the specimen. The observed values for J_n have ranged from a high value of $\sim 10^3$ A m^{-1} ($\equiv$ 1 G) down to 10^{-6} A m^{-1} for sediments rich with organic materials or for wood. The tremendously large (10^9) dynamic range is not easy to cover adequately with equal sensitivity with a single instrument, and below I shall discuss the different types of instruments that are useful for covering the different parts of this large dynamic range. Because of the admitted emphasis on physical principles, the different magnetometers used for measuring NRM have been divided into three classes as follows: (a) those which utilize static samples, (b) those which utilize rotating samples, and (c) those magnetometers whose sensing mechanism is based on the use of a ballistic galvanometer or a resonance phenomenon.

Before describing these instruments, I should remind the reader of the typical information that is required when the NRM of a sample is being measured. The goal of a paleomagnetist is not only to measure J_n, a scalar, but also to determine the attitude of the ancient geomagnetic vector. The following quantities are therefore measured: M_x the northward (geographic) component of **M**; M_y, the eastward component of **M**; and M_z, the vertical (downward) component of **M**. Thus,

$$(M_x^2 + M_y^2 + M_z^2)^{1/2} = |\mathbf{M}| = J_n V$$
$$\arctan(M_y/M_x) = D \text{ (paleodeclination)}$$
$$\arctan M_z/(M_y^2 + M_x^2)^{1/2} = I \text{ (paleoinclination)}$$

Furthermore, using the dipole approximation for the ancient geomagnetic field (geocentric axial dipole hypothesis), it is possible to relate (McElhinny, 1973) the paleoinclination (I) of the specimen to the paleolatitude (λ) of its location through the following formula:

$$\tan I = 2 \tan \lambda$$

The term paleodeclination is a misnomer, however, because except for the last few centuries we have no record of fluctuations in D (the angle between the true north and the geomagnetic north) in the past. What is commonly done, therefore, is to use the present value of the declination at the sampling site to orient the specimen horizontally and thus calculate M_x and M_y. This procedure results in the well-known indeterminacy of paleomagnetism, that of the paleolongitude of the specimen.

2.1. Static Samples

More than forty years ago Johnson and Steiner (1937) designed a magnetometer for measuring the NRM of rock specimens. This magnetometer, meant for static samples, had a detection system based on the astatic principle. Figure 3 shows the basic principle of astatization of a suspended magnet system. The goal is to make the system insensitivie to the extraneous magnetizations (noise) emanating from various other items in the laboratory while retaining a very high degree of astatization (dynamic response) with regard to the magnetization (signal) of the rock specimen alone. The magnitude of astatization is expressed as the ratio $|P|/\langle P \rangle$, where $|P|$ is the modulus of the moment of the magnet closest to the rock specimen and $\langle P \rangle$ is the modulus of the moment of the system as a whole (see Fig. 3). This magnitude can frequently be as high as 5000–10,000.

Figure 3a shows the basic features of the original astatic system, which utilizes two small (approximately 1 cm) permanent magnets and a mirror (m). The whole structure is suspended with a phosphor bronze fiber of high torque sensitivity. This type of a system was developed to perfection by Blackett (1952) and was used by him in the famous negative experiment in which he vainly searched for a purely rotation-induced magnetic moment for macroscopic bodies. The rock specimens commonly used are cylinders of 2.5-cm diameter and 2.5-cm height, although for strong rocks ($J_n \geq 10^2$ A m^{-1}) the appropriate ratio of height/diameter for shape isotropy is ~ 0.9 (Stacey and Banerjee, 1974). Alternative loca-

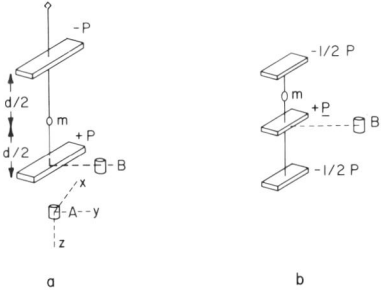

FIG. 3. Construction principle of the classical astatic (a) and parastatic (b) magnet systems. (a) Sketch of two permanent magnets whose magnetic moments are $+\mathbf{P}$ and $-\mathbf{P}$, respectively. The three-magnet parastatic system (b) also has a net magnetic moment, $\langle P \rangle \simeq 0$. The usual locations of a cylindrical rock specimen are indicated as A and B, and the mirror as m.

tions A and B are indicated in Fig. 3a for the specimens, while there is usually only one location used for the three-magnet system (B in Fig. 3b).

An astatic magnetometer is designed to have minimal noise due to magnetic sources other than the rock specimen because the specimen at positions A or B (Fig. 3) is much nearer to the $+\mathbf{P}$ magnet than the other(s) and hence induces a net torque that is measured, for example, by a lamp and scale utilizing the mirror (m) of the suspension. This distance between $+\mathbf{P}$ and the specimen is usually 2–10 cm. In contrast, the extraneous noise sources are usually tens of meters away from both the positively and negatively oriented magnets, and in theory the net extraneous torque on the suspension is zero. In reality, however, this is not so and all sensitive magnetometers (e.g., those that measure the NRMs of sedimentary rocks) must be placed in controlled field-free (i.e., ambient fields less than 10 nT) spaces. It is common, therefore, to build such instruments away from urban environments and install Helmholtz coils with negative electronic feedback to cancel the ambient field and its fluctuating gradients.

Under such conditions, at position A (Fig. 3a) NRM components M_x and M_y can be measured by placing the specimen one way and then rotating it about a vertical axis by 90°. To measure M_z at position A, the specimen must be rotated about a horizontal axis by 180°. At position B the net torque is twice as great as that at position A, although for symmetry reasons error signals due to an inhomogeneous distribution of magnetic carriers in the rock are also greater in position B. Position B in a three-magnet system is most useful at a place where interference from fluctuating field gradients is high. As Pozzi and Thellier (1963) have pointed out, the three-magnet system is highly insensitive to such fluctuations in the ambient field gradient.

Astatic magnetometers used to be direct-reading instruments where a lamp-and-scale method was applied to measure the unrestored torque on the magnet system. In this mode of operation, the critical parameter for increased sensitivity is $|\mathbf{P}|/\langle I \rangle$, where $|\mathbf{P}|$ is the scalar magnitude of the magnetic moment of the magnet nearest to the specimen and $\langle I \rangle$ is the scalar magnitude of the net moment of inertia of the suspension system. Extensive and detailed discussions of the methods for increasing the signal-to-noise ratio for such unrestored torque systems can be found in the work by Collison et al. (1967). Modern astatic magnetometers that utilize negative electronic feedback render such considerations unnecessary. However, the magnitude of $|\mathbf{P}|$ is still crucial for obtaining a high sensitivity. The new generation of rare earth–cobalt magnets with unprecedented values of $|\mathbf{P}|$ and coercivity (H_c, which ensures long time stability of $|\mathbf{P}|$) have already been used to build astatic magnetometers (Deutsch et al., 1967). Examples of astatic magnetometers with negative feedback and the analysis of their signal-to-noise ratio can be found in articles by Roy *et al.* (1972) and deSa *et al.* (1974). The increased dynamic range, short time constants (approximately seconds) at high sensitivity, ability to convert electronically from a dc to an ac output, and choice of analog or digital form of the output—all of these advantages make it unlikely that direct reading instruments of the classical types will ever again be popular. Roy *et al.* (1972) claim a sensitivity of 10^{-5} A m^{-1} for a time constant of 5 sec, which can indeed be achieved in the absence of mechanical vibrations and any other nonelectronic noise sources. As I shall show later, for stronger rocks (10^2 A m^{-1} down to 10^{-3} A m^{-1}) it is not necessary to employ an astatic magnetometer. However, for the weakest sediments and some exotic samples, either an astatic magnetometer or a cryogenic magnetometer is still necessary. The latter will now be described.

The word "cryogenic" to describe this newest arrival in the field of NRM measurement is inappropriate, if not incorrect. This is so because the magnetometer does not, of itself, generate low temperatures, but its sensing mechanism requires an environment maintained close to 4.2°K, i.e., the boiling point of liquid helium at 1 atm. The sole commercial manufacturer of this instrument calls it a superconducting rock magnetometer, but the word cryogenic seems to have caught the imagination of the users of vernacular and I too am forced to use it in this work for the sake of general comprehensibility.

As mentioned in Section 1, the physical principles and diverse applications for the cryogenic magnetometer have been dealt with in detail by Goree and Fuller (1976). I will recapitulate here the basic principles insofar as they throw some light on the signal detection principles and the specific advantages or limitations of the instrument in the measurement of NRM. The heart of a cryogenic magnetometer lies in its sensor—a superconducting quantum interference device (SQUID). A SQUID is based on a wafer in which a dielectric barrier a few hundred angstroms thick separates two superconductors. Josephson (1962) predicted that in such a wafer (now also called a Josephson junction) below the critical temperature for the onset of superconductivity, weakly bound electron pairs would tunnel through the dielectric barrier and give rise to discrete quantized magnetic flux jumps, ϕ_0, whose magnitude is 2.07×10^{-6} A m^{-1} (2.07×10^{-9} G cm^3). It was for this prediction that Brian Josephson was awarded the Nobel Prize in physics in 1973. In the presence of a magnetic field such as that emanating from a rock sample, there exists a quantum mechanical phase difference between the periodically varying supercurrents due to the electron pairs residing in the two superconductors separated by the barrier. As a result there is an interference effect that is the origin of the acronym SQUID. However, the sensitivity of a SQUID used directly as a flux switch is not high enough to allow it to compete with an astatic magnetometer. The commercially available superconducting rock magnetometer employs two additional steps in order to increase the sensitivity. The two steps employ physical concepts of superconductivity and radio-frequency (RF) amplifications that will be briefly introduced here.

First, the Meissner effect in superconductivity is the phenomenon of perfect exclusion of excess ambient magnetic field from the inside of a superconducting ring. When such a ring (or a SQUID) is cooled to its critical temperature for the onset of superconductivity, whatever field was originally present inside the ring is trapped within, and if the ambient field is increased (e.g., due to the presence of the NRM of a rock), a supercurrent of the appropriate sense is produced in the perfectly diamagnetic ring so that the excess flux is completely excluded from entering the ring. However, the critical point of a superconductor is also a function of the ambient magnetic field, and upon increasing the ambient field to a critical value (H_{crit}) the superconducting ring reverts to its normal state. The Meissner effect then vanishes, excess flux enters the ring, and according to Lenz's law, normal current of the appropriate sense (opposite to what was mentioned before) would again flow in the ring so as to decrease the internal flux below H_{crit}, and the ring would again become superconducting. If the excess flux is properly set taking into consideration the geometry and composition of the ring, a sawtooth wave of supercurrent can be produced in such a ring or a SQUID. Second, similar to what is done in fluxgate magnetometers, if the SQUID is driven by a RF (say, 30 MHz), the superimposed sawtooth current due to a dc bias (i.e., the NRM of the rock specimen) produces a series of periodic voltage spikes at the same RF coil, now used as a detector. A feedback loop can then be used to lock the detector to one peak of the periodic response, and a linear output as a function of applied dc bias results. The main difference between the operation of a flux-gate magnetometer and that described above is the a flux-gate sensor senses the dc bias itself, while in the arrangement described here, what is sensed is the (quantized) incremental flux due to the sawtooth current.

Figure 4 is a sketch of the most popular version of the two-axis, 3.8-cm-access diameter cryogenic magnetometer. Although the specimen environment is maintained at close to room temperature, the two SQUID sensors, the two superconducting pickup coils (one set for measuring M_x and the other for M_z), and a superconducting magnetic shield require temperatures $\leq 30°K$ for their operation. A superinsulated (liquid nitrogen-free) Dewar flask is therefore utilized for housing the sensing system, which is maintained at low temperatures by means of the exchange gas

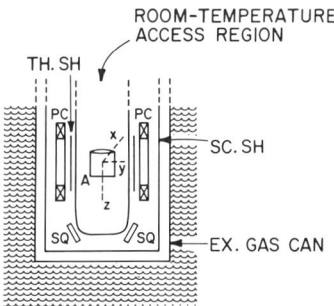

FIG. 4. Diagrammatic sketch of the specimen–sensor region of the superconducting rock (cryogenic) magnetometer. A, specimen; PC, pickup coils for sensing M_z. M_y pickup coils are not shown. SQ, SQUID sensors; TH.SH, thermal reflector shields; SC.SH, internal superconducting shields for excluding external flux from the specimen–sensor region; EX.GAS CAN, exchange gas (helium) can for cooling the pickup coils and the SQUID sensors to temperatures $\leq 30°K$. Liquid helium is indicated with wavy lines. The liquid helium Dewar and the outer Mumetal shield are not shown here.

(helium) can as shown in Fig. 4. Dewar flasks are available with 15- and 30-liter capacities as well as with a coupled helium refrigerator for decreasing helium boil-off.

Figure 5 shows the sensing system. As a specimen is introduced into the detection zone, the magnetic moment components M_x and M_z produce changes in flux in the appropriate pickup coils. These extra fluxes

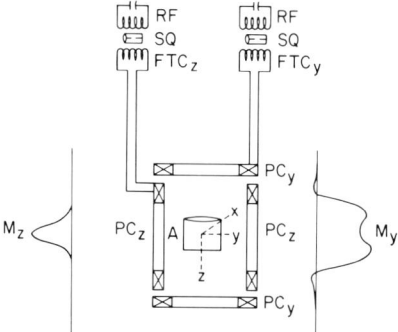

FIG. 5. Schematic illustration of the specimen–sensor geometry and M_z, M_y voltage outputs in a cryogenic magnetometer. PC_z and PC_y are superconducting Helmholtz coils that pick up the specimen flux and deliver it to FTC_z and FTC_y—the field transfer coils for the z and y axes. The use of the SQUID sensors (SQ) and RF circuitry (RF) results in sampling M_z and M_x, shown on either side of the diagram as a function of specimen travel along the z axis.

are delivered to the field transfer coils, which are actually wound on the SQUIDs. An amplification factor of the order of 25 can thus be easily produced in the flux (or incremental dc bias) seen by the SQUIDs. The RF-driven SQUIDs then help deliver a linear output by means of the mechanisms outlined earlier. The sketched voltage outputs for M_z and M_x result when the specimen is slowly traversed along the vertical axis at the center of the Helmholtz pickup coils. M_y is measured by rotating the sample by 90° about the vertical axis. Because of cross-coupling between the Helmholtz coils, it is necessary to devise a measurement scheme so that each component of **M** is measured at least once by each of the two sets of Helmholtz coils. The complete apparatus as shown in Fig. 3 is placed inside a single-layered conventional Mumetal shield so as to decrease further any magnetic field emanating from the laboratory environment.

The claimed sensitivity of the SQUID sensors is 7×10^{-6} A m^{-1} (7×10^{-9} G) but various unavoidable factors reduce it in routine usage. Such factors include instrumental drift, which has been found to be enhanced by a partially filled Dewar flask, undesirable base level signal from the Mylar insert tubes and sample holders (the latter usually made from Delryn plastic or silica), and other causes. A common realistic value for sensitivity is perhaps 2×10^{-5} A m^{-1} (2×10^{-8} G) when a 2.5×2.5 cm cylinder is used as a specimen. One great advantage of the cryogenic magnetometer, however, is its short time constant (1 sec). In this respect, although its ultimate sensitivity is no better than the best astatic magnetometers, measurements are much faster (time constant of 1 sec versus 5–30 sec). As to the dynamic range, the available range in the commercial cryogenic instruments is $10-10^{-5}$ A m^{-1} ($10^{-2}-10^{-8}$ G), approximately two orders of magnitude less than that available from the best astatic magnetometers with negative electronic feedback.

To summarize this section on NRM-measuring magnetometers for static samples, it must be admitted that for short time constant, high sensitivity, and the lack of need for special instrument housing away from urban areas, the cryogenic magnetometer is a much more desirable, if expensive, instrument than the astatic magnetometer. And perhaps it will remain so for some time to come.

2.2. Rotating Samples

A radically different approach to the measurement of NRM is exemplified by magnetometers that use rotating specimens of the rocks. Making the same assumption about homogeneity as in the case of static samples, we consider the net magnetic moment **M** of the rock to reside in its geometrical center. Using Faraday's law of induction, the time derivative of **M** for a rotating sample is related to an observed voltage **E** as follows:

$$\mathbf{M} \cdot C = \mathbf{E}$$

where C is a lumped coefficient of proportionality that depends, among other things, on specimen–sensor distance and symmetry as well as the specific constructional geometry of the sensor(s).

Although magnetometers based on the above principle (and appropriately called "rock generators") were built as early as 1938, the first modern version was built in 1948 (Bruckshaw and Robertson, 1948). In contrast to the direct-reading astatic magnetometers, the rock generators had an immediate advantage in being based on an ac method of detection since electronic noise or unwanted signals from static laboratory objects could be minimized by detecting only the in-phase ac signal. With the passage of time, electronic tube circuitry gave way to the much more sensitive solid-state electronics for signal processing. Hand in hand went the development and diversification of sensors. As a result we now have a family of magnetometers that utilize rotating samples. These have high accuracy if not the highest sensitivity, but much more important is the fact that these instruments can be housed in a conventional laboratory environment with no need for nonmagnetic huts outside the city, which is ideally required for astatic magnetometers (direct-reading or with negative electronic feedback). Although in principle these magnetometers based on the induction law do not need a field-free space for their operation, it has been found to be convenient to cancel the ambient field to, say, 0.2% so that the Earth's field-induced VRM of the rock (or in-phase extraneous signals due to rotation-induced vibration of the sensor) cannot degrade the signal-to-noise ratio. The early versions of the instrument depended heavily on high frequency (100–500 Hz) for reasonable signal-to-noise ratio. Descriptions of these now obsolete instruments can be found in the work by Collinson *et al.* (1967). Excellent phase-sensitive detection by lock-in amplifiers or on-line data processing that uses dedicated minicomputers (and more recently, microprocessors) has made it possible to use highly accurate but low (1–10 Hz) frequencies, and the following discussion will be limited to these. These modern versions can be divided into categories depending on the nature of their sensors: pickup coils or flux-gate magnetometers.

Figure 6 illustrates the basic principle behind these modern rock generators, which are now called spinner magnetometers. Figure 6a shows a rotating sample A situated inside a set of pickup coils (PC). In this illustration the z axis of the Cartesian coordinates has been chosen as the rota-

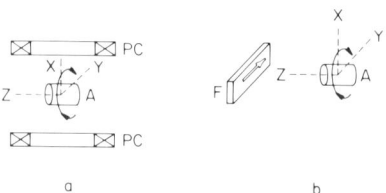

FIG. 6. Schematic diagrams to illustrate the principles of the two common types of spinner magnetometers. (a) Sketch of the pickup coil (PC) sensing system; (b) the fluxgate sensor (F). The sample A and its rotation direction is indicated.

tion axis. As the sample rotates, the NRM components M_x and M_y each alternately go through positions of maximum and minimum coupling with the pickup coil and a periodic ($\sin \theta$) output of **E** results, the period being equal to the rotation period of a motor-driven shaft that houses the rock specimen inside a chuck mounted at one end. At the other end of the shaft a device is mounted that provides a reference signal for the phase comparison of **E**. The device commonly consists of a slotted disk, a small light source, and a photomultiplier tube, although small magnets and associated pickup coils have been used in the past. From a single rotation, single readings for M_x and M_y are obtained using a lock-in amplifier with two analog or digital meters. Using six successive independent orientations of the rock (the so-called six-spin method), four independent measurements for M_x, M_y, and M_z can be obtained, from which the net vector **M** can be accurately determined. More approximate determinations of **M** can be obtained from a four-spin method, but the two-spin method (which is a minimum) is not recommended. The sample and pickup coil geometry shown in Fig. 6a is only one of the many orientations used by different commercial manufacturers and independent investigators. The orientation of Fig. 6a is the best one for canceling errors due to an inhomogeneous distribution of magnetic carriers in a rock specimen.

Figure 6b illustrates the sensing mechanism of a commercially produced spinner magnetometer that employs a RF-driven flux-gate magnetometer (sensitivity = 0.5 nT). Unlike pickup coils, flux-gate magnetometers need cancellation of the Earth's field to approximately 1 part in 1000 for satisfactory operation. Otherwise the dc bias on the flux-gate is so high that it saturates it and discrimination of the signal from the rock specimen cannot be made. Six-layered mumetal shields are used in a commercial instrument, although a carefully designed three-layered shield should be sufficient for most use. An advantage with a flux-gate spinner as shown above is that the dynamic range can be increased by varying the distance from the specimen to the sensor. A dynamic range of $\sim 10^6$, spanning signals as high as 5×10^3 A m^{-1} (5 G) down to 10^{-3} A m^{-1} (10^{-6} G), can be easily obtained in an instrument of the type described here. It is important to point out at this stage that the sensor–specimen geometry of Fig. 6b is much more conducive to the picking up of error signals due to magnetic inhomogeneity than the geometry of Fig. 6a. Several attempts have therefore been made to use more than one flux-gate or ring-shaped flux-gates to decrease this problem, which naturally becomes more severe when the NRM to be measured is weak ($<10^{-2}$ A m^{-1}). Collinson (1975) has given references to such approaches by individual workers.

The detection and display of phase-sensitive signals (e.g., of M_x and M_y in Fig. 6b) in a flux-gate spinner magnetometer is usually the same as that in the pickup coil-using machines. A modern trend, perhaps brought on the the challenge of a new type of rotating sample magnetometer to be described in the next paragraph, is to provide on-line data handling through the use of dedicated minicomputers. Such computer-aided magnetometers can directly apply (field) attitude corrections to the specimen and provide calculated declination, inclination, and intensity of the paleomagnetic vector. It is important to emphasize that although such procedures help remove subjective errors caused by the observer, the results are not immune from electronically produced random errors, and hence, frequent checks should be included in the software used for data reduction.

The final category of rotating-sample magnetometers to be discussed in this section is a fairly new arrival on the scene (Molyneux, 1971), and like the cryogenic magnetometer for static samples, this new magnetometer has made a significant improvement in our NRM-measuring capabilities. It has been called the complete results magnetometer, which is neither a happy nor a specific enough title for this instrument. As shown in Fig. 7a, the rotating rock specimen is completely enclosed by a ring-shaped

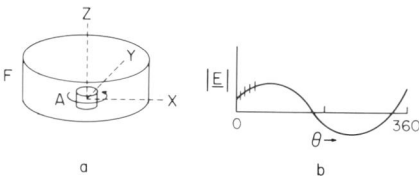

FIG. 7. (a) Sketch of the location of the rotating sample A *inside* a ring-shaped flux-gate magnetometer. For a rotation as indicated about the z axis, M_x and M_y can be measured. (b) The continuous sin θ output of the voltage $|E|$ measured by the flux gate for a complete 360° rotation of the rock specimen. Tick marks on the $|E|$ curve signify how 128 such discrete samplings of $|E|$ are used to digitally process the data for the actual calculation of M_x and M_y for the case shown in (a).

flux-gate magnetometer. From a rotation about the z axis, M_x and M_y can be determined. A four-spin method is commonly used to calculate the average values of M_x, M_y, and M_z, although a six-spin method is preferred so that the associated errors can be properly evaluated. The previously mentioned minicomputer, in association with a phase-sensitive reference signal from the rotation shaft, is fed with 128 discrete samplings of the signal for a full 360° rotation. The data from each rotation are stored until 2^n rotations ($n = 4-11$) have taken place. The minicomputer analyzes the data and yields M_x and M_y values (for the case shown in Fig. 7a). The exponent $n = 4$ corresponds to a spinning time of 2 sec, while $n = 11$ corresponds to 5 min. Since the rotation frequency is low (7 Hz), the specimen can be placed simply on a horizontal table top without the need for a sample holder. This results in reduced noise from magnetic impurities that reside on a sample holder however "clean" it may be. The time required for changing specimen orientation is also reduced since a specimen holder need not be fitted into a chuck for each spin.

The sensitivity, accuracy, and dynamic range for each of the three types of magnetometers described in this section cannot be compared easily and quantitatively. As Collinson (1975) has observed correctly, the manufacturers' claims about sensitivity cannot be compared readily because information about measurement time is usually not provided. Even if a magnetometer is sensitive enough to detect a signal as weak as 10^{-6} A m^{-1} (10^{-9} G), and the time constant for measurement along each axis is high, say, 300 sec, it is not a practicable proposition to attempt the measurement of the average NRM of only one collecting site since it may require as long as 10 hr in all. Another very real problem about long time constants is shot noise, which can render the time for a single spin measurement to three to five times the recommended time constant. Therefore, if one incorporates the relevant information about time constants with the claimed sensitivities, it appears that there is little to choose between the older flux-gate and the modern ring-shaped flux-gate spinner magnetometers. For routine usage with a large number of specimens, their sensitivity is the same, of the order of 10^{-4} A m^{-1} (10^{-7} G). The pickup-coil spinner magnetometer is definitely less sensitive, perhaps by one order of magnitude, suffering as it does from electrostatic charging, mechanical vibrations, etc. If a six-spin method is used at a signal level of about 10^{-2} A m^{-1} (10^{-5} G), all three of the instruments are equally accurate in terms of reproducibility, but the pickup-coil spinner magnetomer fares worse when smaller signals must be measured. The effective dynamic range is thus lower for the pickup-coil spinner magnetomer, about 10^6, while that for the other two is 10^7 or better.

It is important to end this section by emphasizing that even for most igneous rocks that have relatively large NRM values ($\geq 10^{-2}$ A m^{-1}), one requires as much sensitivity and accuracy as when dealing with much weaker rocks. This is where I disagree with Collinson (1975), who wrote: "The ultimate 'noise level' or 'sensitivity' . . . down in the 10^{-9} G range, may not be of much interest to the paleomagnetist. . . ." I do not think this is so because in the present "second generation" period of paleomagnetic and rock magnetic studies involving weak lake sediments, equally weak lunar rocks, and meteorites and the separation or "stripping" of multiple components of NRM in Precambrian rocks, we routinely need magnetometers that can reliably measure *components* of NRM as small as 1% or less of the original as-is NRM. There is no substitute for magnetometers with the highest possible sensitivity and accuracy.

2.3. Ballistic and Resonance Magnetometers

The ballistic and resonance magnetometers are two less common versions of NRM-measuring approaches. In the ballistic magnetometer, the sensing principle is based on Faraday's law of induction, and in the common version of it the sample is rapidly translated from one pickup coil to another. The resonance magnetometer is a hybrid instrument in which amplification of the deflection of a direct-reading astatic magnetometer is achieved by rotating the sample at the natural frequency of the suspended magnet system, and resonance is thereby achieved. Neither of these instruments claim to have sensitivities better than 10^{-1} A m^{-1} (10^{-4} G) and therefore cannot compete with the instruments described in Sections 2.1 and 2.2. They are included here partly for the sake of completeness but partly to being out the unique abilities of the ballistic magnetometer. The value of these unique characteristics will be appreciated more when we discuss artificial or laboratory-imparted remanent magnetization in Section 3.

As already mentioned, the guiding principle of the ballistic magnetometer is Faraday's law of induction, which states

$$\dot{\mathbf{M}} \cdot C = \mathbf{E}$$

which is exactly the same principle utilized for rotating samples. **M** is the net magnetic moment of the sample, C is a coefficient of proportionality depending on the coupling of the magnetic flux lines with the pickup coils and **E** is the induced voltage, commonly measured with a galvanometer (see Fig. 8). For greater sensitivity the pickup coils are wound in series opposition so that the voltages sensed by the two coils are added, resulting in an improvement in sensitivity by a factor of 2.

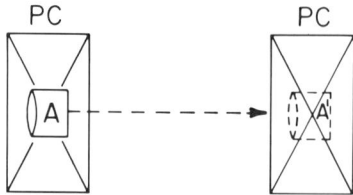

FIG. 8. Basic principle of the ballistic magnetometer. The specimen A is suddenly translated from one pickup coil (PC) to the other, to the new position A'. The rapid translation produces a voltage **E** in the pickup coil system (series wound in opposition) and measured with a ballistic galvanometer.

Thellier (1938) constructed one of the earliest versions of this instruments for paleomagnetic studies involving archaeological samples. Nagata (1967) has provided a detailed account of its various modifications. Dunlop and West (1969) have described a ballistic magnetometer that was very successfully used by them to make a temperature-dependent study of laboratory-imparted remanent magnetizations, e.g., ARM and VRM. It is easy to appreciate the reasons for the relatively poor sensitivity (10^{-1} A m^{-}) of the ballistic magnetometer: the time derivative of **M** cannot be increased indefinitely by simple rapid translation and the number of turns of the pickup coils cannot be increased beyond the level of appreciable Johnson (thermal) noise. We have seen earlier how the spinner magnetometers with lock-in amplifiers or dedicated minicomputers bypass this problem.

The resonance magnetometer was designed to avoid some of the problems faced by the early workers in rock magnetism and paleomagnetism. The most sensitive instrument at that time (during the 1940s and 1950s) was the direct-reading astatic magnetometer. However, theoretical calculations by Blackett (1952) showed that even under the best of circumstances, a direct-reading astatic magnetometer will require 25 min to measure $|\mathbf{M}| = 5 \times 10^{-6}$ A m^{-1} (5×10^{-9} G). This is because the sensitivity is proportional to T^2, where T is the period of oscillation. For a sensitivity of 5×10^{-6} A m^{-1}, Blackett (1952) recommends $T = 30$ sec. Now in a resonant oscillator the deflection (θ_r) at resonance is given by

$$\theta_r = Q\theta_0$$

where θ_0 is the deflection at zero driving frequency and $Q = \langle I \rangle \omega/\lambda$, where $\langle I \rangle$ is the moment of inertia of the system, ω is the natural frequency of the system, and λ is the damping coefficient.

It was, therefore, natural to investigate (Kumagai and Kawai, 1953) whether a sensitivity as high as that of a long-period astatic magnetometer could be achieved in a short-period instrument operated under resonant conditions. The best magnetometer of this type has been described by Farrell (1967), who took pains to control the rotation frequency of the rock specimen as well as use auxiliary techniques to adjust the natural frequency of the magnet system to match the specimen frequency as closely as possible. However, the results were disappointing because the highest sensitivity under optimum conditions turned out to be only about 10^{-3} A m^{-1} versus 10^{-5}–10^{-6} A m^{-1} corresponding to the best direct-reading astatic magnetometers. The reason lies in the noise imposed by mechanical vibration, which according to Farrell (1967) causes noise that is 50 times greater than that due to electronic sources. Another drawback of the resonant magnetometer is that it cannot be operated at its highest sensitivity in an urban laboratory environment. This was unexpected because of the ac detection system, but the instrument is highly sensitive to fluctuating field *gradients,* if not the field itself. Finally, from general experience with claims of instrumental sensitivity, it would perhaps be correct to say that for routine usage, one should expect a sensitivity of about one order of magnitude less than the highest sensitivity (10^{-3} A m^{-1}) claimed by Farrell (1967).

In summary, the ballistic and resonance magnetometers are interesting departures from the more common instruments for measuring NRM of rocks, but in terms of high sensitivity they are much less desirable than the commercially available spinner and astatic magnetometers.

2.4. Selective Demagnetization or "Cleaning" Methods

Up until the latter half of the 1950s, methods for selective demagnetization of unstable (and hence, undesirable) components of NRM were not in use. It was perhaps implicitly believed that there were "good" rocks whose magnetic carriers had preserved stably the direction of the paleomagnetic vector and there were "bad" rocks that, although known to be old from geological information, turned out, upon measurement, to contain a NRM direction parallel to the present-day field. However, this is not so. In general, a rock specimen carries an ensemble of magnetic grains that, even if chemically homogeneous, can have a large range of volumes and magnetic states. Thus, for a uniformly magnetized or single-domain state, the grains can be superparamagnetic when, by virtue of their small volumes, the NRM is randomized by thermal (or Boltzmann) energy kT over laboratory time scales or geological time scales. Such a grain must be large enough to overcome the thermal perturbation in order to preserve stably the orientation of the paleomagnetic vector.

Then there are even larger grains whose instability arises from a change in their magnetic state. Because of the need to minimize the increasing magnetostatic energy, if a single-domain grain grows in volume, it first passes into the pseudo-single-domain state when there are only a few, say, one to four, domains in the grain and then the grain becomes truly multidomain, with a multiplicity of domain walls separating many uniformly magnetized but differently directed domains. A multidomain grain has little magnetic stability since thermally activated movement (usually a translation) of one or more of the domain walls over geological or laboratory time scales is very likely. The single- and pseudo-single-domain grains, on the other hand, have much higher stability, and it is the NRM components contributed by them that are sought out by selective demagnetization techniques. Complete quantitative descriptions of the above phenomenological behavior can be found in textbooks on rock magnetism, (e.g., Stacey and Banerjee, 1974). Following L. Néel, the time-dependent behavior of NRM in a zero ambient field can be expressed as

$$J_n(t) = J_n(0) \exp(-t/\tau)$$

where τ is the relaxation time constant. The relaxation time constant in turn can be expressed as

$$\tau = (1/C) \exp(-E_b/kT)$$

where C is a large frequency constant (10^9 sec^{-1}) that can be taken as invariant with temperature; E_b, energy barrier opposing relaxation; k, Boltzmann's constant; and T, ambient temperature in absolute scale.

The goal of the various selective demagnetization procedures is to preferentially remove the contributions of grains with low values of τ from the net J_n in order to isolate the stable primary component that dates from the genesis of the rock in question. It should not escape the reader's notice that it is implicitly *assumed* that the grains with the highest values of τ also are the ones responsible for the primary component of NRM. Experience tells us that for simple cases (e.g., a primary TRM with an overprint of IRM or partial TRM due to modest heating on burial) this is indeed true, but for old rocks such as those of Precambrian age such an approximation may not be valid. With this warning in mind, we shall turn our attention to the specific demagnetization techniques.

Zero-field demagnetization is none other than an effort to put the exponential time dependence of $J_n(t)$ to practical use. Commonly it is also known as "storage test." The individual specimens are measured for the directions and magnitudes of their natural magnetic moment **M** and then placed in an environment where the ambient laboratory field has been canceled to a very large degree. Using a set of Helmholtz coils and no feedback, the roughly 50,000-nT ambient field can be canceled to 0 ± 100 nT. Although this may be acceptable in many cases, one must manually adjust the currents in the Helmholtz coils from time to time, which is bothersome. Negative electronic feedback can easily reduce the residual field to ±10 nT and expensive commercial instruments are available that can reduce it further to ±0.2 nT. In general, however, Helmholtz coils are not popular because of their large size. Their presence may reduce significantly the working space in a medium-sized laboratory. The alternative is to use Mumetal magnetic shields. A three-layered Mumetal shield can easily reduce the ambient field to ±20 nT, while a commercially available six-layered shield can provide a residual of ±5 nT. Obviously, a completely enclosed shield (i.e., a box with a lid) is preferred to one that has an open end. Mumetal shilds have to be carefully protected from external stress and strain, which can give rise to "hot spots" or regions carrying a net remanence, thus degrading the field cancellation capability.

As mentioned earlier, the effectiveness of zero-field demagnetization can be variable. Thus, if one measures the NRM of a specimen and re-measures it after storage in zero field after 1, 2, 4, 8, . . . days, in geometric progression, the undesirable secondary components of NRM acquired, say, since collection in the field can usually be removed. But for the removal of secondary magnetizations acquired *in situ* in the field, it may become necessary to store the specimen in zero field for periods longer than a few months, which is inconvenient. However, an inspection of the equation governing τ shows that it is exponentially dependent on $-1/T$ and therefore highly sensitive to an increase in ambient temperature. Thus, even a modest increase of 25% in the ambient absolute temperature (300°K) to 375°K (102°C) can make zero-field demagnetization significantly more effective in removing VRM of moderate values of τ. One must keep in mind, however, that not all rocks may resist chemical alteration on heating to 102°C. The most attractive thing about zero-field demagnetization is that the procedure duplicates very closely the very opposite of that that caused the secondary components of NRM. Thus it is a truly selective process that leaves the primary remanence more or less alone, which cannot be said for alternating-field (AF) demagnetization, for example. The chief application of zero-field demagnetization, therefore, is in the case of rocks that carry a small primary component of NRM swamped by a very large value of secondary components with small (days) to moderate (months) values of τ. Otherwise, the procedure can be looked upon as a storage test that shows whether, for example, the rock in question is *likely* to acquire secondary components such as VRM. Such a test should be applied routinely to all or a representative batch of specimens under study.

Thermal demagnetization is considered next since from the point of view of physical principles, it is the natural successor to zero-field demagnetization. Indeed, if one accepts the first-order approximation of the theory of TRM (J_{tr}), i.e., an assembly of noninteracting single domain grains, the total TRM can be expressed as

$$J_{tr} = \sum_i J_{tr}(v_i)$$

where the summation is extended over the whole range of grain volumes (v_i) present. For a given value of v, the magnetic energy barrier E_b can be expressed as

$$E_b = vK_{eff}$$

where K_{eff} is anisotropy energy per unit volume whether the anisotropy has a shape, magnetocrystalline, or magnetostrictive origin. By referring to the equation for relaxation time constant τ, we can see that for a given value of K_{eff} and v, the room temperature value of τ is a constant quantity. Conversely, if we fix the value of τ as the time of formation of the rock specimen, we can calculate a corresponding value of T, which is called the blocking temperature T_b. According to L. Néel, τ depends exponentially on the parameter E_b/kT. As an example, a sphere of magnetite whose volume is 5.86×10^{-17} cm³ (diameter $d \simeq 500$ Å) has a τ of 1 sec at 300°K but an increase in volume by a factor of 3 to 18.34×10^{-17} cm³ increases τ to 3×10^9 years. Again conversely, if this larger grain of $v = 18.34 \times 10^{-17}$ cm³, which obviously can carry a stable TRM ($\tau = 3 \times 10^9$ years), is heated to high temperatures such as two or three times the room temperature, τ will decrease to values that are measured in minutes or seconds.

The above strong dependence of τ on T (for given values of v) is utilized in the following way in thermal demagnetization of rocks. A rock specimen is placed in a noninductively wound furnace, which, in turn, is placed inside a carefully controlled field-free space (Roy *et al.*, 1972). The specimen is heated to a given temperature some 50 or 100°C higher than room temperature (23°C). After holding the specimen at the elevated temperature for 10–15 min in order for the specimen to equilibrate, the heating current in the furnace is turned off, a cooling fluid such as forced air or circulating cold water is turned on, and the specimen is cooled back to room temperature in a field-free environment. Figure 9 is a schematic illustration of a thermal demagnetization apparatus. Under perfect conditions (e.g., no chemical change and resultant creation of a new magnetic phase on heating) in the absence of an appreciable (~ 50 nT or greater) field during cooling, such a step as described above randomizes the components of a natural TRM that were blocked, i.e., acquired in nature

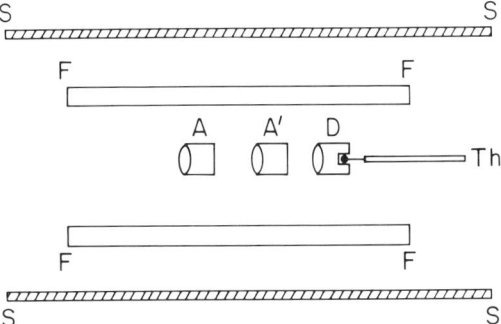

FIG. 9. A set of specimens A, A', etc., suitably separated from one another's magnetic influence, is placed in a water-cooled noninductively wound furnance (FF). A thermocouple (Th) is placed in a hole inside a "dummy" (D) nonmagnetic specimen of thermal conductivity similar to that of the rocks in order to estimate the temperature of A, A'. The magnetic shield SS cancels the Earth's field seen by the rock specimens.

during cooling in the Earth's field in the elevated temperature interval where grains with the smaller volumes and smaller values of τ acquired their remanence. By demagnetizing them preferentially, one removes the components of remanence (TRM) that have been acquired most recently, i.e., the secondary remanences like IRM or VRM. In an actual experiment such stepwise heating is carried out in order of increasing temperatures and the remaining amounts of J_n at room temperature are measured with a magnetometer. The resultant data of intensity (J_n, the scalar magnitude of **M**) and direction of the changing vector can be displayed as in Fig. 10, from which the critical temperature T' is determined at which most of the low-τ and, by inference, secondary components of

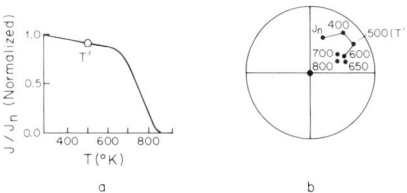

FIG. 10. Changing NRM intensity (a) and direction (b) on thermal demagnetization. The directions are plotted on an equal-area stereonet; the numbers denote heating steps in °K after which remaining NRMs were measured. T' (here equal to 500°K) is the temperature above which most of the remaining signal is primary. In the given example here magnetite must be the carrier of NRM since all the NRM vanishes below 853°K—the Curie point of magnetite.

NRM are remove. Such pilot measurements of complete thermal demagnetizations must be carried out in order to arrive at an optimum T' for demagnetizing a large batch of specimens. In practice, not one value of T' but two or three are selected so that the average direction of the primary paleomagnetic vector at a site can be determined from a tightly clustered group of vector directions on, say, an equal-area stereographic projection. The alternative choices for such projections will be discussed later on in this section after AF demagnetization procedures have been introduced.

It is worth noting that thermal demagnetization, although effective in removing high-coercivity (large τ) components of unwanted remanence, has its drawbacks too. The principal one amongst these is that the observed decrease in J_n on stepwise demagnetization can be due to both physical and chemical processes. The theory outlined above deals only with the physical process of unblocking or randomizing the magnetization of the low coercivity or small grains. Any observed change in J_n that is really due to chemical alteration (e.g., magnetite) will constitute an error and must be recognized as such. Thus the onus is on the experimentalist to prove that there has been *no* chemical alteration on thermal demagnetization. Sometimes this is done by measuring certain intrinsic parameters before and after heating. These include saturation magnetization (J_s), saturation isothermal remanence (J_{rs}), low-field susceptibility (χ), and anhysteretic remanent magnetization (J_{ar}), of which only the last parameter is most closely related to the carriers of stable remanent magnetization. A recent approach has been to control the oxygen fugacity of the specimen environment using a gas buffer system, viz. $CO-CO_2$ or H_2-CO. It should be noted, however, that a mixed-gas system will not be in thermodynamic equilibrium until fairly high temperatures, e.g., $\geq 600°C$, are reached, while most magnetic degradation by chemical alteration starts at temperatures as low as 250°C. Thus the use of controlled oxygen fugacity is no guarantee against chemical alterations due to kinetic processes that occur at temperatures well below that at which thermodynamic equilibrium is achieved. In fact it has been the experience of many an experimentalist that a moderate vacuum ($\sim 10^{-5}$ Torr) combined with *a high pumping rate* is a successful, if qualitative, solution to the problem.

Alternating-field demagnetization is one of the earliest techniques attempted in paleomagnetism to isolate stable components of NRM (As and Zijderveld, 1958; Creer, 1959). Because heating of rocks is not required in this method of demagnetization, it has proved popular with most paleomagnetists who are wary of thermally induced chemical alteration. The magnetic carriers of many naturally oxidized basalts are titanomaghemites ($Fe_{3-x-y}Ti^{\square}_y O_4$), which are metastable compounds. These minerals spontaneously undergo a phase-splitting into a magnetite–ilmenite intergrowth on heating to temperatures as low as 250°C. Neither an oxidizing nor a reducing atmosphere can prevent the phase-splitting and hence AF demagnetization of basalts is a popular choice for removal of secondary low-τ components of NRM. The principle of AF demagnetization can be easily appreciated if we refer to the hysteresis cycles depicted in Fig. 11. If a specimen is placed in an alternating field of peak value H_1, the magnetization of the specimen is described by the cyclic hysteresis loop $+H_1 \rightarrow -H_1 \rightarrow +H_1$. At this stage, single-domain grains whose intrinsic grain coercivities (H_{cr}, also called remanence coercivity) are less than $|H_1|$ will be oriented cyclically and sequentially along the $+H_1$ and $-H_1$ directions. If the field is now *smoothly* decreased to a lower value H_2, the grains with $H_2 < H_{cr} < H_1$ are left magnetized, half along the $+H_1$ and the other half along the $-H_1$ direction. However H_{cr} is given by

$$H_{cr} \propto K_{eff}/J_s$$

and hence a remanence coercivity H_{cr} for grains of a given volume v is related to a unique value of relaxation time constant τ. In other words, just as in thermal demagnetization raising the ambient temperature to a given value of T randomizes the component of remanence controlled by grains with a given value of τ, AF demagnetization achieves the same result by subjecting the specimen to an alternating field of a given peak value H_1. It has been pointed out that the analogy is not perfect because the AF operates along an axis while thermal demagnetization randomizes magnetizations in a spherical symmetry. It is still common in some laboratories to spin the specimen simultaneously about two mutually perpendicular axes. Such "tumbling" AF demagnetization apparatus are now being re-

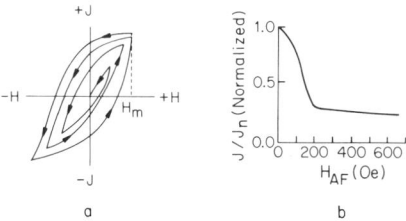

FIG. 11. (a) The traces of the magnetic hysteresis loops when the peak field is smoothly decreased from H_m. The normalized intensities of the demagnetized NRM may follow the behavior shown in (b) when H_m is increased stepwise, e.g., 25, 50, 100 Oe, etc. On the basis of the intensity data alone it would appear that most of the remanence remaining after $H_{AF} = 200$ Oe is primary. It should, however, be confirmed by plotting the directional changes of the NRM vector, as in Fig. 10b.

placed by unidirectional AF demagnetization apparatus based on solid-state components because a sequential demagnetization along three axes in such modern instruments can be shown to produce the same result as in the older tumbling variety.

To return to the complete AF demagnetization process, it was emphasized earlier that the decrease from H_1 to H_2 should be smooth, i.e., slow enough so that there is ample time in between to randomize grains whose coercivities fall between H_1 and H_2. The same process is continued until the AF is reduced to zero (Fig. 11). The procedure is continued in a stepwise fashion to peak fields higher and higher (than H_1) until a complete AF demagnetization curve for intensity as shown in Fig. 11b is obtained. The directions of the remanent vector are also measured and plotted, for example, in a stereographic projection as shown previously for thermal demagnetization in Fig. 10. The importance of using both the magnitude and directional information for isolation of the stable component of NRM in a specimen will be emphasized and clarified in the next paragraph. I shall mention briefly here the sources of error in AF demagnetization. The major source of error is the presence of a dc bias during the decrement of the AF. As it was defined in the Introduction, the simultaneous presence of dc bias and decreasing AF gives rise to anhysteretic remanent magnetization (ARM) in rocks. Thus instead of merely decreasing the total NRM, new and stable components of magnetization (noise) can be added to the rock in the shape of an ARM. The likelihood of ARM pickup can be minimized by careful ambient field cancellation (magnetic shields) and removal of even harmonics from the drive field. Even-harmonic removal is extremely easy with modern solid-state electronics, and commercial instruments are available that can remove even harmonics to 1 part in 10^5. However, if due to parts failure ARM is produced during AF demagnetization, it is easy to detect its presence in the modern single-axis instruments by reversing the orientation of the specimen in the AF solenoid.

We should end the present discussion about the merits of different, commonly employed methods by comparing two of the most popular methods, thermal and AF demagnetization. There exist statements in the literature to the effect that while thermal demagnetization is most effective for red sandstones containing hematite, AF demagnetization is the best one for selective demagnetization of basalts. It is important to appreciate the reasons behind such statements because then one can choose the most effective demagnetization technique for any given rock, be it a nonred sandstone, a shale, or a laterite. Néel (1955) was the first to emphasize that the expressions for unblocking of magnetization are different for thermal and AF demagnetization. Dunlop and West (1969) utilized the

Néel derivations to clarify the difference between the two demagnetization procedures. The two Néel expressions are as follows:

$$vH_{cr} = AT$$
$$[v(H_{cr} - |h|/F')^2)]/H_{cr} = BT$$

where A and B are constants, $|h|$ is the modulus of an applied field, and $F'(T) = H_c/H_{cr}$. H_c, it will be recalled, is the bulk coercivity of an assembly of grains, while H_{cr} refers to an individual grain.

In Fig. 12 the two sets of curves represent the two types of behavior. The solid curves relate to thermal demagnetization and have been drawn for the unblocking condition for given temperatures T. The dotted curves are for constant values of $|h|$, i.e., each curve relates to a given value of peak AF. To understand the demagnetization procedure, one first locates the T curve SP corresponding to the highest blocking temperature (T_b) of the sample above which the magnetic moments are thermally randomized (superparamagnetic), although the actual Curie point (T_c) has yet to be reached. If the remanent magnetizations of a given grain size range are located at AA in the H_{cr}–v space, the purpose of either thermal or AF demagnetization is to translate the curve AA to the right continuously until each point of the whole curve is swept past the T curve SP, corresponding to the highest blocking temperature. It can be seen easily that upon thermal demagnetization to the highest blocking point and beyond, AA will move parallel to SP and when it has gone past SP, *all* the grain sizes (v) will become superparamagentic at the same instant. Now compare the translation to the right of the curve BB with regard to AF demagnetization of a given range of grain sizes in the specimen. Even when a very high peak field (H_1) has been applied and the majority of the grain

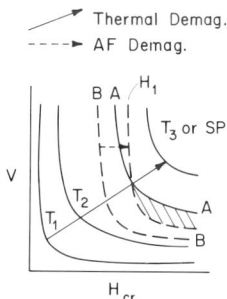

FIG. 12. Representation of thermal and AF demagnetization using (V, H_{cr}) diagrams of L. Néel. The hatched region represents high H_{cr}–low V grains that have not yet been demagnetized by the highest AF used. (↗) Thermal demagnetization; (---→) AF demagnetization.

size ranges are superparamagnetic, there exists a region (shaded in Fig. 12) where the remanence is unaffected and is retained by grains with high values of H_{cr} (remanence coercivity) and small values of v. To an experimentalist, therefore, it would appear on an intensity demagnetization plot that although AF demagnetization to a fairly high value removes the majority of the room temperature NRM, there is a tenacious fraction of remanence carriers that are still unaffected by the demagnetization step. Since hematite grains often have the combination of large H_{cr} and small v, the red sandstones containing fine-grained hematite do show persistent stable remanence on AF demagnetization as compared to thermal demagnetization. The reasons why basalts are usually demagnetized by AF are as follows: (1) magnetite does not have such high H_{cr} values for small v and (2) basalts are liable to oxidation and/or other chemical alteration processes on thermal demagnetization. Behavior such as that of red sandstones mentioned here has been seen for aluminum-substituted maghemite grains in baked laterite. The clue to the observation of such behavior is not the rock type or the chemistry of the grains per se but whether or not they display a high H_{cr}–small v combination that may, in the long run, have a dependence on chemistry.

Since the three previously discussed methods of selective demagnetization constitute the most common ones, it is appropriate at this point to describe how the demagnetization data can be analyzed to provide the most insight. In Fig. 13 I have indicated how the vector additions of NRM components of differing relaxation time constants (τ) result in a net NRM whose stable component can often account for only 1–10% of the net NRM. Each of the components S_1, S_2, and S_3 may be carried by a range of grain sizes, with a distribution of remanence coercivities (or blocking temperatures). If these distributions do not overlap one another,

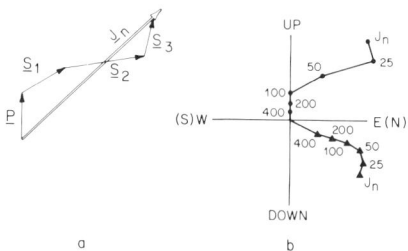

FIG. 13. Changes in the complete NRM vector on AF demagnetization. (a) The primary (**P**) and secondary (S_i) components of NRM (J_n) in one plane, say, up–down and east–west. (b) Diagram showing how the projection of NRM on this plane (solid points) changes with increasing peak fields of demagnetization. The numbers refer to the steps of the peak fields used for AF demagnetization.

only then is it possible to separate the different components by selective demagnetization to increasingly larger values of either temperature (thermal demagnetization) or peak AF (AF demagnetization). Assuming this is so, it is possible to use plotting techniques as in Fig. 10 to isolate the primary (i.e., highest-τ) component of NRM. Both intensity and directional variations on demagnetization need to be studied for clear isolation. If, for example, only stereographic projection of directional changes is used and all the vectors S_1, S_2, and P are coincident, the point at which the secondary components S_1 and S_2 have been reduced to zero and the primary component P is being attacked cannot be identified. A similar case can be made for not just using the intensity variation either. Zijderveld (1967) has suggested a different way of plotting so as to provide the full vector information on demagnetization. This is shown in Fig. 13b. If one plots the projections of the vector on two planes simultaneously, one can, at any state of demagnetization, read off the remaining intensity as well as the direction of the NRM vector. Furthermore, it is easier in a diagram of this type to recognize the direction(s) along which the secondary component(s) had resided. Such information helps not only to separate the secondary components but also to try and identify their ages and origins.

If, however, the τ-spectra of the different components do overlap to some extent, the breaks in the Zijderveld plots (Fig. 13b) become less sharp. Still, as long as the majority of the secondary components have distinctly lower values of H_{cr} or T_b, *and* the primary (higher H_c or T_b) component is not very small to begin with, it should be possible to recognize the primary component as the vector that appears to point to the origin (i.e., the direction remains the same on demagnetization but the intensity decreases monotonically). Halls (1976) has described a method called the "remagnetization circle" in which only stereographic projections of directions are used to separate the stable direction. Hoffman and Day (1978) have described a slightly different technique that utilizes vector differences between two demagnetization steps rather than the diminishing vector itself. Another type of analysis which can be helpful as a complementary tool is the plotting of a directional scatter parameter (k, the precision parameter of McElhinny, 1973) as a function of demagnetization steps. The rationale is that in the early stages of demagnetization the primary component (by virtue of its high H_{cr} or T_b) is least affected and the variously directed secondary (i.e., lower H_{cr} and T_b) components are destroyed, resulting in a minimum scatter of remanence directions from a given site. The hypothesis further assumes that with even higher values of AF or temperature step the primary component itself is attacked and gets weaker, resulting in a renewed increase in the scatter parameter k. In this type of analysis a minimum in k versus H_{AF} or T plots is sought, and once located, this value of H_{AF} or T is then used as a blanket value for demagnetizing a whole batch of specimens from one site. Another approach has been to determine some sort of a generally applicable paleomagnetic stability index based on hypothetical projections of demagnetization trends of secondary components (Symons and Stupavsky, 1974). Such indices are attractive to users of large numbers of samples, but because of the assumptions made in constructing the indices, the indices are qualitative at best. Whether one uses a large or small number of samples, there is no real alternative to point-by-point analysis of demagnetization data of many pilot specimens after plotting the data so as to display both the intensity and directional variation information. In many ways even this is a crude approach to the problem. In modern studies of multicomponent NRM, single grains or grain types have been used and the ages of the different components determined from high-sensitivity radiometric dating techniques. Some examples of this approach will be discussed in the last section of this article.

We shall now turn our attention to some other techniques of selective demagnetization. These techniques may be less common but are by no means esoteric, and they hold promise in cases where the three conventional demagnetization methods have failed. They are chemical demagnetization, low-temperature demagnetization, and reversed direct-field demagnetization methods.

In chemical demagnetization the goal is to use a dilute reagent that will selectively attack certain chemical compounds or grain sizes, thereby permanently removing the contribution to the total NRM of a certain component or components. The reader should realize that for chemical dissolution to be effective the specimen should be porous, which means that of all the different rock types used in paleomagnetism and rock magnetism, only coarse-grained sandstones are likely to be the best candidates for chemical demagnetization. On the other hand, it has been the experience of most workers in this field that coarse-grained sandstones (which often carry large multidomained magnetite) are less likely to carry stable (large τ) components of remanence than are fine-grained siltstones and shale, which are less permeable to chemical reagents. Burek (1969, 1971) improved upon the usual technique of soaking samples in dilute acid solutions by using a high-pressure bomb to force the acid through the interconnecting pores and claimed success with medium- to fine-grained sandstones. Roy and Park (1974), on the other hand, have used weak solutions (8–10 N) of HCl and very long solution times (4500 hr) to treat red sandstones that contain very fine-grained cement of hematite. Anyone planning to use the chemical demagnetization method should, however, be prepared to use a trial-and-error approach before anything suitable can be found for a specific rock type. On the other hand, it should be pointed out that virtually nothing except dilute HCl has been tried as a reagent up to now and the way is open for experimentation.

Low-temperature demagnetization is another technique that has been little used in routine studies. The principle of this method is based on the two isotropic points of hematite and magnetite. For hematite the critical temperature is 263°K and for magnetite it is 130°K. At these respective temperatures there is a change in sign of the magnetocrystalline anisotropy constant of the minerals. Thus if K_{eff} of a grain is controlled by magnetocrystalline anisotropy alone at the isotropic point $K_{eff} = 0$, and if the cooling and warming are done in zero ambient, the directions of remanent magnetization will become randomized in the absence of anisotropy energy. The procedure does not necessarily guarantee removal of secondary magnetization but it does ensure that those grains whose remanences are controlled by magnetocrystalline anisotropy alone will, in fact, lose their remanence. What are these grains? They are the large multidomain (md) grains whose domain walls are controlled by magnetocrystalline anisotropy and magnetic exchange interaction. Since these are also the same grains that have low values of τ, they are indeed more likely to carry less stable components of secondary magnetization. Their demagnetization will selectively enhance the contribution to NRM of the more stable single-domain (sd) and pseudo-single-domain (psd) grains. The method has been applied successfully by Merrill (1970). However, it *is* of limited application since the isotropic point is a function of grain size and impurity atom content in magnetite and hematite. First, for submicron grains of magnetite and hematite, the isotropic point is lower than the liquid nitrogen temperature (77°K), and it is not so convenient or cheap to carry out the experiment. Second, because of the presence of common impurities such as titanium, the isotropic point may be completely suppressed and the method becomes inapplicable. For a detailed discussion of the phenomenology of isotropic points the reader should consult the book by Stacey and Banerjee (1974).

The final method is an old one—the application of a gradually increasing (but reversed in sense) steady field and its subsequent withdrawal. As shown in Fig. 1 in the Introduction, such an application of slowly increasing steady field will result in a slow decrease of the NRM. But those grains will be demagnetized first whose H_{cr} (and T_b) values are the lowest; thus a preferential demagnetization of low τ grains will result. The truth, however, is that when such an application of a steady field reaches a moderate magnitude, the primary component will also be partially affected and not-so-clear isolation will result. It is not, therefore, a highly recommended approach and is resorted to mainly in the absence of the more sophisticated techniques for selective demagnetization.

3. Artificial or Laboratory-Imparted Remanent Magnetizations

3.1. Viscous Remanent Magnetization

Of all the different types of remanent magnetizations that can be applied in the laboratory for diagnostic purposes, viscous remanent magnetization (VRM) will be discussed first. Usually it is also the first test recommended for newly collected paleomagnetic samples because the experimenter needs to know not only the magnitude of the NRM and its time stability (as measured by AF and thermal demagnetizations) but also whether the rocks sampled contain either primary or secondary grains that can carry large amounts of secondary VRM so as to mask the original primary component of the NRM. A commonly used method is to measure the as-is NRM and then store the specimen in zero field for a length of time, e.g., 1 month or more. During this month, the previously mentioned "storage test" can be applied, i.e., the specimen is withdrawn at predetermined intervals in a geometric progression such as 1, 2, 4, . . . days and the NRM is remeasured. An obvious decay of NRM with time will indicate the presence of soft (i.e., low H_{cr}, low τ) components of VRM and the experimenter normally waits until the decay stabilizes to get on with selective demagnetization or "cleaning" of the rock. However, there is a disadvantage in this method because if the rock has a very hard (high H_{cr}, large τ) component of VRM, it may not show up in such a month-long storage test. It is therefore also necessary to impart VRM to a specimen and observe the acquisition and decay parameters of such artificial VRMs.

The apparatus for measuring VRM at room temperature is quite simple. One can set up a solenoid with a very stable dc power source to generate a small constant field (e.g., 0.05 mT $\equiv$ 0.5 Oe) to house the specimen. After suitable exposures to this field, the specimen can be carried in a Mumetal box (to protect it from the Earth's field) to a magnetometer that will measure the acquired VRM. A better approach is to house the solenoid as an integral part of the magnetometer so that any stray VRM is not acquired during the transfer of the specimen. For example, it is possible to buy a commerical spinner magnetometer that has a built-in Helmholtz coil for producing such small fields. The astatic magnetometer is well suited for adaption to measure acquisition and decay of VRM, particularly if it is desired to also raise or lower the specimen temperature during the measurements. It should be appreciated that in a VRM experiment such as the ones described above, the observed remanence is a sum of the actual VRM (which is, by definition, time-dependent) and an IRM is acquired by the specimen at $t = 0$. It is usual, therefore, to subtract the IRM from the total remanence for comparison with theory. It was observed quite early (Ewing, 1885) that VRM (J_{vr}) has a logarithmic time dependence:

$$J_{vr} = J_{ro}(1 + \log t_a)$$

where J_{ro} is the initial remanent magnetization, and t_a is the acquisition time, i.e., the time during which the field is left on. Also, S_a is the viscous acquisition coefficient:

$$S_a = \partial J_{vr}/\partial(\log t_a)$$

In a converse experiment where an acquired VRM is allowed to decay in zero field, the relationship becomes

$$J_{vr} = J_0(1 - \log t_d)$$

Here t_d is the decay time during which the field was switched off. As before,

$$|S_d| = \partial J_{vr}/\partial(\log t_d)$$

The log t dependence of VRM acquisition and decay can be predicted theoretically for both md and interacting ensembles of sd magnetic grains (Stacey and Banerjee, 1974). For a recent review of the theory of VRM see Dunlop (1973). He has stated that for experimental times (t) that are less than the characteristic relaxation times for a complete acquisition (τ_a) and a complete decay (τ_d),

$$S_a = |S_d| \quad \text{for} \quad t < \tau_a, \tau_d$$

S_a and S_d depend on, among other things, the initial magnetic state of the specimen, its ambient temperature, and the applied field. Complete theoretical descriptions of such dependences are not yet exact enough to describe the experimentally observed dependences although attempts have been made (Walton, 1980). One reason for this is the inability of the present theoretical models to incorporate a realistic model of the specimen, but another important reason is that there have not yet been many experiments in which the experimental variables have been controlled carefully and separately. Also, too few synthetic specimens with known concentrations of sized magnetic carriers have been used. As a result, the literature is replete with contradictory claims regarding S_a and $|S_d|$: while some find that $S_a = |S_d|$, others find that they are not equal; again some claim that S_a is higher for an AF demagnetized specimen than for a NRM-carrying specimen, and others find the opposite to be true. So far as temperature dependences of S_a and $|S_d|$ are concerned, a corroboration of the phenomenological theory that the two coefficients should increase with increasing temperature has indeed been obtained, but there is no pretension that the magnitude of the increase is thoroughly understood.

As a result, most published work has tended to deal with the gross description of thermal and AF demagnetizations of laboratory-imparted VRM and the conclusions that can be drawn from such experiments with regard to the removal of the secondary remanence acquired by the same rock under natural conditions. Instead of concentrating on the median demagnetizing field (MDF) or blocking temperatures (T_b) obtained by such experiments, it is much more instructive to use the graphical approach pioneered by Néel and successfully used by Biquand and Prévot (1971). Here I shall follow the approach outlined by Dunlop (1973). Figure 14 shows the two different situations obtained when a previously acquired VRM (dashed curves) is demagnetized by thermal or AF demagnetization. Figure 14a shows that on heating to higher temperature ($T_3 > T_2 > T_1 > T_0$) the blocking curves (or more accurately, the unblocking curves) move diagonally along the NE direction, "sweeping up" and demagnetizing the previously acquired VRM (area to the left of the dashed VRM acquisition curve). Note two things: (1) the same movement of the blocking curves could be achieved by storage in ambient temperature T_0 for long durations but the required lengths of time would be prohibitively long; (2) the blocking curve at $T = T_3$ coincides nearly, but not exactly, with the VRM acquisition curve. This is because the acquired remanence has a small component of IRM that tilts the acquisition curve to the right, exposing the shaded region of as yet undemagnetized VRM

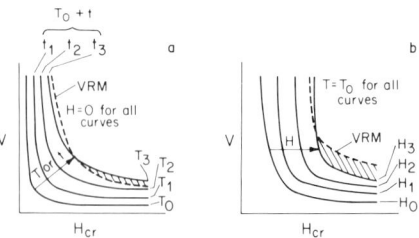

FIG. 14. (V, H_{cr}) diagrams of constant τ to illustrate the effects of thermal (a) and AF demagnetization (b) processes on a previously acquired VRM (dashed curves). The shaded area in each diagram denotes the as-yet-undemagnetized portion of the VRM.

carried by the appropriate grain sizes v_i with their corresponding H_{cri} values. Obviously a temperature slightly higher than T_3 will be required to demagnetize completely all the VRM.

In Fig. 14b, we see that the blocking curves under AF demagnetization are almost parallel to the v axis and especially so for high values of AF ($H_3 > H_2 > H_1 > 0$). Under the influence of increasing peak values of AF the blocking curves move to the right subparallel to the v axis, and at a peak AF value of H_3 there is a shaded region of undemagnetized VRM that is larger than that shown in Fig. 14a (thermally demagnetized). Consequently, an AF peak value that is even larger than H_4 will be necessary to demagnetize completely all the acquired VRM. As mentioned in the previous paragraph, Biquand and Prévot (1971) used this approach to explain how even a short-term (hours) exposure to a weak steady field could result in a VRM that needed a peak AF of more than 1000 Oe for complete removal. Instead, thermal demagnetization would be much more effective for rocks carrying grains with small values of v and high values of H_{cr}, as is the case, for example, with nearly superparamagnetic hematite grains that have a sd magnetic state.

An interesting example of the application of VRM in the laboratory has been given by Pullaiah et al. (1975). Because of the time–temperature equivalence in VRM acquisition and decay in the Néel formalism, it should be possible to observe the motion of the blocking curves in the laboratory time scales at much elevated temperatures to make conclusions about the past natural residence time of the same rock at a lower temperature. If successful, such experiments can lead to extremely useful geological information about the temperature or the duration of past metamorphisms. In order to apply the formulas properly, however, a complete knowledge is required of the temperature variations of spontaneous magnetization (J_s) and coercivity of remanence (H_{cr}) of the magnetic species, and these data are often not available explicitly. Pullaiah et al. (1975) made an approximation for the $H_{cr}(T)$ variation and used the metamorphic grade and estimated temperature of hematite-bearing rocks to conclude that a residence time of a million years will completely erase the primary TRM record if the rock has suffered greenschist facies metamorphism. However, they also carried out a calibration experiment at two temperatures for a rock and found that the Néel formula was not able to describe exactly the thermally induced relaxation of remanence. Much more work in this area is obviously needed.

Heller and Markert (1973) have used a similar approach for a room-temperature problem. They assumed that the basalts used in building Hadrian's Wall in North England had acquired a steady VRM in the present field since their emplacement according to a log t formula and a constant value of S_a. Therefore, they determined S_a from short-term laboratory tests and separated the magnitude of the VRM component in the rocks to calculate a magnetic age of the emplacement of the wall that agreed with historical information. For such "dating" methods to succeed, however, the specimens should have seen only the present sense (i.e., normal) of the geomagnetic field. Thus rocks older than 700,000 years (the Brunhés epoch) cannot be used, and if the allegedly more recent geomagnetic events or short reversals are real, the magnetically determined "age" would be incorrect by an unknown amount.

It is perhaps useful to draw the reader's attention here to what are perceived to be the problems in VRM studies today. One problem is that although we have a phenomenological understanding about the VRM process in sd grains, especially as to the different sensitivities to thermal and AF demagnetization, we do not have an equivalent understanding of the reasons for the observed high AF stability of the VRM in some md grains. Nor do we understand why the absolute value of S_a for md grains can often be as high as that of sd grains because the spontaneous magnetization of md grains is only a percent of that of sd grains. Another effect that is being observed more and more is that the magnitude of S_a (and $|S_d|$) is dependent on the relative value of the duration of the experiment. For example, it has been observed that S_a increases sharply after especially long times such as 10^4 or 10^5 sec. It is the hope of many rock magnetists that in the future, controlled VRM experiments will throw some light on this behavior. For the geologist and the geophysicist alike the importance of such experiments lies in our ability to extract paleomagnetic information from a much larger number of rocks of Paleozoic or older age for global tectonic studies. At present these rocks are often considered useless paleomagnetically because of their highly stable VRM components, which mask successfully the primary components of NRM.

3.2. Isothermal Remanent Magnetization

As defined in Section 1, isothermal remanent magnetization (IRM) is applied to a rock specimen by exposing it deliberately to a steady field at a given temperature and then reducing the field to Zero. The resultant magnetization is an IRM. Most commonly, However, the given temperature is room temperature, and more than 90% of the IRM data quoted in the literature for given rocks refer really to room-temperature IRM values. The range of field values used in IRM experiments is from one millitesla (10 Oe) to hundreds of millitesla ($\times 10^3$ Oe). Since it is obvious that imparting of IRM inexorably alters the NRM record in a rock, such experiments are conducted usually after the NRM has been measured and demagnetized selectively in order to obtain information about magnetic stability. The thermally or AF demagnetized specimen is then given an IRM to extract information about the intrinsic remanence-carrying abilities of the specimen. For a most profitable outcome the demagnetized sample is given an IRM in slowly increasing steps of steady field in order to construct an acquisition curve of IRM as shown in Fig. 15. In an ensemble of sd grains such a procedure will magnetize sequentially the softest (i.e., lowest H_{cr}), the median-range, and finally the hardest (i.e., highest H_{cr}) grains. The energy barriers against which such magnetization takes place depend on the chemical composition, grain size, and the degree of dispersion of the magnetic carrier involved. Also the magnetizing mechanisms at different levels of the applied steady field are different. For md grains the mechanisms are domain wall translation and domain wall rotation in increasingly higher fields, while for sd grains the mechanisms are coherent and incoherent rotations. In spite of such a variety of parameters, however, it is possible to observe in nature composition-dependent IRM acquisition curves as shown in Fig. 15. Such an approach was taken by Dunlop (1971) to recognize the presence of very small amounts of hematite in specimens of sandstones heated in the laboratory. Instead of determining carefully the full IRM acquisition curve, some workers have merely applied a fairly large steady field, anywhere between 200 and 1000 mT (2000–10,000 Oe), and then withdrawn the field to measure the magnitude of saturation IRM (J_{rs}). Unless the acquisition curve as shown in Fig. 15 has been measured, however, it is not possible to conclude that the observed remanence is indeed the saturation value. In addition, J_{rs} by itself is a less instructive parameter than is the complete acquisition curve.

The next type of measurement done with IRM is to construct thermal or

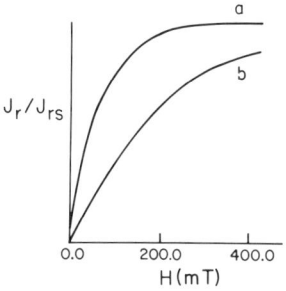

FIG. 15. Acquisition curves of IRM as a function of applied steady field H. The observed remanence J_r has been normalized by the saturation value (J_{rs}) of IRM at fields much higher than shown here. Curves: (a) for the magnetite-bearing rock; (b) for a hematite-bearing rock.

AF demagnetization curves with J_{rs} as the starting point. Alterating-field demagnetization is more common than thermal demagnetization, and the goal of this kind of a demagnetization curve is not so much to simulate the stable NRM characteristics but to characterize the total spectrum of magnetic carriers in the rock. The reason for the above statement is the following: in ARM or TRM the steady field bias used is quite small, on the order of the Earth's field (0.05 mT ≡ 0.5 Oe). When superposed on the large-amplitude peak AF or on high temperatures near the highest blocking temperature, such a small biasing field is most effective in magnetizing the highest H_{cr} grains and the resulting remanences have high values of median destructive field (MDF). By comparison, IRM is more of a brute force method of acquiring a remanence. As the magnitude of the steady field increases domain wall translation and rotation processes occur, starting with the softest (low H_{cr}) fraction and at saturation ending with the hardest (highest H_{cr}). The resultant MDF reflects the complete spectrum of H_{cr} values present in the rock and is invariably lower than the MDF of low (steady)-field ARM or TRM. However, it is the latter two types of artificial remanence that simulate most closely the stable components of NRM. Thus it is not likely that the AF demagnetization curve of a saturation IRM (J_{rs}) would be characteristic of the stable fraction of the NRM. On the other hand, if the two are indeed similar, the conclusion can be made that the NRM is carried mostly by the less stable md grains, as has been the case with the majority of iron-bearing lunar samples. Also if the AF demagnetization curve of IRM acquired at a low value of steady field (e.g., 10 mT ≡ 100 Oe) appears to be similar to that of the NRM, it can be concluded that the NRM is mostly carried by less stable grains and such rocks should not be used for detailed studies.

As mentioned earlier, the chief use of the demagnetization curve of a saturation IRM is in characterizing the complete range of remanence coercivities (H_{cr}) present in the rock. For example, if thermal demagnetization has been carried out on a rock and there is some question as to whether a new magnetic mineral has been formed unintentionally as a result of the heating, IRM demagnetization curves could be employed to distinguish the "before heating" sample from the "after heating" sample. Some workers have used only the total value of J_{rs} as a parameter to infer thermal alteration, e.g., during a paleointensity experiment performed by the method of Thellier and Thellier (1959). It is perhaps inappropriate to do so since a comparison of the demagnetization curves (AF or thermal) of the two values of J_{rs} would lead to a more complete appreciation of the *reason* behind the observed change in J_{rs}. Carmichael (1968) has gone further in that he has used the ratio of J_{rs} values before and after heating to provide a "correction factor" to the inferred paleointensity. It is difficult to defend such a "correction factor" since the onus is on the experimenter to show that the J_{rs} increase or decrease upon heating in fact pertains to the sole stable component of NRM in which we are interested.

A somewhat similar use of IRM was made by Fuller (1974) in which he normalized the observed NRM of lunar rocks with the IRM in order to correct for possible fluctuations in the iron content from rock to rock. After such normalization, the ratio was used as a relative measure of paleointensities at the lunar surface corresponding to the radiometric ages of the given lunar rocks. When such studies were made, enough data were not available on the comparative demagnetization curves of the NRM and IRM of lunar rocks. Hence such relative paleointensity values were not wholly defensible as accurate. However, the large amount of data on demagnetization of NRM of lunar rocks since accumulated indeed show that for the most part the NRM of lunar rocks can be simulated by moderate-field IRM. As a result, Fuller's (1974) method for obtaining relative paleointensity values for the lunar surface is acceptable. An unacceptable case is that of Kawai et al. (1975), who gave a similar set of relative paleointensities obtained by normalization with moderate-field IRM for wet lake sediments. Since no comparative demagnetization curves for NRM and IRM have yet been provided, it is not possible to estimate the accuracy of the obtained data.

3.3. Anhysteretic Remanent Magnetization

Anhysteretic remanent magnetization (ARM) is imparted usually in the following manner: the specimen is first placed in a small solenoid, which is then placed inside a larger AF solenoid, which, in turn, is placed in a field-free space. The small solenoid, connected to a stable dc source, produces a constant weak field, e.g., 0.05 mT (0.5 Oe). After this solenoid is energized, the current of a given peak value in the AF solenoid is switched on—the direct field and the AF are usually kept parallel to each other, although experiments have been reported (Rimbert, 1959) where a 90° orientation led to a smaller value of ARM. The AF is smoothly decreased, for example, with the aid of an electronic ramp function, to zero with the weak direct field left on. Finally, the latter is switched off and the specimen is extracted to measure the acquired ARM. In general, there are two types of uses of ARM measurements. First, ARM, as an analog of the stable component of NRM, is used to confirm that the characteristics observed in the total NRM are indeed those of the more stable and hopefully primary component and not of any subsequent secondary imprinting of magnetization. Second, ARM can be used as a rock magnetic tool to study the characteristics of the magnetic carriers per se. Of course, the goal of such a study is also tied to a proper understanding of the origins of NRM.

In her classic study of ARM versus other kinds of artificial remanences, Rimbert (1959) showed that ARM is a close, if not perfect, analog of TRM. Thus rocks that normally carry a TRM, e.g., highly oxidized basaltic lava flows, could be studied for their original versus present NRM characteristics using ARM. For example, a comparison of the AF demagnetization of the ARM with that of the NRM might reveal either a loss of certain remanence coercivity (H_{cr}) fraction due to low-temperature hydrothermal alteration or a gain of certain H_{cr} fraction due to a higher-temperature oxidation and the consequent formation of new magnetic carriers at the expense of primary but previously nonmagnetic minerals. Petrological observations alone do not constitute sufficient evidence for claiming that the magnetic carriers themselves have been altered, although a combination of petrological and geochemical observations plus ARM studies can constitute a powerful tool in discovering the alteration history of a rock. Nor should the ARM characteristics be regarded as good analogs of TRM alone—ARM is often found to be an excellent analog of stable depositional remanent magnetization (DRM) or postdepositional remanent magnetization (PDRM) in sediments (Levi and Banerjee, 1976) as well as some types of chemical remanent magnetization (CRM) (Hoye and Evans, 1975). In both cases, the analogy stems from the fact that alternating field appears to be a good, if not a perfect, simulator of temperature (Levi and Merrill, 1976). In DRM and PDRM the stable component is due to the oriented small sd or psd grains that normally carry an original TRM. In the case of CRM, the blocking of the remanence occurs because of a rapid increase in the term $K_{eff}v/kT$ due to an increase in volume v. This increase in the magnitude of the term is, of course, mathematically indistinguishable from that due to a decrease in T, which is what occurs in the case of TRM. And if ARM can simulate TRM, it can also simulate CRM in specific cases. An example would be CRM in grains that are large enough to be sd or psd in size but less than md in size. The attraction for the ARM simulation of the stable NRM of a rock lies in the thermally nondestructive nature of ARM application. This property is even more appreciated when it is known that the rock specimen in question has been altered in nature and is at a thermodynamically metastable state. Such was the case studied by Johnson and Merrill (1972, 1973), in which it was necessary to inquire into the comparative AF demagnetization behavior of TRM and CRM in maghemite (low-temperature-oxidized magnetite) and titanomaghemite (similarly oxidized titanomagnetite, $Fe_{3-x}Ti_xO_4$). Both maghemite and titanomaghemite are thermally metastable compounds, and temperatures of the order of 250°C or above are able to produce a kinetic alteration in composition in laboratory time scales. In the case of maghemite the final stable product is hematite (Fe_2O_3), and in the case of titanomaghemite, although the end products can vary (Readman and O'Reilly, 1972) depending on the degree of prior oxidation, usually the stable products are magnetite and ilmenite ($FeTiO_3$). Johnson and Merrill (1972, 1973) had wanted to investigate how stable the CRMs in maghemite and titanomaghemite are compared to TRMs in the same compounds. It has been argued that in nature titanomaghemites may be kinetically stable for a duration that is long enough for them to acquire a TRM. In the laboratory, however, when they are reheated in the absence of natural gases and other volatile species, the titanomaghemite exsolves into magnetite and ilmenite before a TRM can be imparted to them. Johnson and Merrill (1972, 1973) observed that CRM in maghemite was less stable than ARM, while in titanomaghemite it was the reverse. Drawing upon the analogous AF stability of TRM and ARM, these authors have claimed that when titanomagnetites are chemically altered in nature at low temperatures to produce titanomaghemites, they will carry a CRM that will be more stable to AF demagnetization than the TRM carried by these minerals.

As mentioned earlier, the second area in which ARM has found most applications is rock magnetic studies. A major hope has been to understand thoroughly the similarities and dissimilarities between ARM and TRM, and between ARM and DRM so that ARM can be used to extract information about absolute or relative paleointensities. Some rocks such as lunar basalts or metabreccias contain very fine-grained ($\sim 10^2$ Å) iron that was formed in the extremely low oxygen fugacity present at the lunar surface. When such rocks are heated under terrestrial conditions to impart a laboratory TRM, the iron grains invariably oxidize to produce magnetite or other oxide minerals. The kinetic breakdown of the other minerals on heating at moderate temperatures is so prevalent that having a low-oxygen-fugacity environment cannot guarantee the integrity of the original iron grains. In such a situation it is desirable to replace the laboratory heating step (and the imparting of TRM) by the measurement of ARM, provided one knows the correct value of the ratio R of ARM and TRM susceptibilities:

$$R = \chi_{ARM}/\chi_{TRM}$$

Stephenson (1971) obtained $R \approx 0.8$ for one lunar rock and a synthetic sample containing iron of unknown grain size. He then applied this value of R to a larger number of lunar samples. Banerjee and Mellema (1974) measured R for highly elongated sd grains of CrO_2 at room temperature and at elevated temperatures all the way up to the highest blocking temperature on the ensemble. It was argued that since χ_{ARM} must approach χ_{TRM} at the blocking temperature, a knowledge of the temperature-dependent behavior of R will lead to a better simulation of TRM (which is acquired at elevated temperatures) with ARM (which is obtained at room temperature). These workers obtained a room-temperature value of $R \approx$ 0.2. Levi and Merrill (1976) have obtained a range of room-temperature R values between 0.7 and 0.1 for powdered magnetite specimens whose grain size ranged between 3 and 0.04 μm (400 Å). Although these workers obtained R greater than 1.0 for two magnetite chips of size 1–10 mm, this result should be treated with caution because, unlike the other specimens, these were not dispersed powders and it is well-known that remanences in single chips are notoriously prone to secondary influences arising from their irregular shapes. It appears, therefore, that both experimental and theoretical determinations of R have not yet reached the stage at which empirical relationships between ARM and TRM can be used for a quantitative simulation by the former of the latter. This is also the conclusion reached by Dunlop *et al.* (1975). Future investigations may succeed in resolving the problem.

Even though a *quantitative* simulation of TRM with ARM is not yet possible, a *qualitative* simulation of the stable component of any NRM with ARM is possible. The stable remanences in TRM, DRM, or CRM often arise from the same (v, H_{cr}) fraction of grain sizes as those that contribute to ARM. The relative susceptibilities of these remanences are, however, not only different but even incalculable, as described earlier. In this situation one can resort to a qualitative simulation with the premise that although one does not know the absolute contributions to ARM and a stable NRM of the different (v, H_{cr}) grain size fractions, the *identical* grain size fractions carry *both* types of remanences. The one sure way to ascertain this is to compare the thermal or AF demagnetization curves of the NRM with a laboratory-imparted ARM. Figure 16 shows such a comparison of demagnetization curves of ARM, IRM, and NRM (Levi and Banerjee, 1976). The comparison could be made quantitative by normalizing the NRM with ARM (or IRM or some other laboratory remanence) and plotting the normalized data as a function of peak values of AF or temperature step (Fig. 16). A "flat" behavior with regard to AF or temperature would indicate excellent agreement. As shown in Fig. 16 taken from Levi and Banerjee (1976), such an agreement was seen for ARM simulation of NRM but not for IRM simulation for the same sediment samples. It should be emphasized that because of the grain size dependence of ARM, IRM, etc., no single artificial method of NRM simulation can be generally valid for all types of rocks and sediments. The onus is upon the inves-

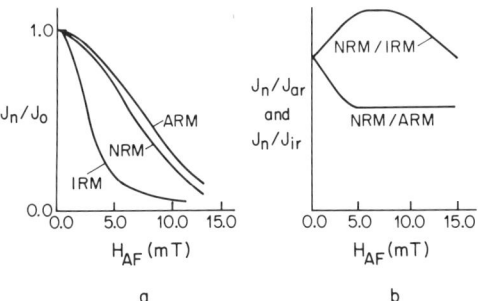

FIG. 16. (a) AF demagnetization of ARM, NRM, and IRM for a natural sediment from a lake. (b) The same data replotted in a normalized fashion. After demagnetization to about 5 mT (50 Oe) the NRM seems to be very well simulated by ARM in terms of demagnetization characteristics. (From Levi and Banerjee, 1976.)

tigator to present the relative demagnetization data first before selecting a specific type of artificial remanence such as ARM for NRM simulation. One should also study the applicability of such simulations for a representative number of specimens from the suite before comparing data of one set with that of another. In general, however, it has been seen empirically that for the stable component of NRM carried by fine grains, ARM is more likely to be a good simulation than the other types of artificial remanence.

With these provisos fully satisfied it is possible to get relative values of paleointensity from a given suite of rocks and sediments. Levi and Banerjee (1976) and Lund et al. (1977) have obtained relative paleointensity values of the geomagnetic field for the past 10,000 years from wet lake sediments by the ARM simulation method.

Anhysteretic remanent magnetization has been used as a rock magnetic tool to confirm the presence or absence of md grains in a rock (Johnson et al., 1975). Originally it was Lowrie and Fuller (1971) who suggested this way of utilizing the often reported behavior that sd grains carry a TRM that becomes less stable to AF demagnetization as the magnitude of the weak TRM-giving field increases. Lowrie and Fuller found that for md grains the trend of AF stability is the opposite and they therefore suggested that the magnetic state of a rock (i.e., sd grains versus md grains) could be recognized from such studies. In order to make the test fully nondestructive by eliminating the heating procedure, Lowrie and Fuller suggested that the TRM in a large steady field be replaced by a saturation IRM in a large (1000 mT ≡ 10,000 Oe) steady field. Dunlop et al. (1973) found that the test was equivocal, when applied to natural rocks and Johnson *et al.* (1975) recommended that when a saturation IRM is replaced by an ARM in a small steady field, the latter is indeed a better simulator of a TRM given in a similar field. However, Johnson *et al.* (1975) also found that the modified Lowrie–Fuller test cannot distinguish between sd and psd grains and thus its chief use should be limited to a search for the presence of md grains. It is, however, a perfectly satisfactory alternative since both sd and psd grains give rise to NRMs with similar stabilities. It is the md grains that have lower stabilities, and rocks carrying NRM due to md grains should be recognized quickly and removed from a paleomagnetic study of old rocks.

3.4. Thermoremanent Magnetization

This process is one of the most studied in the short history of rock magnetism. The procedure consists of placing a rock specimen inside a furnace that is water-cooled and noninductively wound in order not to have an associated steady field when the furnace is on. The controlled steady field source is either a solenoid with a stable power source or the uncanceled ambient Earth's field. If it is the former, the solenoid is usually slipped over the water-cooled furnace, the TRM is given by heating the specimen to a temperature slightly above the Curie point (T_c) of its magnetic carrier, and then, turning the steady field on and the furnace off in order to cool the specimen back to room temperature, the field can be switched off. No magnetic shielding is necessary if either the ambient field is used as the steady field or the required steady field is much greater (say, 5×) than the ambient field in the laboratory. I use the word "ambient" as opposed to "the Earth's field" because in most laboratories the ambient field direction can be quite different in direction and magnitude from the expected reference geomagnetic field at the location.

Thermoremanent magnetization has the distinction of being the one process of nature that can be simulated nearly perfectly in the laboratory. The same cannot be said, for example, for DRM and CRM. The only problem that may arise in the simulation is the control of ambient atmosphere such as oxygen fugacity or sulfur fugacity. In the previous section I have mentioned the example of lunar rocks whose ambient fugacities are difficult to reproduce in the laboratory. A similar problem is faced with dredged or drilled basalt samples from the upper oceanic crust. For most other terrestrial samples, however, sample integrity during imparting of TRM can be preserved with use of a vacuum of the order of 10^{-5} Torr or, more exactly, using a solid-state or mixed-gas buffer system to control the oxygen fugacity of the specimen environment.

The first use of TRM is to convince the experimentalist that the observed NRM in, say, a basalt is a true TRM acquired in nature at an instant in time and not due to a subsolidus, sub-T_c occurrence of CRM. Another secondary source can be thermochemical remanent magnetization (TCRM) in which the specimen acquires a secondary TRM due to thermal alteration and creation of a new magnetic phase at a temperature above its T_c. A third kind of secondary remanence in a basalt can be partial thermoremanent magnetization (PTRM) acquired by moderate heating to, say, 200°C after the initial cooling of the basalt body. In the most general case, one can assume that the TRM will be carried by the hardest (highest H_{cr}, large τ) volume fraction, whereas the secondary remanences of thermal origin will affect other (v, H_{cr}) fractions. Therefore, a simple test of the primary nature of the NRM will be to demagnetize it thermally or with an AF and then compare this demagnetization curve with that of a TRM given in the laboratory in a known field. For reasons enumerated in the earlier sections, thermal demagnetization is preferred over AF demagnetization if the specimen is not thermally degradable. A perfect overlap between the NRM and TRM demagnetization curves will help strengthen the conclusion that the NRM is indeed a primary TRM. Some adjustments must be made depending on the particular situation. For example, the specimen may have acquired a secondary, but soft, VRM in the Earth's field. This component of the NRM will likely have low H_{cr} values and correspondingly low values of blocking temperatures (T_b). Figure 17 shows a possible situation where the stable component of the NRM appears to be very similar to the laboratory TRM although there is a superimposed low-T_b secondary component. Any part of the NRM that survives these

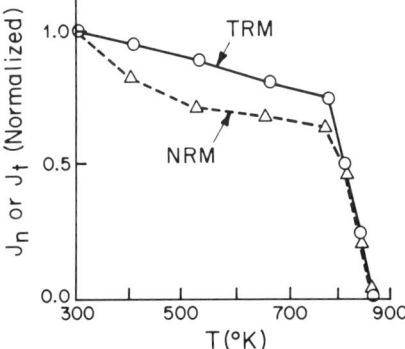

FIG. 17. Comparative thermal demagnetization of a NRM and a laboratory TRM shows that most of the high-T_b NRM was probably acquired by a thermoremanent process. At lower temperatures the NRM indicates secondary components due to VRM, PTRM, etc.

lower temperatures can be regarded as originating from a thermoremanent process similar to the one used in the laboratory.

There are a couple of by-products of the above type of measurements. The magnitude J_{tr} of the TRM may be compared with the magnitude J_n of the NRM of a given igneous rock and if the two are similar, one can make a likelihood argument about the NRM of the given igneous rock being a TRM. The other type of information becomes available on demagnetization of the given TRM. If AF demagnetization has been used, the resulting coercivity spectrum provides an idea of the H_{cr} distribution in the given specimen. If thermal demagnetization has been used, as in the example above, the result is information about the distribution of blocking temperatures (T_b) in the specimen. Dunlop and West (1969) showed that the AF demagnetization of a TRM and the thermal demagnetization of an ARM of the same specimen can be used to obtain a distribution curve of (v, H_{cr}) of a given specimen. This is a rather laborious approach, however, and Dunlop (1976) has recently suggested an alternative but less time-consuming approach that will be discussed in the next section.

An important diagnostic use of laboratory TRM is in the recognition of reproducible self-reversal of NRM. Self-reversal is an unusual phenomenon in nature but when it does occur, it results in a rock specimen acquiring a NRM antiparallel to the magnetizing field. The reason for this lies in intragrain or intergrain negative magnetic interactions (electronic superexchange). The various types of reversals possible in nature have been enumerated in textbooks on rock magnetism. Here I deal with an example that was discovered by Nagata and Uyeda (1959) in the now-famous Haruna dacite. The rock had a reversed sense of NRM for the present Brunhés epoch, and when samples of it were heated to the Curie point and cooled in a weak field in the laboratory, the TRM was also found to be reversed with regard to the applied field. This was ample proof that the NRM, which was of TRM origin, had not recorded a reversed Earth's field but a normal one. By separating out the active magnetic carrier, Nagata and Uyeda were able to show that the negative superexchange was an intrinsic property of a single-phase rhombohedral mineral, of composition $\sim Fe_{2.5}Ti_{0.5}O_3$. A thorough explanation of this unusual behavior was offered later by Hoffman (1975). Unfortunately, there are a large number of rocks with reversed NRM that when heated to their Curie points suffer physiochemical breakdown. Hence these rocks cannot be shown to acquire a reversed TRM in the laboratory. In any case it is a good practice to attempt to give a laboratory TRM to an igneous rock that has been found to contain reversed magnetization.

The major use of a laboratory TRM is in the determination of paleointensities of the geomagnetic field. The intensity of the geomagnetic vector is, of course, as important a component as its direction. However, the extraction of the information about past intensities is fraught with many complexities of interpretation and sources of error. In order to determine the "calibration constant" of a NRM it should be possible to impart the same type of remanence in the laboratory without altering the chemical composition and microstructure of the specimen. Of the NRMs of TRM, DRM, and CRM origin, only TRM can be reproduced reasonably well in the laboratory. [For the other types of remanences the best that can be achieved is a simulation of the demagnetization curve, as has been discussed in the previous section.] But even for TRM duplication there exist some severe problems that will be discussed later. I will first deal with the basic principles behind paleointensity determination methods for TRM-bearing rocks.

The most well-known method is named after Thellier and Thellier (1959), who applied it, however, not to rocks but to archaeological samples like bricks and pottery. These commonly contain hematite ($\alpha - Fe_2O_3$), the highest oxide of iron, and therefore can be reheated in the laboratory without fear of thermally induced alteration as would be the case with maghemite, magnetite, and titanomagnetite. Thellier and Thellier (1959) first convinced themselves that partial TRMs (or PTRMs) acquired by a specimen in different high-temperature intervals could be added to equal the total TRM that would have been obtained had the specimen cooled from the highest temperature to room temperature. Therefore, given the proviso that all of the NRM is a TRM, these workers recommended a method of partial demagnetization and remagnetization in a known field so as to lead to a determination of paleointensity with a given uncertainty value. The apparatus used is a thermal demagnetizer (Fig. 9) with the added convenience of a steady field source (a dc solenoid) that could be switched on during cooling if needed. The common version for paleointensity determination by the Thelliers' method is as follows: the room-temperature value of NRM (J_n) is measured and plotted as shown in Fig. 18. The specimen is then thermally demagnetized by heating to, say, 50°C in zero field. The loss of remanence (ΔJ_n) for 50°C is recorded. Then the specimen is reheated to exactly the same temperature but cooled in a given field, say 0.1 mT (1 Oe). The gain in remanence (ΔJ_t) due to this step is also noted. The point marked 50 is then plotted in Fig. 18, whose ordinate and abscissa correspond to the above ΔJ_n and ΔJ_t values for 50°C. The paired heatings are repeated for other chosen values of temperature so as to uniformly cover the complete blocking temperature (T_b) spectrum of the given specimen. Thus a prior determination of the T_b spectrum using a second specimen is clearly desirable. The genius of Thellier and Thellier (1959) lay in the fact that instead of assuming that the

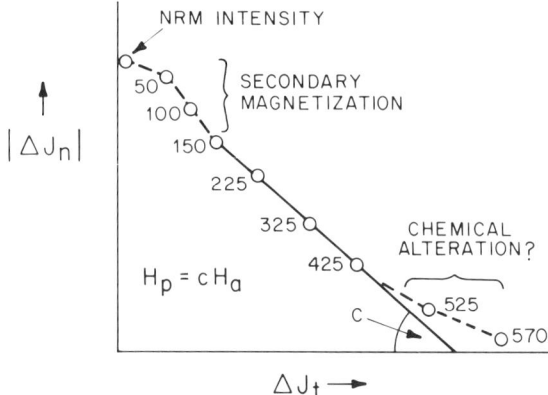

FIG. 18. A commonly used version of the Thelliers' (1959) method of paleointensity determination by imparting TRM (J_t) in a field H_a in stepwise fashion and then comparing it with the loss of NRM (J_n) at the same temperature steps (50, 100, 150°C, etc.). Note departure from linearity of (ΔJ_n, ΔJ_t) data at low temperatures due to the presence of secondary magnetizations and a different slope of the highest temperatures due to a possible role of thermally induced chemical alteration. The paleointensity H_p is given by cH_a, where c is the best-fit slope of ΔJ_n, ΔJ_t for the most reliable temperature interval.

paleointensity coefficient c ($\simeq \Delta J_n / \Delta J_t$) of a specimen was constant over the whole T_b spectrum, they decided to determine experimentally if indeed this was so. Furthermore, for a rock that does not alter chemically on heating, it is possible to put a greater confidence on c determined at the highest T_b values since those blocking temperatures correspond to the most stable (large τ) grains. If in an experiment c is indeed observed to be constant over all temperature studied, one, of course, has great confidence in the result. If, on the other hand, the lower T_b values of c are significantly different from the higher ones, as shown in Fig. 18, it is reasonable to consider the low-temperature values as due to secondary VRM or PTRM in a different magnetic fraction and to put one's faith only in the higher-temperature data for the calculation of paleointensity.

There is, however, a second possible source of error. In Fig. 18 I have indicated a slightly different value of c for the highest-temperature intervals. This could indeed be due to a *physical* reason, i.e., unblocking of a different (v, H_{cr}) fraction in the rock. On the other hand, it could likely be due to a different rate of thermal decay of J_n caused by a *chemical* breakdown of the mineral carrying the NRM or heat-induced microstructural alterations of the NRM carriers. Shaw (1974) has recommended that if such problems are anticipated, after the final heating step the specimen should be AF-demagnetized to a large peak value (e.g., 100 mT or 1000

Oe), then given an ARM in a weak field, and the AF demagnetization of this "after heating" ARM should be compared with a "before heating" ARM demagnetization curve obtained from a neighboring virgin specimen. Shaw's point is that if indeed the heating has caused irreversible changes in the ARM-carrying capability of the specimen, by analogy the TRM-carrying capability has also been affected. I will describe below in detail the recommendation of Shaw (1974) for such a situation. The approach of Thellier and Thellier (1959) to this problem is more conservative. They would use only the linear high-temperature part of the ΔJ_n versus ΔJ_t plot (Fig. 18) and if there is the slightest doubt about chemical alteration, they would recommend discarding the specimen. The paleointensity (H_p) is calculated in the Thelliers' method by fitting a least squares line through the reliable points. The slope (c) of the line and the knowledge of the applied field (H_a) in the laboratory lead to a determination of the paleointensity since

$$H_p = [\langle \Delta J_n \rangle / \langle \Delta J_t \rangle] H_a$$
$$= cH_a$$

Coe (1967) and Coe and Grommé (1973) have discussed the various sources of error in the Thelliers' method and the reader is referred to these articles for the details. The fact remains that if the rejection criteria of Thellier and Thellier (1959) are used, one obtains a few but very reliable paleointensities, and as a by-product of such an experiment one gets a very clear idea of possible thermally induced alterations. Levi (1976) has suggested that some of the observed departures from linearity at low and intermediate temperatures may be intrinsic (physically) to a multidomained specimen.

It is probably clear to the reader that one major problem with the Thelliers' method of paleointensity determination is the immense amount of time necessary to perform the complete experiment on a single rock specimen. Even if only 6 temperature steps are chosen per specimen, it automatically means 12 heatings and coolings (6 for demagnetization and the other 6 for remagnetization in a known field). Furthermore, with successively higher values of temperature it takes longer and longer times to cool the rock back to room temperature for the measurement of remanence. An alternative approach was suggested by Van Zijl et al. (1962), who replaced thermal demagnetization with the much faster AF demagnetization. In this approach the NRM is first demagnetized and the intensity plotted as a function of H_{AF} (Fig. 19). Then the rock is given a total TRM in a given field by heating it to a temperature higher than the Curie point. This TRM is then AF-demagnetized and plotted as shown in Fig. 19. If the two demagnetization curves are similar in shape for the higher values of

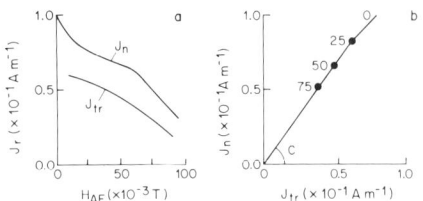

FIG. 19. (a) The AF demagnetization of NRM (J_n) and TRM (J_{tr}). (b) The same data are replotted where the numbers denote steps of AF demagnetization. The slope c is multiplied by the TRM-giving steady field (H_a) to yield the paleointensity by the method of Van Zijl *et al.*

H_{AF} it is concluded that the most stable remanence carriers contribute both to the NRM and to the laboratory TRM. The NRM versus TRM plot for this coercivity window can then be plotted (Fig. 19). Analogous to the case for the Thelliers' method, the slope of the best-fitting straight line is multiplied by the magnitude of the laboratory field to yield the paleointensity. Because only one heating is required in this method, the time required to study a single specimen can be as little as $\frac{1}{10}$ of what is required in a paleointensity experiment of the Thellier type. It should, however, be pointed out again that AF demagnetization is not the exact parallel of thermal demagnetization, and since the NRM was acquired as a TRM, the best way to study the conjugate process of acquisition is by thermal demagnetization. When AF demagnetization replaces thermal demagnetization, the observed remanent state after demagnetization is a complex one, being due to both the unblocking of discrete grains with discrete H_{cr} values equal to the H_{AF} applied as well as an effectively unblocked state arrived at by some other grains in an interactive mode. The individual H_{cr} values of these latter grains may be higher than the specific value of the H_{AF} used to achieve this remanent state. The error produced may not be large, but the difficulty lies in its intractability.

One other problem with the method of Van Zijl *et al.* (1962) is that the specimen must be given a total TRM by heating to the Curie point. Unlike the Thelliers' method, here there is no opportunity to monitor possible thermal alterations which (in the Thelliers' method) may appear as slight but distinctly increasing departures from linearity on heating to increasingly higher temperatures. Thus, in the Thelliers' method there is still a possibility of salvaging the lower-temperature part of the data, while in the method of Van Zijl *et al.* the physiochemical integrity may well be lost during the single high-temperature heating step. It may fortuitously happen that the second AF demagnetization curve is not drastically different in shape, and the experimenter may not recognize the amount and/or character of the changed magnetic carriers in the rock. This problem is much more serious than the problem of nonequivalence of thermal and AF demagnetization. To overcome this, Shaw (1974) has suggested a monitoring mechanism that was alluded to in Section 3.3. Shaw's method of paleointensity determination is basically the same as that of Van Zijl *et al.* (1962) except for the added steps of AF-demagnetizing an ARM given to the sample before and after the single-step heating. If no thermal alteration took place, the "before" and "after" curves should overlap. When they do not, Shaw concludes that only the overlapping H_{cr} values should be used in the final comparison of J_n versus J_t plot (Fig. 20), leading to a more reliable paleointensity determination. Indeed such selective use of data has allowed him to retrieve historically known paleointensity values from basalts with an accuracy of ±10%. In spite of the fact that the monitoring tests need extra time, Shaw's modification of the method of Van Zijl *et al.* still takes less time than Thelliers' method, and this fact alone should make it more popular with paleomagnetists in the future. There is one problem with Shaw's method that I have discovered. When the stable remanence is *not* due to sd or psd grains, the monitoring test in the form of a low-field ARM seems to fail. Specifically, what happens is that for md grains even if the "before" and "after" curves of AF demagnetizing of TRM are the same, the two ARM demagnetization curves look dissimilar and may lead the experimentalist to conclude wrongly that a part of the relevant H_{cr} spectrum has changed while what it really means is simply that low-field ARM is *not* a good measure of md TRM and should not be used as such.

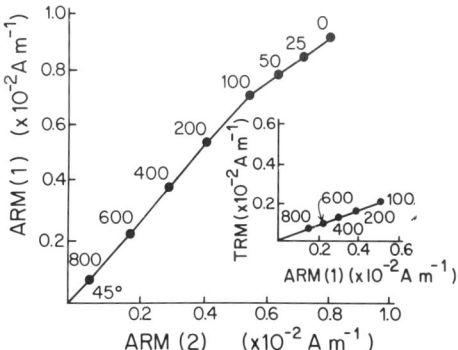

FIG. 20. Shaw's method of paleointensity determination. ARM(1) applied to the sample before heating is AF-demagnetized at discrete steps (e.g., 25, 50, 100 Oe, etc.) and the remaining ARMs after each step are compared with those pertaining to ARM(2), imparted after heating. The 45° slope between points marked 800 and 100 delimit the steps within which TRM and ARM(1) may be compared against each other (see inset) to yield an acceptable paleointensity.

4. Parameters Measured in the Presence of a Field

4.1. Magnetic Susceptibility

While the previous two sections of this chapter have dealt with the measurement of remanent magnetizations that require a strict exclusion of an applied field, the present section will deal with the determination of various intrinsic magnetic parameters in the presence of a field. Therefore, the magnetic measurements are not different in character from those made commonly by pure and applied physicists; the only difference in our case is the special emphasis put on their geological and geophysical aspects.

Magnetic susceptibility (χ) could perhaps be termed the zeroth-order magnetic property—it is a tensor that measures how magnetic a certain sample is in the presence of an applied magnetic field. When one refers to low-field susceptibility measurements, the field is of the order of the Earth's field (0.05 mT) or as much as 0.1–1 mT (1–10 Oe). High-field susceptibilities, on the other hand, are measured at fields as high as 1000 mT (10,000 Oe). The distinction between the two types of measurements rests on the dominant magnetic processes in the two field regimes—in the low field region, the processes are reversible domain wall translation, reversible domain wall rotation, and reversible rotation of sd magnetization. These are nonhysteretic processes and the magnetization state returns to the initial state after application of the field. If the field is low enough (say, 0.01 mT or 0.1 Oe) it can be assumed with confidence that the stable NRM of the sample has not been affected by the susceptibility determination. It is, however, a good practice to measure the NRM of a sample and record it before measuring low-field susceptibility. An ac inductance bridge is commonly used to determine the out-of-balance signal when the specimen is introduced in one of the matched coils. The sensitivity of such a bridge is of the order of 1.2×10^{-7} (1×10^6 G/Oe). An older approach is to switch off the vertical field cancellation coils in an astatic magnetometer and measure the induced magnetization (**J**) of the specimen in the presence of the known vertical component of the Earth's magnetic field (**H**). The low-field susceptibility tensor (χ) is then given by

$$\mathbf{J} = \chi \cdot \mathbf{H}$$

In order to simplify the measurement procedure, the tensor relationship is relaxed usually by making measurements along principal axes defined for this purpose and the three vector components χ_1, χ_2, χ_3 are determined according to the relationship

$$\mathbf{J}_i = \chi \cdot \mathbf{H}$$

In a high-field measurement the apparatus is usually a susceptibility balance or a vibrating-sample magnetometer. The most popular susceptibility balance utilizes a commercially available electrobalance and an electromagnet with shaped ("Faraday") pole caps that produce a constant product $H \cdot (dH/dZ)$ inside the measurement region. The force (**F**) exerted on a vertically hanging specimen is measured and susceptibility (χ) determined from the relationship

$$\mathbf{F} = \chi \cdot \mathbf{H} \cdot (d\mathbf{H}/dZ)$$

The applied field is in the range 200–1000 mT (2000–10,000 Oe) and the sensitivity is better, down to 1.2×10^{-8} (1×10^{-7} G/Oe), than that in an ac bridge. In a vibrating-sample magnetometer, one simply measures the slope of the J–H curves in low or high fields and the susceptibility is equal to the slope. In terms of sensitivity, the vibrating-sample magnetometer is about the same as or one order of magnitude worse than the susceptibility balance.

There are two broad areas of applicability for susceptibility data. The first is in characterizing the NRM of a rock and the second is that of characterizing the magnetic carriers in a given rock. For the first group of applications, a knowledge of low-field susceptibility allows one to calculate the induced magnetization in a rock in the Earth's field and then compare it with the NRM of the rock. The usual procedure is to divide the NRM intensity (J_n) by the induced magnetization (χH, where H is the Earth's field) to obtain the so-called Koenigsberger ratio (Q_n):

$$Q_n = J_n/\chi H$$

Q_n values of 1 or greater show that the NRM is as strong as or greater than the induced magnetization, and in order to calculate the magnetic anomaly over such a body, the two must be separately determined before combining so as to obtain a true measure of the anomaly signal. Q_n values of 10 or greater suggest that the NRM is the predominant source for the magnetic anomaly. When Vine and Matthews (1963) offered their now-famous explanation for the linear marine magnetic anomalies, they made the assumption that for the magnetic source layer under the ocean $Q_n \gg 1$, and hence the normal and reversed anomaly signals were produced by antiparallel NRMs recorded in the oceanic crust. Conversely, when a paleomagnetic collection yields samples with $Q_n \ll 1$, it is fair to conclude that unless the NRM has very much higher stability to AF or thermal demagnetization than that of the induced magnetization, the samples will not yield reliable information about ancient directions of magnetization. It is therefore customary to measure NRMs of a newly collected suite of rocks, and then test for viscous magnetization (described earlier) and measure the low-field susceptibility.

The second major application of low-field susceptibility to remanence studies is related to rocks with considerable magnetic anisotropy. Uyeda et al. (1963) showed that for a variety of rocks—igneous, metamorphic, or sedimentary—the presence of magnetic anisotropy can deflect the acquired NRM direction by as much as 60°. One way to approach the problem is to avoid the use of rocks with pronounced petrofabric, but often the inherent fabric may not be apparent as in some sedimentary rocks. In those cases it is of advantage to measure the anisotropy of susceptibility and to plot the data on a stereonet (Fig. 21) to see if (a) the magnitude of the anisotropy is large (>10%) and (b) whether the NRM directions are strongly influenced by the location of the anisotropy maximum. Kent and Lowrie (1977) have pointed out that specimen shape and sensor coil geometry can very strongly influence the observed anisotropies, and they recommend the use of a low-field torque magnetometer (Stone, 1967) for accurate determinations of anisotropic susceptibility. The shape of the susceptibility ellipsoid commonly found in rocks is that of an oblate ellipsoid with the maximum and intermediate directions lying in the equatorial plane. The determination of the amount of anisotropy plus a reference to the charts of remanence deflection in Uyeda et al. (1963) can lead to an es-

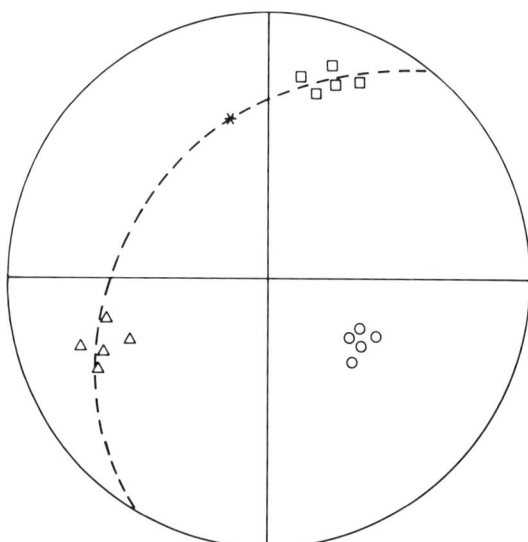

FIG. 21. Equal-area stereonet projections of maximum, minimum, and intermediate susceptibility axes determined from five samples. The average NRM direction seems to be an artifact of susceptibility anisotropy. (□) Maximum χ; (○) minimum χ; (△) intermediate χ; (*) NRM direction.

timate of the error caused by the deflection of the NRM vector from the true ancient field direction.

I shall now discuss a few cases where susceptibility data have helped to characterize the carriers of remanence or simply to identify the mineral present and provide an estimate of its grain size. The ac bridge method for low-field susceptibility measurement can be easily adapted for the measurement of long cores (Radhakrishnamurthy *et al.*, 1970). In such an instrument, the whole core is translated slowly through one of the matched coils and a large Helmholtz coil system provides a low-amplitude alternating drive field. The variations of observed susceptibility are related to changes in both chemistry (e.g., magnetite versus hematite) and grain size. As a first approximation, however, the chemically produced variations are the dominant ones, especially if magnetite has been identified as the magnetic mineral present. A low-field susceptibility log in such a situation can be a reliable measure of the fluctuations in magnetite content with time (Thompson, 1975). For a more detailed analysis, Fuller (1974) has suggested a graphical approach in which the ratio of IRM (J_r) to saturation magnetization (J_s) is plotted against the ratio of low-field susceptibility (χ) to saturation magnetization (J_s). When the data for a large number of lunar samples (with iron as the magnetic carrier) were plotted in this manner, a cluster corresponding to md grains could be distinguished from another one containing sd and psd grains. In the Section 4.2 suggestions will be given for a few other graphical plots. One should again emphasize here the difference between low-field susceptibility of md grains, on one the hand, and sd and psd grains, on the other. In the first group the chief physical process responsible for the observed susceptibility is reversible domain wall translation, leading to fairly large values. In the second group, i.e., sd and psd grains, magnetization increase can take place chiefly by coherent rotation and a low field is not very effective. Therefore low-field susceptibility of sd and psd grains is usually much lower than that of md grains. The situation is made complex if superparamagnetic grains are also present. They help to increase the susceptibility of sd and psd grains to such an extent that they appear to be md grains.

High-field susceptibility of whole rock specimens is due to a variety of reasons, the most common of which is the nonsaturable (at room temperature) contribution due to iron impurities (Fe^{2+}, Fe^{3+}) in silicate grains, hematite (Fe_2O_3) grains, and superparamagnetic ultrafine grains, if any. The iron impurities are paramagnetic in nature. Therefore both they and the superparamagnetic grains can be identified if high-field susceptibility is measured at increasingly lower temperatures. Susceptibilities measured at different values of temperature (T) and applied field (H) can be superposed on one another to yield a unique χ versus H/T plot, where T is the ambient temperature in degrees Kelvin. This is just a special case of the general susceptibility versus temperature plots, which can help to identify the types of interatomic magnetic ordering responsible for ferromagnetism, ferrimagnetism, and antiferromagnetism. Since the nature of magnetic ordering is not normally a subject of inquiry in paleomagnetism and rock magnetism, it will not be dealt with here any further. Interested readers are referred to the textbooks on rock magnetism by Nagata (1961) and Stacey and Banerjee (1974).

There is another type of experiment on temperature dependence of susceptibility that is very useful in paleomagnetism and rock magnetism. The two most common magnetic minerals, hematite and magnetite, possess isotropic points below room temperature where the intrinsic magnetocrystalline anisotropy goes to zero before changing sign. Therefore, if low-field susceptibility is measured as a function of temperature, anomalous peaks in susceptibility are observed at these characteristic temperatures. In the case of pure hematite the transition is called the Morin transition and is located at 263°K. For magnetite the isotropic point is at 130°K. Thus, by simply measuring low-field susceptibility down to the liquid-nitrogen temperature (77°K), it is possible to distinguish between the presence of pure magnetite and pure hematite in a rock. Impurities and fine grain size can shift the isotropic points and sometimes suppress them completely. Stacey and Banerjee (1974) have described these variations.

The final effect to be discussed here is the Hopkinson effect in low-field susceptibility. When the temperature of a magnetic compound is slowly raised, a point is reached immediately below the Curie temperature where the spontaneous magnetization (J_s) is still present, but most of the anisotropies that control the direction of magnetization are vanishingly low. This is because other parameters like magnetocrystalline anisotropy, shape anisotropy, magnetostriction vary as J_s^n, where n varies from 2 to 9. Therefore, as temperature is increased, the restrictions on the orientation of the magnetization decrease, and close to but below the Curie temperature it becomes possible to orient the magnetization with great ease using a very weak applied field. This is the Hopkinson effect of an anomalous low-field susceptibility rise below the Curie temperature. [This effect should be distinguished from the phenomenon of blocking temperature, which is a grain size-dependent effect (for sd grains).] The Hopkinson effect is more dramatic if the low-field susceptibility at room temperature is low—the rise is then much more remarkable. As previously explained in this section, sd and psd grains have lower susceptibility than do md grains. Therefore, the Hopkinson effect is more obvious if sd or psd grains are present in the rock.

4.2. Hysteresis Parameters

As explained in the Introduction, all magnetic materials that exhibit remanence also display magnetic hysteresis in a cyclic field. The hysteresis is due to irreversible magnetization changes, is proportional to the area of the J versus H loop (see Fig. 1), and shows up as heat in the specimen. The applied steady magnetic field H is varied from a peak value (say, positive sign) to the peak value of the opposite sign (negative) and then back to the initial peak value. The steady field can be replaced by an AF-driven solenoid, but in that case, because of the high inductance of multiple-turn solenoids, the peak value is limited usually to about 200 mT (2000 Oe). When a steady field is used, the peak value can be raised routinely to 1500 mT (15,000 Oe), and with special cobalt–iron pole pieces of an electromagnet, to as much as 2000 mT (20,000 Oe). The most convenient apparatus for measuring high-field hystersis loops is a vibrating-sample magnetometer combined with an electromagnet with a programmed electronic sweep. The working volume is usually small because of the requirement of field homogeneity. Thus specimens are only a few hundred milligrams in weight and are packed into a holder so that they cannot move when vibrated. Low-temperature Dewar flasks and high-temperature furnaces extend the temperature range of observation from 4.2 to 1000°K.

It should be appreciated that the magnetic parameters extracted from a hysteresis loop display effects of strong intergrain interaction because of the presence of a large ambient magnetic field. The parameter that suffers least from such interaction effects is the saturation magnetization (J_s) measured usually at fields in excess of 1000 mT (10,000 Oe). There are two precautions, however, that should be kept in mind. The first precaution applies to all the parameters measured from a hysteresis loop. Before reading off any data the specimen should be cycled from $+H_{max}$ to $-H_{max}$ and back to $+H_{max}$ at least three or four times until the loop looks obviously symmetric (e.g., $|J_{max}|$ and $|J_{min}|$ should be the same within experimental error). The second is that if J_s is being determined from $\frac{1}{2}(|J_{max}| + |J_{min}|)$, in some cases at least there is a large paramagnetic contribution (χH_{max}) arising from the unsaturable iron impurity ions in the silicate minerals. A quick way to subtract this contribution is to extrapolate backward the high-field part of the loop to determine the value of J_s from the intercept on the J axis at $H = 0$ (see Fig. 1). In a mixture of magnetic and nonmagnetic materials, if the magnetic material is homogeneous in terms of chemistry, measurement of J_s leads to an estimate of the fluctuations in the content of the magnetic fraction. Thus room-temperature J_s values of a sediment core as a function of depth can lead to an estimate of variations in magnetite content if it is the sole magnetic carrier. Saturation magnetization measured at low temperatures (e.g., 77 or 4.2°K) in a high field (1500 mT or 15,000 Oe) can lead to an estimate of cation distribution in a pure compound as was done for the titanomagnetite series by O'Reilly and Banerjee (1965). However, this is a nonunique approach and must be supplemented by other types of measurements such as electrical conductivity (O'Reilly and Banerjee, 1965) or Mössbauer effect (Jensen and Shive, 1973).

The remanence ratio J_{rs}/J_s is the next important parameter that can be determined from a hysteresis loop. J_{rs}, the saturation isothermal remanent magnetization, is usually determined from the intersection of the descending branch of the hysteresis loop with the magnetization (J) axis (see Fig. 1). However, if the specimen is strongly magnetic and its shape departs from that of a sphere by as much as 5–10%, it is necessary to apply a correction to the observed J_{rs} value (Stacey and Banerjee, 1974). The correction is based on the inequality of the demagnetization factors (N) in the different directions in the sample. The magnitude of remanence ratio is a sensitive indicator of the magnetization state of a specimen. Values less than 0.1 indicate that the specimen is composed of large md grains. Alternatively, if we have reason to believe that there is a large fraction of fine grains in the specimen, $J_r/J_s < 0.1$ indicates that an appreciable fraction of these grains are superparamagnetic and thus responsible for the low remanence ratio. Remanence ratios between 0.1 and 0.4 indicate that the grains are most likely in a psd state. Values between 0.4 and 0.5 indicate sd state; ideally the ratio should be exactly 0.5 for uniaxial magnetic anisotropy but it is usually lower in a natural specimen because of dilution with psd, md, or superparamagnetic grains. The assumption of predominant uniaxial magnetic anisotropy has been borne out by practice for most natural compounds even though crystallographically some are cubic (magnetitie) and some trigonal (hematite). It should be noted that the above approximate method of identifying the magnetic state of an ensemble of grains is no less accurate than the methods based on the comparison of AF demagnetization curves of artificial remanences (Lowrie and Fuller, 1971; Johnson *et al.*, 1975). The main problem with both of the methods is that in nature the grain ensembles are rarely found to be homogeneous in magnetic state and therefore it is difficult to identify whether they are *mostly* sd, psd, or md.

The third parameter measured from a hysteresis loop is coercive force H_c, obtained from the second (or "northwest") quadrant of a hysteresis plot (Fig. 1). After a specimen has seen the largest field $\mathbf{H}_{max}$ in the positive direction, the field is decreased until $\mathbf{H} = 0$. At this point there is left a remanent magnetization J_r due to irreversible hysteretic processes. Coercive force H_c is the magnitude of the reverse or negative field necessary to reduce J_r to zero while the negative field is still present. If, however, the field is withdrawn at this point, the specimen will have a fraction of the original J_r still present. Thus an even large negative field (H_{cr}) will have to be applied if the remanence J_r is to go to zero when the negative field is withdrawn completely. This value, H_{cr}, is called the coercivity of remanence or remanence coercivity. The term coercivity replaces coercive force when it has been assured that the applied $\mathbf{H}_{max}$ was large enough to have saturated the specimen. The evidence of saturation is that the hysteresis loop has closed to form a line, and this line has a fairly small slope with the field axis. The symbol for coercivity is the same as that of coercive force H_c. As has been mentioned earlier, both coercive force and the coercivity of remanence of an ensemble are influenced by magnetostatic grain interactions. The interactions lower the observed H_c drastically if even a part of the ensemble is composed of md, paramagnetic, or superparamagnetic grains, i.e., those that have either extremely small intrinsic H_c or zero value of H_c. For an estimate of the stability of remanence, H_{cr} is a much more reliable measure. It can be determined without recourse to a hysteresis loop simply with the use of an electromagnet and, say, a spinner magnetometer. First, the specimen is saturated in a given direction, and as soon as the specimen is placed in the zero-field region of the spinner magnetometer, the specimen reaches the remanent state J_r. After the measurement of J_r, by trial and error a set of negative fields are applied and the resulting remanences are measured until J_r reaches zero or acquires a negative value. If the latter is the case, H_{cr} can be determined by linear extrapolation between the field values that gave rise to small but positive J_r and those that were small but negative (Fig. 1). There are two useful relationships involving the various hysteresis parameters so far described:

$$J_{rs} = H_c/N$$

and

$$\chi = (H_{cr} - H_c)/H_{cr}N$$

These formulas have been derived by Stacey and Banerjee (1974) and their applications have also been indicated there. They are highly suitable for determining graphically H_{cr} or N values. In the same work the dependence of coercivity on grain size has been discussed and graphical approaches have been outlined that lead to a determination of the type of crystallographic defects present in a specimen.

4.3. Thermomagnetic Properties

Thermomagnetic properties refer to those magnetic properties of a specimen that are determined from temperature-dependent studies. The simplest and most commonly measured parameters are the Curie temperature and the blocking temperature. In a susceptibility-measuring device the specimen is heated until susceptibility shows a sharp rise (Hopkinson effect) immediately followed by steady low values indicating the paramagnetic state. The high-temperature side of the Hopkinson peak is the Curie temperature defined as the temperature above which the specimen is paramagnetic. In a vibrating-sample magnetometer, induced magnetization in constant fields >200 mT (2000 Oe) is measured as a function of temperature until the magnetization shows a rapid decrease and above the Curie temperature acquires steady low values due to paramagnetism. If the field is fairly high, e.g., 1000 mT (10,000 Oe), the paramagnetic contribution is also high and it is difficult to discern clearly the Curie temperature. In such a situation the inflection point in the magnetization–temperature $(J-T)$ plot is identified as the Curie temperature. The applied field may be decreased somewhat in order to alleviate the problem of finding the inflection point, but the field should not be lower than the "knee" of the hysteresis loop (Fig. 1) in order to ensure that the observed Curie temperature will not vary with the applied field. Unlike the Curie temperature, the blocking temperature (or temperatures) are determined by heating the specimen in zero applied field so that the critical temperature or temperatures are identified at which the remanent magnetization drops sharply. Above the highest blocking temperature the remanence should be zero. The usual practice is to use a noninductively wound furnace (in a field-free space) and a temperature controller to heat the specimen to given temperatures steps, to hold the specimen for 5–15 min at the set temperature in order to equilibrate, and then to turn the furnace off and cool the specimen with blown cold air so that it reaches room temperature again. It is crucial that there be no field present during the cooling, otherwise the specimen will acquire a new TRM instead of being thermally demagnetized. This problem can be particularly acute if the specimen contains thermally unstable nonmagnetic oxides, hydroxides, or sulfides, which on heating may form new, magnetic compounds. To prevent these secondary magnetic compounds from acquiring a TRM, the ambient field should be canceled to less than ±5 nT. Blocking temperatures can be determined more quickly with the use of an astatic, a spinner, or a ballistic magnetometer with a built-in furnace.

The importance of determining both the Curie temperature(s) and the blocking temperatures lies in the ability to identify correctly all the magnetic phases contributing to an observed remanent magnetization. This is particularly true if more than one blocking temperature has been found; if the highest one appears to be close to the Curie point of a known compound (e.g., magnetite), it is tempting to attribute a lower blocking temperature to magnetite grains also, albeit with much smaller grain sizes. Only a Curie temperature measurement will tell, however, whether this latter blocking temperature really corresponds to a discrete magnetic compound (e.g., pyrrhotite) with its distinctive lower Curie temperature. For those workers interested in self-reversal or remanence due to physiochemical mechanisms inherent in a specimen or those interested in the nature of magnetic exchange interactions amongst the magnetic ions in a compound, it should be emphasized that continuous determinations of induced magnetization versus temperature ($J-T$) and/or thermoremanent magnetization versus temperature ($J_{tr}-T$) are essential. Examples of applications may be found in Stacey and Banerjee (1974).

The last technique to be discussed under thermomagnetic properties is thermal fluctuation analysis—a new technique based on Néel (1949) theory of thermal fluctuations as recently developed by Dunlop (1976). Néel advanced the concept of thermal fluctuations of magnetization and relaxation time constant (τ) in order to provide a theory for TRM of sd grains. As discussed earlier in this article, thermal fluctuations arising from the Boltzmann energy (kT) are responsible for a finite probability of reversal in sign of remanent magnetization. The same physical process, however, can produce an effective thermal fluctuation field (H_q) that is related to the observed coercive force (H_c) in the following manner:

$$H_c = H_{cr} - H_q$$

where

$$H_q = [2kTn(f_0 t)H_{cr}/vJ_s]^{1/2}$$

and k is Boltzmann's constant, T is ambient temperature (°K), $f_0 = 10^9$ sec^{-1}, t is experimental time, say, 1 sec, H_{cr} is the intrinsic coercive force necessary to reverse the remanence of a grain, v is grain volume, and J_s is saturation magnetization at ambient temperature. Dunlop (1976) has pointed out that this equation could form the basis of a method ("thermal fluctuation analysis") that would allow the determination of the average $H_{cr}(\langle H_{cr}\rangle)$ and the average $v(\langle v\rangle)$ of a given ensemble. Of course, the method is exact only for sd grains with predominantly uniaxial shape anisotropy for which the Néel formalism was derived, but Dunlop claims that md and psd grains could also be approached by this method as long as it is borne in mind that $\langle v \rangle$ in these latter cases refers not to the physical volume but to an "effective volume" that must be activated in order to cause magnetization changes. Dunlop's derivation utilizes the temperature variations of H_c and J_s and is as follows:

$$\frac{H_c(T)}{j_s(T)} = \langle H_{cr}\rangle - \left[\frac{2k \ln(f_0 t)}{J_s}\right]^{1/2} \langle H_{cr}\rangle^{1/2} \frac{1}{\langle v\rangle^{1/2}} \frac{T^{1/2}}{j_s(T)}$$

where $j_s(T) = J_s(T)/J_s$ (room temperature). The experimental approach is to measure J_s versus T and H_c versus T from the temperature dependence of hysteresis loops in order to fit the above equation and determine the two unknowns, $\langle H_{cr} \rangle$ and $\langle v \rangle$. However, as we have mentioned earlier, the observed H_c values do not reflect the true coercive force of the sd grains because of admixture with and influence of md, paramagnetic, or superparamagnetic grains in an ensemble. Dunlop (1976) therefore recommends the determination of the coercivity of remanence (H_{cr}) of the whole ensemble as a function of temperature and either replacement $H_c(T)$ by multiplying with measured $H_{cr}(T)$ or correction of $H_c(T)$ by multiplying by a factor equal to the ratio H_{cr}/H_c as measured at room temperature. Figure 22 shows a fit of the theoretical equation with the experimentally observed parameters for three sd ensembles with different $\langle H_{cr} \rangle$ and $\langle v \rangle$ values (Dunlop, 1976). $\langle H_{cr} \rangle$ is obtained from the intercept on the ordinate axis and then utilized in the expression for the slope of the straight line [the second term in Dunlop's equation for $H_c(T)/j_s(T)$] to calculate $\langle v \rangle$. As is clearly seen from Fig. 22, the approach is most successful for rod-shaped grains with strong uniaxial anistropy and is less successful for

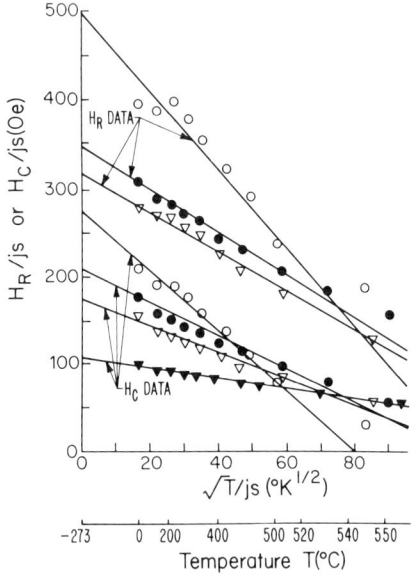

FIG. 22. Dunlop's method of thermal fluctuation analysis. H_{cr} is obtained from the intercept and the slope leads to the determination of V. As the data shows, closest agreement with theory is shown here for the 2200-Å-long rod-shaped grains. (○) 370 Å; (●) 760; (▽) 1000; (▼) 2200.

the cube-shaped grains, where the effective anisotropy is of magnetocrystalline origin. The actual value of $\langle v \rangle$ was known for only one of the rod-shaped grains, and there was good agreement between the value of $\langle v \rangle$ deduced from thermal fluctuation analysis and that known from observation. Dunlop (1976) has pointed out that the thermal fluctuation analysis is most successful when the $H_c(T)$ and $J_s(T)$ data are obtained far from the blocking temperatures. Hence ensembles with a single sharp and high blocking temperature are most suitable for this kind of analysis, and not those with a broadly distributed range of blocking temperature, when single values of $\langle v \rangle$ and $\langle H_{cr} \rangle$ fail to describe the ensemble.

4.4. Magnetocrystalline Anisotropy and Magnetostriction

The two critical parameters magnetocrystalline anisotropy and magnetostriction are intrinsic, dependent on the chemistry and crystallographic structure of a magnetic mineral and independent of their grain size. A knowledge of these two parameters and the shape anisotropy of a given grain enables one to estimate the magnitude of the energy barrier to magnetization reversal. It is true that in pure magnetite the shape anisotropy is the predominant contributor to the energy barrier, but this is not so for titanomagnetites ($Fe_{3-x}Ti_xO_4$), hematite, and pyrrhotite, to name only a few of the other magnetic minerals. The determination of the critical grain size thresholds (d_s) for thermally stable behavior of this last group of compounds requires a knowledge of magnetocrystalline anisotropy and magnetostriction. Magnetocrystalline anisotropy arises from the spin–orbit coupling of ionic magnetic moments, resulting in crystallographically controlled "easy" and "hard" directions of magnetization, even in a perfect sphere (Stacey and Banerjee, 1974). The magnetocrystalline energy (E_K) for a cubic crystal is given by

$$E_K = K_1(\alpha_1^2\alpha_2^2 + \alpha_2^2\alpha_3^2 + \alpha_3^2\alpha_1^2) + K_2(\alpha_1^2\alpha_2^2\alpha_3^2)$$

where K_1 and K_2 are empirical anisotropy constants; α_1, α_2, and α_3 are direction cosines. For a given composition such as iron or magnetite there are characteristic values of K_1 and K_2. Of course, K_1 and K_2 vary as a function of temperature. For K_1 the classical theory of magnetic interactions predicts a dependence on saturation magnetization (J_s) such that

$$K_1(T) \propto J_s^n(T)$$

where $n \approx 10$ for cubic structure.

Values for K_1 and K_2 may be determined in one of two common ways: either induced magnetization (J) is measured as a function of applied field

(H) along three principal directions of a single crystal sphere, or the sphere is suspended in a high-field torque magnetometer (Banerjee and Stacey, 1967) and the observed magnetic torque as a function of orientation of the sphere is measured in a given crystallographic plane. In both cases an electromagnet capable of producing very high fields (e.g., 1000–1500 mT) must be used since the specimen must be truly saturated. The specimen is ground and polished to make a perfect sphere. Even a 1% variation in the diameter can cause considerable error in strongly magnetic materials such as magnetite or iron. In addition, the specimen should not be under a nonuniform stress while the measurements are conducted because it will lead to spurious contributions to K_1 and K_2 values. The measurement techniques become slightly more complex when the ambient temperature must be varied. Fletcher *et al.* (1969) have described an automatic torque magnetometer for high-temperature measurements. Figure 23 shows the two methods for determining K_1 and K_2 for magnetite.

Magnetostriction has been defined as the strain dependence of magnetocrystalline anisotropy. Its microscopic origin, therefore, lies in the same

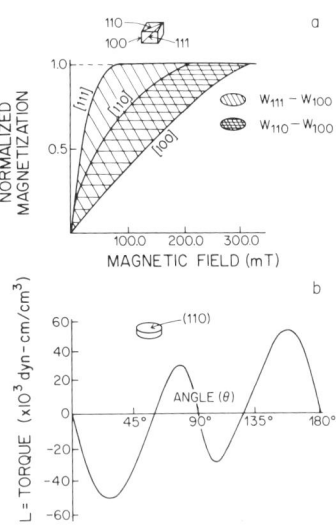

Fig. 23. Two methods for determining K_1 and K_2 of magnetite. (a) Curves are determined from field dependence of magnetization along different crystallographic axes of a cubic specimen. $K_1 = 4(W_{110} - W_{100})$; $K_2 = 27(W_{111} - W_{100}) - 36(W_{110} - W_{100})$. (b) Curve is obtained from the angular dependence of magnetic torque (L) experienced along different directions in the (110) plane of a suspended disk-shaped specimen. $L = -(K_1/4 + K_2/64) \sin 2\theta - (3K_2/8 + K_2/16) \sin 4\theta$.

spin–orbit coupling of ionic magnetic moments that is responsible for magnetocrystalline anisotropy. If a specimen is magnetized along the different principal crystallographic axes, magnetostriction can be observed as a linear strain ($\pm \Delta l/l$) along those directions. The converse of magnetostriction is piezomagnetism, where changes in magnetization can be observed when a specimen is nonuniformly stressed. Piezomagnetism forms the basis of the seismomagnetic effect that is being eagerly sought as a precursor to major earthquakes. For cubic crystals the magnetostrictive strain ($\Delta l/l$) can be expressed as

$$\frac{\Delta l}{l} = \tfrac{3}{2}\lambda_{100}(\alpha_1^2\beta_1^2 + \alpha_2^2\beta_2^2 + \alpha_3^2\beta_3^2 - \tfrac{1}{3}) + 3\lambda_{111}(\alpha_1\alpha_2\beta_1\beta_2 + \alpha_2\alpha_3\beta_2\beta_3 + \alpha_3\alpha_1\beta_3\beta_1)$$

where λ_{100} and λ_{111} are characteristic constants that are saturation values of $\Delta l/l$ along crystallographic axes [100] and [111]; α_1, α_2, and α_3 are direction cosines of the magnetization; and β_1, β_2, and β_3 are direction cosines of the direction of measurement of $\Delta l/l$.

For polycrystalline assemblages with no preferred orientation, as is common in most rocks used in paleomagnetism, an average magnetostriction constant (λ_s) may be used, given by

$$\lambda_s = 2\lambda_{111} + \lambda_{100}$$

For magnetite λ_s is approximately equal to $+40 \times 10^{-6}$, which must be multiplied by the magnitude of the prevailing nonuniform stress to give the amount of magnetostrictive anisotropy energy per unit volume in a given magnetite grain. The usual method for the determination of magnetostriction constants λ_{111} and λ_{100} is to attach strain gauges of the resistance type along the principal directions [111] and [100], and then measure the saturation values of $\Delta l/l$ along the given directions when the applied field is also applied in parallel. Strain gauges must be carefully chosen so that they do not contribute significant errors due to temperature dependence of resistivity or due to intrinsic magnetoresistance, a change in the value of resistance due to the effect of the applied magnetic field. Care must also be exerted to bond the gauges to the specimen as securely as possible and to make sure that the specimen has been thoroughly demagnetized in all directions prior to measurement. The latter step is essential for making sure that the observed saturation value of $\Delta l/l$ in a given direction is indeed the maximum possible length change. The actual measurement of the change in resistance is made by making the sensing strain gauge part of a four-arm resistance bridge and using a sensitive potentiometer. Syono (1965) has described the techniques for measuring both magnetocrystalline anisotropy constants for magnetite and titanomagnetite. A table collating the values of Curie temperature, saturation magnetization, anisotropy constants, and magnetostriction constants for magnetite and titanomagnetite has been given by Stacey and Banerjee (1974). Measurements of magnetostriction constants at elevated temperatures are required for understanding the origin of TRM. However, the measurements are extremely sensitive to fluctuations in temperature of the sensing gauge and precautions must be taken to isolate the specimen in a thermally insulated chamber. Klapel and Shive (1974) have measured the magnetostriction constants λ_{111} and λ_{100} for magnetite from room temperature to near the Curie temperature. Although the higher temperature data had a large amount of scatter, it appears likely that the temperature (T) dependence in the interval studied is linear, contrary to the classical theory, which predicts a T^3 law.

So far as piezomagnetism and the seismomagnetic effect are concerned, there have been a large number of experiments to determine the approximate magnitude of the seismomagnetic anomaly that should accompany a stress release of, say, 10^5 Pa (100 bar). Unfortunately, most of the experiments to date have been performed on rock specimens that were not studied thoroughly for their petrology and microstructure. Thus it is not quite clear how reliable are the determined magnitudes of the seismomagnetic effect. It is known, for example, that the application of stress will produce two types of changes—the change in the NRM and the production of a low-field susceptibility anisotropy that will cause an observed anomaly in magnetization induced by the geomagnetic field. The relative contributions of these two effects and their relative directions will influence strongly the observed seismomagnetic effect, and that would depend among other things on the composition, size, shape, and texture of the magnetic minerals. Kean *et al.* (1976) have made a first attempt at this kind of study, which has been followed up by Revol *et al.* (1977). Field studies have been generally inconclusive up to now, although there are indications in some recent experiments that there has been seismomagnetic effects of a few nanotesla (i.e., 10^{-4} × geomagnetic field) preceding some earthquakes along the San Andreas fault in California (Johnston *et al.*, 1976). The field observations are carried out with proton precession magnetometers, and it is necessary to make simultaneous or near-simultaneous measurements at several length scales so as to subtract the base level variations produced by fluctuations in the geomagnetic field that manifest themselves on a large length scale, while the seismomagnetic effect should have a smaller length scale near the epicentral region.

4.5. Magnetic Domain Structure

Magnetic domains, i.e., zones of uniform magnetization in a grain separated by $\sim 10^3$-Å-wide domain walls, exist in the presence of low (e.g., ~ 1 mT or 10 Oe) or zero ambient field in magnetic grains whose effective diameter (d) is greater than d_c, the critical diameter above which nonuniform magnetization becomes an energetically favorable state. Domain walls move rapidly in response to fields as low as 1–10 mT (10–100 Oe), and it is easy to saturate magnetically the larger grains (~ 100 μm) with fairly low fields. This is the experimentally observed phenomenon that was explained by postulating the existence of domains and domain walls. Now, however, their reality has been proven beyond doubt with the aid of magnetic colloids, which when applied to the surface of a grain delineate the locations of the domain walls. To study the domain structure of the magnetic grains in a rock specimen, it is first necessary to polish the surface with fine-grained alumina. The polishing process creates a strained surface layer that masks the intrinsic domain structure. Therefore etching of the surface layer is necessary. This may be done by chemical methods such as HCl–HF solutions or by sputtering with an inert gas such as argon in a vacuum. The colloid solution is prepared by precipitating colloidal magnetite grains of extremely small size (10–100 Å) and then "stabilizing" them in a soap solution so that they do not coagulate to form grains larger than the domain wall thickness. In that case, the resolving power of the technique is vastly diminished. This method of observing magnetic domains has been perfected for minerals and rocks by Soffel (1969) and the reader is referred to his articles for technical details.

There are two principal uses of magnetic domain structure studies. The first is simply to identify which of the opaque grains observed under a petrological microscope truly contributes to the observed magnetization. In particular cases when a grain is only partially magnetic (because of compositional variations etc.), this technique can be very useful (Soffel, 1977). There is also potential for such a technique to be applied to altered material such as submarine pillow lavas with variable zones of low-temperature oxidation. The second application is more important for identifying which grains in a given mineral ensemble are likely to carry the most stable components of NRM. If the magnetic colloids show that the most easily visible, i.e., the larger, grains contain a multiplicity of domain walls, it can be concluded that in the first order these grains are the least likely to possess stable remanence, since even a small field fluctuation will alter their magnetic remanent state. On the other hand, if it can be shown that some of the very same grains have domain structure that is very resistant to change on the application of laboratory fields, it may be concluded that they are psd grains and may be responsible for the stable remanence. In that case, a petrological study of their type of occurrence may lead to a clearer and direct idea of the the origin of the stable remanence (Soffel, 1969; Halgedahl and Fuller, 1980).

Temperature-dependent studies of domain structure are difficult because most fluids used in colloid preparation can only withstand a small range of temperature variation on either side of room temperature. Transmission or scanning electron microscopy can then be attempted as has been done for ferromagnetic metals, but as yet no such thorough application has been made for minerals and rocks.

5. Future Trends

The above discussion of the various methods in paleomagnetism and rock magnetism has mainly dealt with the more widely known and extensively applied approaches. I would now like to describe the future trends in the field. The field of paleomagnetism, i.e., applied rock magnetism, has indeed come of age. Its most well-known contribution has been a quantitative proof of the phenomenon of continental drift through the determination of past latitudes of different continental masses of Paleozoic age. In doing so, the paleomagnetists have relied solely on a "cut-and-dry" approach whereby only those rocks that carry mostly a single stable component of NRM have been used, the others have been discarded, and the identification of an average paleomagnetic pole from a given number of virtual geomagnetic poles contributed by different formations has been achieved by the sheer strength of statistics, i.e., increasing the sampling size until the signal-to-noise ratio is satisfactorily high. In the statistical analysis (McElhinny, 1973) there has been no attempt at weighting the information obtained from the different samples or at analyzing the information content of the so-called secondary components of NRM, i.e., those that display lower stability to thermal and AF demagnetization. Now, however, the time has come for attacking the "second generation" problems (definitely not "second order") in paleomagnetism and we find that the previous "cut-and-dry" approach is unsuitable for the present types of problems. An example that comes to mind immediately is the problem of Precambrian paleomagnetism. The Precambrian period constitutes the major part of the Earth's history—the first 4 billion years. Admittedly, the nonmagnetic problem of obtaining accurate dates for rocks of this age is a severe one, but to my mind, there has recently been more progress in the age determination of these rocks than in deciphering the history of multiple components found in the NRM of Precambrian rocks. With a few notable exceptions, the general approach has been that of selective demagnetization (by thermal, AF, or chemical technique) of the observed NRM of a suite of rocks, and as soon as there emerges an average direction with an acceptably low scatter (or a small cone of confidence at the 95% level), this direction has been dubbed the stable NRM direction. Roy and Lapointe (1978) have recently criticized this kind of an approach and pointed out that even after thermal or AF demagnetization has been able to isolate a component of high stability, the use of another demagnetization technique such as chemical leaching can produce a completely new (sometimes 180° reversed) stable NRM direction. Although this observation supports my thesis, it is worth pointing out that these different NRM directions have no age-labeling, especially in sediments, and subjective judgments must be exercised in order to assign the "primary" label to one of the many apparently stable components. In the examples provided by Roy and Lapointe (1978), for example, it is by no means certain that even the direction isolated after chemical leaching is, in fact, the primary stable direction. Who knows what new direction would have emerged if after chemically leaching the specimen was resubjected to AF demagnetization? The point worth making is that for older rocks particular attention must be paid to *all* the components separated by selective demagnetization techniques, and each one of these should be treated as a likely candidate for primary component unless petrology or isotope geochemical evidence is conflicting in nature. With this example in mind, I shall now indicate some of the future trends in the methodology of paleomagnetism and rock magnetism.

5.1. *Studies on Single Grains*

If the analysis of the multiple components of NRM in a given rock specimen is the problem of the future, a direct way to approach it would be to study the individual contributions to the NRM by the individual mineral grains. After all, such grains are separated by hand for radiometric age determination and it would make sense if such "labeled" grains could also be utilized for NRM measurement. Before the advent of the superconducting or cryogenic magnetometer such talk was indeed fanciful because even for strongly magnetized grains the total signal is small, of the order of 10^{-5} A m^{-1} (10^{-8} G). In a narrow (1 cm) bore cryogenic magnetometer, however, it is possible to measure the signal from such small grains, and the true physical sites of stable (or unstable) components of NRM can be identified (Wu *et al.*, 1974). By carrying out radiometric dating on such grains, it may further be possible to assign ages to such components of magnetization (Wu *et al.*, 1976).

Hysteresis parameters of single grains have also begun to be studied, although not among paleomagnetists and rock magnetists. Zijlstra (1967) has developed a vibrating reed magnetometer that is capable of measuring induced magnetization (J) versus field (H) of single grains of rare earth–cobalt alloy grains, which are of great interest to researchers in the permanent magnet industry. It would seem possible to employ such a magnetometer to study the grain size-dependent magnetic properties of important minerals in order to resolve the question about the exact size range of psd grains.

5.2. Comprehensive Instruments

At the beginning of this section I emphasized the need for separation of stable but younger chemical remanent magnetization (CRM) and partial thermoremanent magnetization (PTRM) from the NRM of very old rocks. However, it has become clear recently that both old and young rocks can be victims of extremely stable viscous remanent magnetizations (VRM) whose coercivity spectra overlap the coercivity spectra of the signal, i.e., the stable NRM. In such cases demagnetization can hardly be "selective," sometimes resulting in such a drastic diminution of the stable NRM that no significant information about past magnetization can be deciphered. A second problem associated with VRM is that as demagnetization (particularly of the AF type) proceeds, some rocks become very prone to acquiring new strong VRM in the laboratory field. An attractive solution to both of these problems may lie in designing magnetometers inside of which one can house different types of demagnetization setups. Since the whole instrument will have to be located in a field-free space, the laboratory field cannot have any influence on the measurements. Additionally, the removal of a very stable VRM due to, for example, ultrafine grains can be attempted with an incorporated furnace (i.e., thermal demagnetization). The specimen magnetization can be continuously followed and, as soon as most of the unwanted VRM is removed, the demagnetization stopped. Such comprehensive instruments can also be built to include equipment to measure low-field susceptibility and even saturation magnetization. One magnetometer that would seem to be capable of these modifications is the cryogenic magnetometer if its sensors could be insulated from the thermal and AF fluctuations. Preliminary attempts have been made in this regard but serious problems still remain.

5.3. Continuous Measurement of Long Cores

When a long sediment core from a lake or a drill core from hard rocks must be studied thoroughly, it is much more attractive to be able to do so without having to first cut the core into numerous hand samples and then measure each of them. Dodson *et al.* (1977) have designed a cryogenic magnetometer with a pull-through access hole (and hence, an annular helium Dewar). The instrument has three sensing coils for measuring the three components of magnetization, and a dipole function is used to deconvolve the data and remove the error introduced by the sensing of an extended region over and above that that is immediately within the sensing coil region. This instrument is much superior to the instrument in which only two azimuthal components (i.e., declination) can be measured (Molyneux *et al.*, 1972), resulting in a loss of inclination data.

5.4. On-Line Signal Processing

Like most other geophysical fields, paleomagnetism and rock magnetism have also discovered the advantages of on-line signal processing or data reduction, and at least two commercial instruments are now available that gather the data of three components of magnetization and, given the information about the attitude of the sample horizon in the field, can provide the corrected absolute directions of magnetization. With more input data, the on-line minicomputer can also provide the latitude and longitude of the virtual geomagnetic pole as well as the confidence statistics for a set of specimens. The rapid progress in the electronic microprocessor field, however, has already made some of these approaches obsolete. For example, an on-line microprocessor and a graphics terminal (TV screen) can handle the "first cut" of a data set and alert the investigator as to whether one should proceed at all with a thorough statistical analysis of the rest of the data set.

Acknowledgments

The preparation of the first draft of this article was begun when I was a Visiting Professor at Stanford University and later, at the University of California, Berkeley. I thank both the institutions and my colleagues there for their hospitality. The work was supported, in part, by NSF grant EAR 75-21796. I am greatly indebted to Timothy Canaday and Kathy Ohler for their help in the final stages of the preparation of this article.

References

As, J. A., and Zijderveld, J. D. A. (1958). Magnetic cleaning of rocks in paleomagnetic research. *Geophys. J. R. Astron. Soc.* **1**, 308–319.
Banerjee, S. K., and Mellema, J. P. (1974). A new method for the determination of paleointensity from the ARM properties of rocks. *Earth Planet. Sci. Lett.* **23**, 177–184.
Banerjee, S. K., and Stacey, F. D. (1967). The high-field torque-meter method of measuring magnetic anisotropy in rocks. *In* "Method in Paleomagnetism" (D. W. Collinson, K. M. Creer, and S. K. Runcorn, eds.), pp. 470–476. Elsevier, Amsterdam.

Biquand, D., and Prévot, M. (1971). AF demagnetization of viscous remanent magnetization of rocks. *Z. Geophys.* **37**, 471–485.

Blackett, P. M. S. (1952). A negative experiment relating to magnetism and the earth's rotation. *Philos. Trans. R. Soc. London, Ser. A* **245**, 309–370.

Bruckshaw, J. M., and Robertson, E. I. (1948). The measurement of magnetic properties of rocks. *J. Sci. Instrum.* **25**, 444–446.

Burek, P. J. (1969). Device for chemical demagnetization of red beds. *J. Geophys. Res.* **74**, 6710–6712.

Burek, P. J. (1971). An advanced device for chemical demagnetization of red beds. *Z. Geophys.* **37**, 493–498.

Carmichael, C. M. (1968). An outline of the intensity of the paleomagnetic field of the earth. *Earth Planet. Sci. Lett.* **3**, 351–354.

Coe, R. S. (1967). The determination of paleointensities of the earth's magnetic field with emphasis on mechanisms which could cause non-ideal behavior in Thellier's method. *J. Geomagn. Geoelectr.* **19**, 157–179.

Coe, R. S., and Grommé, C. S. (1973). A comparison of three methods of determining geomagnetic paleointensities. *J. Geomagn. Geoelectr.* **25**, 415–435.

Collinson, D. W. (1967). The variation of magnetic properties among red sandstones. *Geophys. J.* **12**, 197–207.

Collinson, D. W. (1975). Instruments and techniques in paleomagnetism and rock magnetism. *Rev. Geophys. Space Phys.* **13**, 659–686.

Collinson, D. W., Creer, K. M., and Runcorn, S. K., eds. (1967). "Methods in Paleomagnetism." Elsevier, Amsterdam.

Creer, K. M. (1959). AC demagnetization of unstable triassic Keuper marls from S. W. England. *Geophys. J.* **2**, 261–275.

deSa, A., Widdowson, J. W., and Collinson, D. W. (1974). The signal to noise ratio of static magnetometers with negative feedback. *J. Phys. E* **7**, 1015–1019.

Deutsch, E. R., Roy, J. L., and Murthy, G. S. (1967). An improve astatic magnetometer for paleomagnetism. *Can. J. Earth Sci.* **5**, 1270–1273.

Dodson, R. E., Fuller, M. D., and Kean, W. F. (1977). Paleomagnetic records of secular variations from Lake Michigan sediment cores. *Earth Planet. Sci. Lett.* **34**, 387–395.

Dunlop, D. J. (1971). Magnetic properties of fine-particle hematite. *Ann. Geophys.* **27**, 269–293.

Dunlop, D. J. (1973). Theory of the magnetic viscosity of lunar and terrestrial rocks. *Rev. Geophys. Space Phys.* **11**, 855–901.

Dunlop, D. J. (1976). Thermal fluctuation analysis: A new technique in rock magnetism. *J. Geophys. Res.* **81**, 3511–3517.

Dunlop, D. J., and West, G. F. (1969). An experimental evaluation of single domain theories. *Rev. Geophys.* **7**, 709–757.

Dunlop, D. J., Hanes, J. A., and Buchan, K. L. (1973). Indices of multidomain magnetic behavior in basic igneous rocks: Alternating-field demagnetization, hysteresis, and oxide petrology. *J. Geophys. Res.* **78**, 1387–1393.

Dunlop, D. J., Bailey, M. E., and Westcott-Lewis, M. F. (1975). Lunar paleointensity determination using anhysteretic remanence (ARM): A critique. *Geochim. Cosmochim. Acta, Suppl.* **6**, 3063–3069.

Ewing, J. A. (1885). Experimental researches in magnetism. *Philos. Trans. R. Soc. London* **176**, 523–640.

Farrell, W. E. (1967). The resonance magnetometer. *In* "Methods in Paleomagnetism" (D. W. Collinson, K. M. Creer, and S. K. Runcorn, eds.), pp. 100–103. Elsevier, Amsterdam.

Fletcher, E. J., deSa, A., O'Reilly, W., and Banerjee, S. K. (1969). A digital vacuum torque magnetometer for the temperature range 300–1000°K. *J. Sci. Instrum.* [2] **2**, 311–314.

Fuller, M. (1974). Lunar magnetism. *Rev. Geophys. Space Phys.* **12**, 23–70.

Goree, W. S., and Fuller, M. D. (1976). Magnetometers using R–F driven squids and their application in rock magnetism and paleomagnetism. *Rev. Geophys. Space Phys.* **14**, 591–608.

Haalck, H. (1942). "Der Gesteinsmagnetismus." Becker and Erler Kom.-Ges., Leipzig.

Halgedahl, S., and Fuller, M. (1980). Magnetic domain observations of nucleation processes in fine particles of intermediate titanomagnetite. *Nature (London)* **288**, 70–72.

Halls, H. C. (1976). A least-squares method to find a remanence direction from converging remagnetization circles. *Geophys. J. R. Astron. Soc.* **45**, 297–304.

Heller, F., and Markert, H. (1973). The age of viscous remanent magnetization of Hadrian's Wall (northern England). *Geophys. J.* **31**, 395–406.

Hoffman, K. A. (1975). Cation diffusion processes and self-reversal of thermoremanent magnetization in the ilmenite-hematite solid solution series. *Geophys. J. R. Astron. Soc.* **41**, 65–80.

Hoffman, K. A., and Day, R. (1978). Separation of multi-component NRM: A general method. *Earth Planet. Sci. Lett.* **40**, 433–438.

Hoye, G. S., and Evans, M. E. (1975). Remanent magnetizations in oxidized olivines. *Geophys. J. R. Astron. Soc.* **41**, 139–151.

Irving, E. (1964). "Paleomagnetism and its Application to Geological and Geophysical Problems." Wiley (Interscience), New York.

Jensen, S. D., and Shive, P. N. (1973). Cation distribution in sintered titanomagnetites. *J. Geophys. Res.* **78**, 8474–8480.

Johnson, E. A., and Steiner, W. F. (1937). An astatic magnetometer for measuring susceptibility. *Rev. Sci. Instrum.* **8**, 236–238.

Johnson, H. P., and Merrill, R. T. (1972). Magnetic and mineralogical changes associated with low-temperature oxidation of magnetite. *J. Geophys. Res.* **7**, 334–341.

Johnson, H. P., and Merrill, R. T. (1973). Low-temperature oxidation of titanomagnetite and the implications for paleomagnetism. *J. Geophys. Res.* **78**, 4938–4949.

Johnson, H. P., Lowrie, W., and Kent, D. V. (1975). Stability of anhysteretic remanent magnetization in fine and coarse magnetite and maghemite particles. *Geophys. J. R. Astron. Soc.* **41**, 1–10.

Johnston, M. J. S., Smith, B. E., and Mueller, R. (1976). Tectonomagnetic experiments and observations in western U.S.A. *J. Geomagn. Geoelectr.* **28**, 85–97.

Josephson, B. D. (1962). Possible new effects in superconductive tunnelling. *Phys. Lett.* **1**, 251–253.

Kawai, N., Yaskawa, K., Nakajima, T., Torii, M., and Natsuhara, N. (1975). *Paleolimnol. Lake Biwa Jpn. Pleistocene* **3**, 143–160.

Kean, W. F., Day, R., Fuller, M., and Schmidt, V. A. (1976). The effect of uniaxial compression on the initial susceptibility of rocks as a function of grain size and composition of their constituent titanomagnetites. *J. Geophys. Res.* **81**, 861–872.

Kent, D. V., and Lowrie, W. (1977). VRM studies in Leg 37 igneous rocks. *In* "Initial Reports of the Deep Sea Drilling Project" (F. Aumento *et al.*, eds.), pp. 525–529, US Govt. Printing Office, Washington, D.C.

Klapel, G. D., and Shive, P. N. (1974). High-temperature magnetostriction of magnetite. *J. Geophs. Res.* **79**, 2629–2633.

Kumagai, N., and Kawai, N. (1953). A resonance type magnetometer. *Mem. Coll. Sci. Univ. Kyoto, Ser. A* **20**, 306–309.

Levi, S. (1976). The effect of magnetite particle size on paleointensity determinations of the geomagnetic field. *Phys. Earth Planet. Int.* **13**, 245–259.

Levi, S., and Banerjee, S. K. (1976). On the possibility of obtaining relative paleointensities from lake sediments. *Earth Planet. Sci. Lett.* **29**, 219–226.

Levi, S., and Merrill, R. T. (1976). A comparison of ARM and TRM in magnetite. *Earth Planet. Sci. Lett.* **32**, 171–184.

Lowrie, W., and Fuller, M. (1971). On the alternating field demagnetization characteristics of multidomain thermoremanent magnetization in magnetite. *J. Geophys. Res.* **76**, 6339–6349.

Lund, S., Banerjee, S. K., Levi, S., Eyster-Smith, N., Wright, H. E., Jr., and Long, A. (1977). High resolution paleomagnetic fluctuations from Minnesota-correlations with paleoclimatic data. *EOS, Trans. Am. Geophys. Union*, **58**, 708 (abstr. only).

McElhinny, M. W. (1973). "Paleomagnetism and Plate Tectonics." Cambridge Univ. Press, London and New York.

Merrill, R. T. (1970). Low-temperature treatments of magnetite and magnetite-bearing rocks. *J. Geophys. Res.* **75**, 3343–3349.

Molyneux, L. (1971). A complete result magnetometer for measuring the remanent magnetization of rocks. *Geophys. J. R. Astron. Soc.* **24**, 429–433.

Molyneux, L., Thompson, R., Oldfield, F., and McCallan, M. E. (1972). Rapid measurement of the remanent magnetization of long cores of sediment. *Nature (London) Phys. Sci.* **237**, 42.

Nagata, T. (1961). "Rock Magnetism," 2nd ed. Maruzen, Tokyo.

Nagata, T. (1967). Principles of the ballistic magnetometer for the measurements of remanence. *In* "Methods in Paleomagnetism" (D. W. Collinson, K. M. Creer, and S. K. Runncorn, eds.), pp. 105–114. Elsevier, Amsterdam.

Nagata, T., and Uyeda, S. (1959). Exchange interaction as a cause of reverse thermoremanent magnetism. *Nature (London)* **184**, 890.

Néel, L. (1949). Théorie du trainage magnétique des ferromagnétiques en grains fin avec applications aux terres cuites. *Ann. Geophys.* **5**, 99–136.

Néel, L. (1955). L'inversion de l'aimantation permanente des roches. *Ann. Geophys.* **7**, 90.

O'Reilly, W., and Banerjee, S. K. (1965). Cation distribution in titanomagnetites $(1 - x)Fe_3TiO_4$. *Phys. Lett.* **17**, 237–238.

Pozzi, J. P., and Thellier, E. (1963). Perfectionnements récents apportés aux magnétomètres de trés haute sensibilité utilisés en minéralogie magnétique. *C. R. Hebd. Seances Acad. Sci.* **257**, 1037–1041.

Pullaiah, G., Irving, E., Buchan, K. L., and Dunlop, D. J. (1975). Magnetization changes caused by burial and uplift. *Earth Planet. Sci. Lett.* **28**, 133–143.

Radhakrishnamurty, C., Likhite, S. D., and Sahasrabudhe, P. W. (1970). Some curious magnetic properties of rocks. *In* "Paleogeophysics" (S. K. Runcorn, ed.), pp. 223–234. Academic Press, New York.

Readman, P. W., and O'Reilly, W. (1972). Magnetic properties of oxidized (cation-deficient) titanomagnetites (Fe, Ti, □)304. *J. Geomagn. Geoelectr.* **24**, 69–90.

Revol, J., Day, R., and Fuller, M. D. (1977). Magnetic behavior of magnetite and rocks stressed to failure—relation to earthquake prediction. *Earth Planet. Sci. Lett.* **37**, 296–306.

Rimbert, F. (1959). Contribution à l'étude de l'action de champs alternatifs sur les aimantations rémanentes des roches. Applications géophysiques. *Rev. Inst. Fr. Pet.* **14**, 17–54, 123–155.

Roy, J. L., and Lapointe, P. L. (1978). Multiphase magnetizations: Problems and implications. *Phys. Earth Planet. Int.* **16**, 20–37.

Roy, J. L., and Park, J. K. (1974). The magnetization process of certain red beds: Vector analysis of chemical and thermal results. *Can. J. Earth Sci.* **11**, 437–471.

Roy, J. L., Reynolds, J., and Sanders, E. (1972). An astatic magnetometer with negative feedback. *Publ. Earth Phys. Br. (Can.)* **42,** 166–182.

Shaw, J. (1974). A new method of determining the magnitude of the paleomagnetic field: Application to five historic lavas and five archaeological samples. *Geophys. J. R. Astron. Soc.* **39,** 133–141.

Soffel, H. C. (1969). The origin of thermoremanent magnetization of two basalts containing homogeneous single phase titanomagnetite. *Earth Planet. Sci. Lett.* **7,** 201–208.

Soffel, H. C. (1977). Domain structure of titanomagnetites and its variation with temperature. *J. Geomagn. Geoelectr.* **29,** 277–284.

Stacey, F. D., and Banerjee, S. K. (1974). "The Physical Principles of Rock Magnetism." Elsevier, Amsterdam.

Stephenson, A. (1971). Single domain grain distributions. II. The distribution of single domain iron grains in Apollo 11 lunar dust. *Phys. Earth Planet. Int.* **4,** 361–369.

Stone, D. B. (1967). Torsion-balance method of measuring anisotropic susceptibility. *In* "Methods in Paleomagnetism" (D. W. Collinson, K. M. Creer, and S. K. Runcorn, eds.), pp. 381–386. Elsevier, Amsterdam.

Symons, D. T. A., and Stupavsky, M. (1974). A rational paleomagnetic stability index. *J. Geophys. Res.* **79,** 1718–1720.

Syono, Y. (1965). Magnetocrystalline anisotropy and magnetostriction of $Fe_3O_4-Fe_2TiO_3$ series with special application to rock magnetism. *Jpn J. Geophys.* **4,** 71.

Theillier, E. (1938). Sur l'aimantation des terres cuites et ses applications géophysiques. *Ann. Inst. Phys. Globe Univ. Paris Bur. Cent. Magn. Terr.* **16,** 157–302.

Thellier, E., and Thellier, O. (1959). Sur l'inténsité de champ magnétique térrestre dans le passé historique et géologique. *Ann. Geophys.* **15,** 285–376.

Thompson, R. (1975). Long period European geomagnetic secular variation confirmed. *Geophys. J. R. Astron. Soc.* **43,** 847–859.

Uyeda, S., Fuller, M. D., Belshé, J. C., and Girdler, R. W. (1963). Anisotropy of magnetic susceptibility of rocks and minerals. *J. Geophys. Res.* **68,** 279–291.

Van Zijl, J. S. V., Graham, K. W. T., and Hales, A. L. (1962). The paleomagnetism of the Stromberg lavas of South Africa. *Geophys. J.* **7,** 23–39, 169–182.

Vine, F. J., and Matthews, D. H. (1963). Magnetic anomalies over ocean ridges. *Nature (London)* **199,** 947–949.

Walton, D. (1980). Time and temperature relationships in the magnetization of assemblies of single domain grains. *Nature (London)* **286,** 245–294.

Wu, Y., Pearce, G. W., Jowett, C. E. and Beales, F. W. (1976). Application of paleomagnetism to the study of Mississippi Valley-type ore deposits. *EOS, Trans. Am. Geophys. Union* **57,** 903 (abstr. only).

Wu, Y. T., Fuller, M., and Schmidt, V. A. (1974). Microanalysis of NRM in a granodiorite intrusion. *Earth Planet. Sci. Lett.* **23,** 275–285.

Zijderveld, J. D. A. (1967). AC demagnetization of rocks: analysis of results. *In* "Methods in Paleomagnetism" (D. W. Collinson, K. M. Creer, and S. K. Runcorn, eds.), pp. 254–286. Elsevier, Amsterdam.

Zijlstra, H. (1967). "Experimental Methods in Magnetism." Elsevier, Amsterdam.

CIRCULATION IN THE COASTAL OCEAN*

G. T. Csanady

Woods Hole Oceanographic Institution
Woods Hole, Massachusetts

1. Introduction 101
2. Wind-Driven Transient Currents 103
   2.1. Quasi-Geostrophic Model 104
   2.2. Coastal Constraint 105
   2.3. Velocity Distribution 106
   2.4. Longshore Pressure Gradients 108
   2.5. Comparison with Observation—Great Lakes 110
   2.6. Pacific Type Continental Shelves 115
3. Upwelling, Downwelling, and Coastal Jets 119
   3.1. Response of Stratified Water Column to Wind 119
   3.2. Coastal Jet Generation 121
   3.3. Large Pycnocline Movements 122
   3.4. Comparison with Observation 124
4. Trapped Waves and Propagating Fronts 126
   4.1. Linear Theory Models 127
   4.2. Surface Fronts 130
   4.3. Observational Evidence on Wave and Front Propagation 133
5. Steady Currents 139
   5.1. Frictional Adjustment 141
   5.2. Vorticity Tendencies 143
   5.3. Steady Parallel Flow over a Straight Continental Shelf 147
   5.4. Shelf Circulation as a Boundary-Layer Problem 149
   5.5. Thermohaline Circulation 155
   5.6. Mean Circulation of a Stratified Fluid 159
   5.7. Mean Circulation of the Mid-Atlantic Bight 162
   5.8. Storm Currents over Atlantic Type Shelves 166
   5.9. Mean Circulation in Lake Ontario 170
   5.10. Mean Summer Circulation over the Oregon Shelf 174
6. Conclusion 176
   References 177

1. Introduction

Dynamic processes in shallow seas and over continental shelves differ markedly from those in the deep ocean for several reasons. The horizontal scales of motion are much smaller and the presence of coasts is a strong constraining influence in most locations. The depths involved are only of the order of 100 m, so that surface effects such as wind stress, or surface cooling or heating extend to a larger fraction of the water column, sometimes to all of it, whereas in the deep ocean the same influences reach what amounts to only a thin skin at the surface. At the same time,

---

* Woods Hole Oceanographic Institution Contribution Number 4660.

basins or continental shelves with characteristic horizontal dimensions of the order of 100 km behave in an "oceanic" manner in the sense that motions in them are strongly affected by the Earth's rotation. Seas of this size and larger, with depth ranges of up to a few hundred meters, will be taken to constitute the coastal ocean. This definition includes enclosed shallow seas such as the North American Great Lakes, open seas such as the broad and flat continental shelves of "Atlantic" type, or the narrow and steep shelves of "Pacific" type, as well as semienclosed bodies of water such as the Gulf of Maine or the North Sea.

The dominant observable motions in shallow seas are rotary currents, associated with tides over continental shelves, with inertial oscillations in stratified, enclosed seas. Such motions in what one might call a pure form are characterized by the rotation of the current vector through 360° in a period not very different from the Earth's rotation rate, and illustrate the dynamic importance of rotation. Water particle motions during a full tidal or inertial cycle are along a closed ellipse with a typical longer axis length of a few kilometers, there being no net displacement in an idealized pure tidal or inertial oscillation. In reality, there is of course always some residual motion, which adds up cycle after cycle and produces fluid particle displacements over the longer term that are much larger than the diameter of the tidal or inertial ellipse. The problem of "circulation" is to describe and understand the pattern of these longer-term water particle displacements. The distribution of important water properties, such as temperature, salinity, or the concentration of heavy metals or nutrients, and the transport of these properties or of life forms incapable of locomotion depends critically on the pattern of circulation, but not very much on the oscillatory water motions, at least not in a direct way (indirect effects include, for example, turbulence and mixing produced by tidal currents).

The present review is thought to be timely because of vigorous recent development of our ideas on coastal circulation, stimulated partly by environmental concerns, partly by the widespread recent use of sophisticated modern instrumentation (such as the moored current meter) and other instruments deployed at fixed points in space (recorders of temperature, salinity, and bottom pressure). It is worthy of note that most of the modern instruments supply information relevant to what in fluid dynamics is known as the "Eulerian" description of fluid motion. The problem of circulation, or longer-term fluid particle motion is essentially "Lagrangian," on the other hand. A month-long (Eulerian) average of a current velocity component at a fixed point does not necessarily bear any relationship to month-long average water particle (Lagrangian) velocities. Eulerian data must therefore be treated with caution in any deductions relating to circulation. The pioneers of coastal oceanography pieced together patterns of circulation in a few shallow seas from the distribution of properties (Bigelow and Sears, 1935) or from the motion of surface and bottom drifters (Harrington, 1895; Bumpus, 1973). While these methods had their shortcomings, it is important to bear in mind that they related more or less directly to the displacement of water particles or parcels over periods long compared to the tidal or inertial cycle, i.e., to circulation. One should not lightly reject this earlier evidence.

Much of the observational evidence to be quoted comes from three shallow seas: Lake Ontario, the Mid-Atlantic Bight, and the Oregon shelf. These are coastal oceans the circulation of which has recently been explored through major cooperative experiments: the International Field Year on the Great Lakes (IFYGL), the Marine Eco-Systems Analysis (MESA) project and other environmentally oriented projects in the Mid-Atlantic Bight, and the Coastal Upwelling Experiment (CUE) and other intensive observations off Oregon. Varying amounts of information are available on other shallow seas, but, in general, this cannot be interpreted with the same degree of confidence that is possible in the well-explored cases. Three recent review articles have summarized some of the evidence available on the three shallow seas mentioned. Beardsley and Boicourt (1980) have described the circulation of the Mid-Atlantic Bight, focusing on the history of research and on the observational evidence. Allen (1980) gave an account of some shelf circulation models, comparing them mainly with observations on the Oregon shelf. Elsewhere (Csanady, 1978a), I have reviewed transient currents in the Great Lakes, relating observations to a series of conceptual models. The present article aims at unifying this evidence from three rather different shallow seas, because a comparison between them is likely to bring the unifying fundamental principles into sharper focus. The emphasis therefore will be on the dynamic principles that underlie the conceptual models found to be successful in quantitatively accounting for observations, i.e., on the fundamental physics of flow phenomena in the coastal ocean.

2. Wind-Driven Transient Currents

The prime driving force of circulation in the coastal ocean is the wind. This is not always obvious in tidal waters, but people living along the shores of the Great Lakes, for example, where tides are practically nonexistent, are well aware that wind action on coastal waters rapidly generates flow predominantly parallel to the coast. The coast prevents perpendicular movement, but longshore motion is unhindered and the longshore component of the wind is particularly effective in generating longshore currents and corresponding long particle displacements.

One of the fruits of recent field studies and associated theoretical work has been an understanding of the structure and dynamics of such wind-driven nearshore currents. Earlier ideas came mainly from pioneering theoretical studies, such as those of Ekman (1905) or Freeman *et al.* (1957), which were based on various idealizations and usually applied only to steady-state frictional equilibrium flow. However, winds at midlatitudes are variable, rarely remaining constant for more than a day. Under these circumstances the transient properties of coastal currents are often of greater practical importance than their asymptotic steady state for constant wind. These transient properties depend more on inertial than on frictional forces, a fact that makes the Ekman type models of limited use.

2.1. Quasi-Geostrophic Model

Inertial effects may be simply understood with the aid of simple models in which bottom friction is supposed absent and a longshore wind stress is suddenly imposed at the surface. Any persistent longshore motion that arises must somehow adjust to geostrophic equilibrium, i.e., the Coriolis force associated with longshore motion must eventually be balanced by an appropriate pressure field. Rossby (1938) first discussed such problems of "geostrophic adjustment," and Charney (1955) extended Rossby's work to coastal current generation in a two-layer ocean of constant depth. In Charney's quasi-geostrophic model the accelerating longshore current is postulated to adjust continually to geostrophic equilibrium. In reality, this may be expected to be true for periods of order f^{-1} and longer (f is the Coriolis parameter). More complete calculations for simple, constant-depth closed basin or coastal zone models confirm that the response of a modest-size sea ($f \cong$ constant) to sudden wind stress can be regarded as a superposition of a quasi-geostrophic (developing) coastal current and various long waves (Crépon, 1967, 1969; Csanady, 1968b; Birchfield, 1969). Essentially the same developing flow model may be applied to a coastal ocean with a more realistic depth distribution (sloping plane beach or even an arbitrary depth distribution as a function of the cross-shore coordinate), containing a homogeneous or a two-layer fluid (Bennett, 1974; Csanady, 1973, 1974a, 1977a; Birchfield and Hickie, 1977) to highlight important effects associated with the changing depth of a coastal zone.

The dynamic principles involved in quasi-geostrophic longshore current generation, elucidated by these theoretical studies, are illustrated in

Fig. 1. The surface level perturbation and the longshore velocity increase hand in hand, maintaining geostrophic balance. In the longshore direction, the wind-stress impulse equals the depth-integrated momentum of the water column, as long as bottom friction is negligible:

(1) $$V = I$$

where

$$V = \int_{-H}^{0} v \, dz \quad \text{and} \quad I = \int_{0}^{t} \frac{\tau_y}{\rho} \, dt$$

v is the longshore velocity; H is the total depth; τ_y is the longshore component of wind stress; and ρ is the water density. Note that V and I are "kinematic" momentum and impulse, i.e., they are divided by water density. The depth-integrated velocity will for simplicity be referred to as "transport."

2.2. Coastal Constraint

The simple longshore momentum balance in Eq. (1) holds provided that the longshore pressure gradient, the bottom shear stress, and the depth-integrated Coriolis force associated with cross-shore flow all

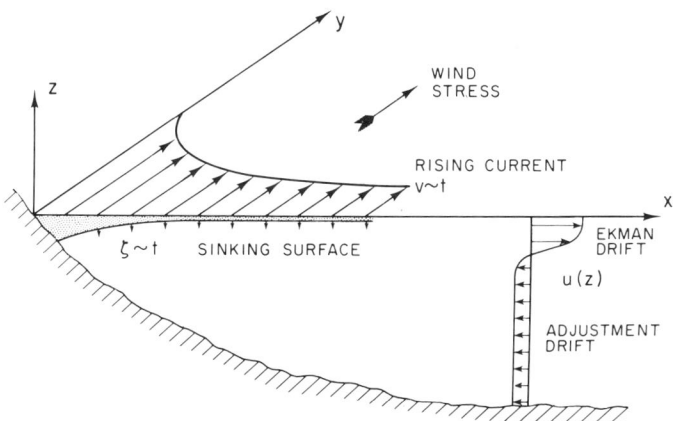

FIG. 1. Idealized transient wind-driven current generated by longshore wind stress, with negligible bottom friction. Current is quasigeostrophic, $v = (g/f) \, \partial \zeta/\partial x$, both v and ζ increasing linearly in time. Below surface Ekman layer longshore acceleration is due to the Coriolis force of adjustment drift, $\partial v/\partial t = -fu$.

vanish. Strictly speaking, this is almost never the case, the more realistic question being under what conditions Eq. (1) is a useful approximation.

The depth-integrated Coriolis force is proportional to transport, the cross-shore component of which is certainly zero at the coast and may be expected to be vanishingly small at suitably short distances from the coast:

$$(2) \qquad U = \int_{-H}^{0} u\, dz \cong 0$$

This "coastal constraint" holds for some distance range from the coast, the magnitude of which is an important characteristic of the coastal flow regime. Frictionless, constant-depth models of the kind investigated by Rossby (1938) and Charney (1955) yield for the scale of the coastal constraint the "radius of deformation":

$$(3) \qquad R = f^{-1}(gH)^{1/2}$$

where g is the acceleration of gravity. If the coastal zone is modeled by an inclined plane beach of slope $s = dH/dx = $ constant, the horizontal scale of variation is (Csanady, 1974a):

$$(4) \qquad L = gs/f^2$$

Typical characteristics of shallow seas at midlatitudes are $f = 10^{-4}$ sec^{-1}, $H = 100$ m, $s = 10^{-3}$, giving $R \cong 300$ km, $L \cong 1000$ km or larger than the typical width of shallow seas. Over distances of this order models assuming a straight coast and constant wind stress are unrealistic, and the results merely show that the coastal constraint is limited in reality by some factor not taken into account in these models. Nevertheless, such model results suggest that Eq. (2) should hold to a considerable distance from the coast.

2.3. Velocity Distribution

At such larger distances from the coast, i.e., in relatively deep water, the force of the wind affects directly only a thin layer at the surface, the rest of the water column responding indirectly, through pressure forces generated by the displacement of water masses. Within the surface shear layer subject to direct wind action, turbulence governs the distribution of wind-imparted momentum, while the Coriolis force acts as an important modifier of the flow. A "turbulent Ekman layer" develops, the depth of which depends only on wind stress and Coriolis parameter (see, e.g., Monin and Yaglom, 1971). The e-folding depth of this layer is empirically found to be

(5) $$D_E \approx 0.1 u_*/f$$

where $u_* = (\tau/\rho)^{1/2}$ is the friction velocity associated with the wind stress magnitude τ. The observable shear (vertical gradient of horizontal velocity) extends over a depth of about $2.5D_E$. For the typical wind stress magnitude of $\tau = 0.1$ Pa (1 dyne cm^{-2}), $u_* = 0.01$ m sec^{-1}. This wind stress is generated by a wind of about 7 m sec^{-1} velocity. At midlatitudes ($f = 10^{-4}$ sec^{-1}) the corresponding Ekman depth D_E is 10 m.

Within the Ekman layer, a longshore wind stress causes a transport to the right of the wind:

(6) $$U_E = \tau_y/\rho f$$

To satisfy the coastal constraint, Eq. (2), a compensating return transport occurs, evenly distributed over the water column. If, as supposed in the simple models, the longshore pressure gradient vanishes, this return flow gives rise to an unbalanced Coriolis force, which accelerates the water alongshore. Similar cross-stream displacements are deduced in other geostrophic adjustment problems, and the phenomenon will be referred to as "adjustment drift."

By Eq. (2) the adjustment drift totals U_E, which distributed evenly over the water column, gives rise to a cross-shore velocity of

(7) $$u = \tau_y/\rho g H$$

in a direction opposite to the Ekman drift. For the typical value of the wind stress of 0.1 Pa, $f = 10^{-4}$ sec^{-1}, and $H = 100$ m, one finds a cross-shore adjustment velocity of $u = 0.01$ m sec^{-1}. The longshore Coriolis acceleration associated with the adjustment drift is fu, or $\tau_y/\rho H$, which is exactly the same as if the force of the wind were evenly distributed over the water column by vigorous stirring.

The longshore velocity so generated by the Coriolis force of cross-shore adjustment drift is constant with depth and has a magnitude

(8) $$v = \int_0^t \frac{\tau_y}{\rho H} dt$$

a result in agreement with Eq. (1). This component of the velocity is geostrophically balanced, so that the cross-shore surface elevation gradient is

(9) $$\frac{\partial \zeta}{\partial x} = \frac{fv}{g} = \int_0^t \frac{f\tau_y}{\rho g H} dt$$

For constant wind stress both v and ζ increase linearly with time, a particularly simple model result.

Results of Crépon (1967) place the predictions of the quasi-geostrophic model further in perspective. Crépon calculated the full response of a semi-infinite sheet of water to suddenly imposed longshore wind. The important point is that, in spite of the apparent complexity, the quasi-geostrophic theory gives excellent approximations to the longshore particle displacement for periods $t \gg f^{-1}$, at $x \ll R$.

A longshore wind stress of 0.1 Pa lasting for 10^5 sec (a little over one day) exerts an impulse I of 10 m² sec^{-1}. Where the depth H is 100 m, this produces a longshore velocity V/H of 0.1 m sec^{-1}. Below the Ekman layer, the fluid acceleration is due to the cross-shore adjustment drift of $u = 0.01$ m sec^{-1} calculated above, $fu = 10^{-6}$ m sec^{-1}, acting for $t = 10^5$ sec.

2.4. Longshore Pressure Gradients

In the above discussion of the quasi-geostrophic model, sea level was supposed independent of the longshore coordinate for simplicity. However, sea level gradients are generally not negligible in the longshore balance of forces along a real coastline. For example, in a closed basin wind "setup" is a well-known effect. When wind blows along the longer axis of a long and narrow basin, such as Lake Erie or Lake Ontario, the level at the downwind end of the basin rises appreciably. In typical cases in Lake Erie the level rise is of the order of a meter, a sufficient amount to affect the output of the hydroelectric power plant on the Niagara River (Platzman, 1963). As is shown in standard texts (e.g., Defant, 1961), the surface level gradient characterizing wind setup is given by

$$(10) \qquad \frac{\partial \zeta}{\partial y} = \frac{\tau_y}{\rho g H_a}$$

where H_a is the average depth of a basin cross section perpendicular to the wind. If basin length is b, the level gradient of Eq. (10) is established in a period of order $b/2c$, where $c = (gH_a)^{1/2}$ is the celerity of long gravity waves. Given $H_a = 100$ m and $b = 300$ km, this period is 10^4 sec (2.8 hr), or about the same as is necessary for the appearance of rotational effects. For periods of the order of a day it is therefore realistic to consider the problem of longshore current generation with a pressure gradient added, opposing the wind stress, of a magnitude according to Eq. (10).

A quasi-geostrophic coastal current model of this kind has been explored in detail by Bennett (1974), and was further discussed and related to observations in Lake Ontario by Csanady (1973). In the case without bottom friction, the force of the wind is balanced by the pressure gradient of Eq. (10) at the locus of the cross-sectional average depth. Shoreward from this locus, in shallower water, wind stress dominates and accelerates the coastal water mass downwind. In water much shallower than the average depth H_a the pressure gradient force is negligible compared to the wind force and the previous results for flow without longshore gradient are recovered. In deep water, the pressure gradient dominates and causes return flow. For a long and narrow basin the calculations are particularly simple and result in a longshore transport distribution $V(x)$ proportional to the difference between the local and cross-sectional average depths (Fig. 2).

Calculated interior velocities of the quasi-geostrophic flow, in a cross-shore transect, given a pressure gradient opposing the wind stress, are similar to those shown in Fig. 1. However, the cross-shore motion below the Ekman layer, which compensates for the Ekman drift, is now partly geostrophic flow associated with the longshore pressure gradient. Where the depth is equal to the section-average depth H_a, geostrophic cross-shore flow exactly compensates for Ekman drift. In much shallower water the compensation (in the transient case) is mostly through adjustment drift. Longshore acceleration is only produced by the adjustment-drift component, so that $\partial v/\partial t$ is generally less than $-fu$, the difference being equal to the pressure gradient force, $-g(\partial \zeta/\partial y)$. The net result is that a transient, wind-driven longshore current develops only in shallow water, within some range l of the coast, where l is roughly the distance to the section-average depth in a closed basin.

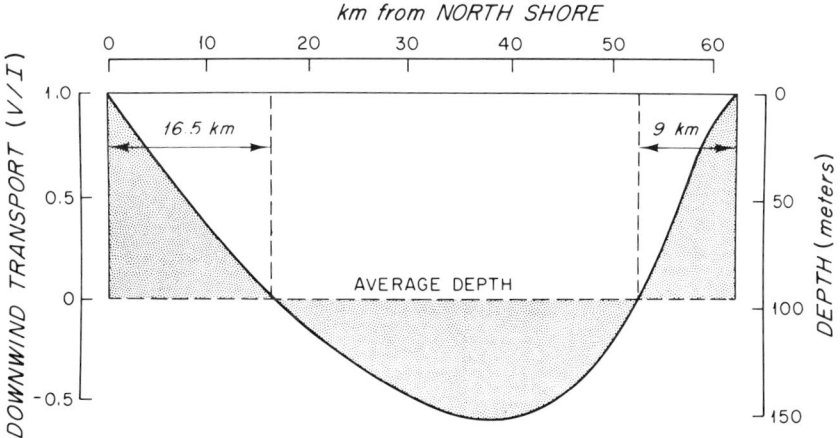

FIG. 2. Depth distribution of Lake Ontario along Oshawa–Olcott transect. Idealized transient transport is given by shaded diagram, pointing downwind in shallow water, upwind in deep water. Transport scale on left, nondimensionalized with wind stress impulse I.

2.5. Comparison with Observation—Great Lakes

Well-documented examples of clear-cut, isolated wind-stress episodes are rare, many complicating factors usually being present in the coastal ocean. In order to interpret the usual noisy observational evidence, it is helpful to distill the results of the quasi-geostrophic model into the following three propositions, amenable to direct and indirect comparison with observation:

(1) Longshore wind stress generates a transient longshore current within some range l of the coast, more or less in accord with $V = I$ [Eq. (1)].

(2) The longshore current so generated is in geostrophic equilibrium, so that a coastal sea level rise or fall accompanies the flow according to Eq. (9), or

$$(11) \qquad \zeta_0 = - \int_0^l \frac{fV}{gH}\, dx$$

where the integration is to extend over the range of the coastal current, l.

(3) Below the surface Ekman layer, much of the longshore acceleration is due to the Coriolis force of cross-shore adjustment drift, i.e., fu is comparable to dv/dt.

The above theoretical framework of quasi-geostrophic current generation has been successfully related to observations carried out in Lake Ontario during the International Field Year on the Great Lakes (IFYGL, carried out 1972–1973). One component of this large-scale experiment has been a detailed survey of five coastal transects around the lake, in sufficient spatial detail and with a frequency of once per day, weather permitting, to allow calculations of the depth-integrated transports and their variation with distance from shore (Csanady, 1973, 1974b, 1976a; Csanady and Scott, 1974, 1980). Other aspects of the IFYGL results have been discussed by Bennett (1977, 1978), Birchfield and Hickie (1977), Blanton (1974, 1975), Boyce (1974, 1977), Marmorino (1978, 1979), Simons (1973, 1974, 1975, 1976), and others (see references in the cited articles).

To illustrate the coastal transect results obtained during IFYGL, distributions of longshore transport along the north and south shores of Lake Ontario, in the same section as shown in Fig. 2, observed on Aug. 8 and 10, 1972, are shown in Fig. 3. The dates were selected because the surveys of the coastal zone velocities on these days followed wind-stress episodes preceded by a quiescent period. The wind stress was directed along the EW axis of the lake: during the first episode, on Aug. 6, in the eastward, and during the second, on Aug. 9, in the westward direction.

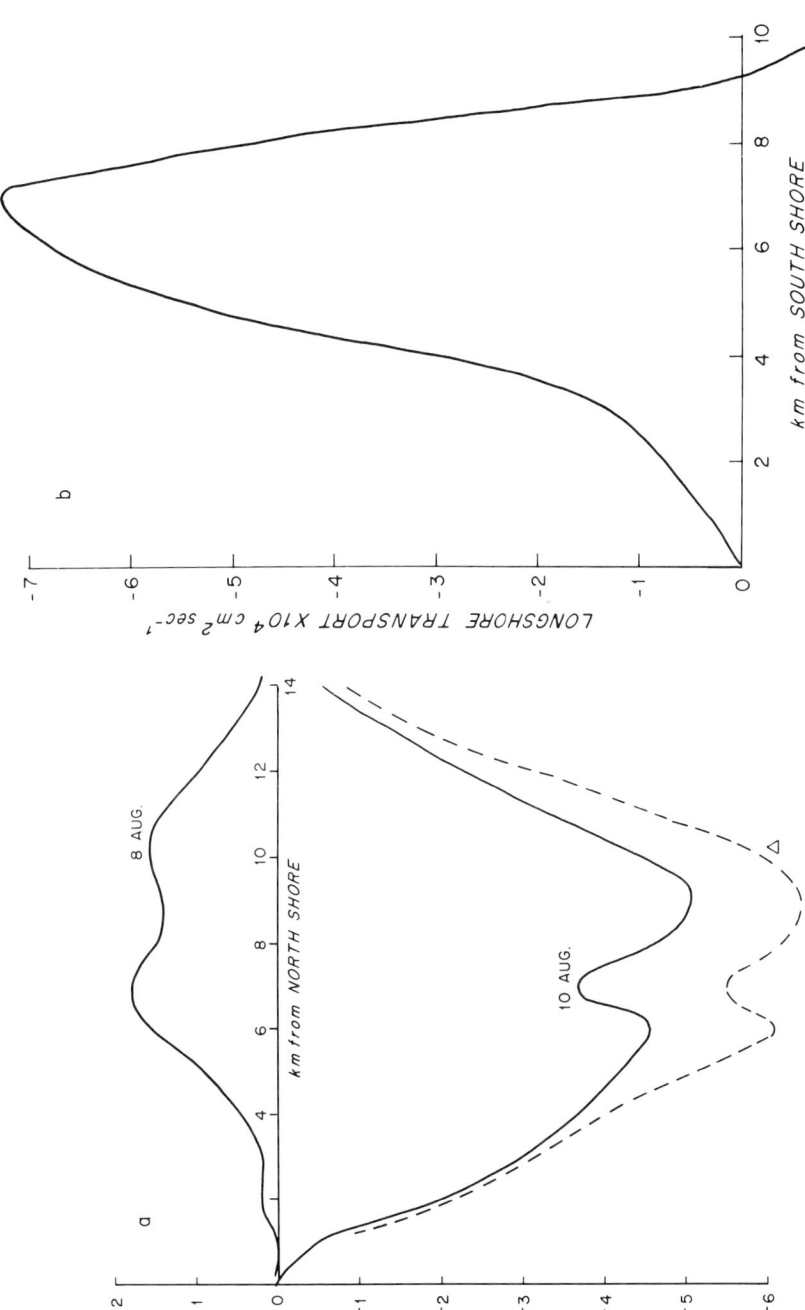

FIG. 3. (a) Depth-integrated longshore transport following eastward (positive) wind impulse on Aug. 10, 1972, off Oshawa, Ontario. Dashed line shows difference. (b) Longshore transport following westward (negative) wind impulse on Aug. 10, 1972, off Olcott, New York.

The wind-stress impulse of the first storm was estimated at $I = 2.5$ m^2 sec^{-1}, the second at $I = -9$ m^2 sec^{-1}, the easterly direction being taken as positive. The surveys were taken after the storms had died down, so that currents in very shallow water had time to decay. Along the south shore at about 7 km from shore and along the north shore at about 9 km, the longshore transport V peaked fairly close to the value predicted by the theory ($V \cong I$). It also dropped to zero at about the locus of section-average depth, 9 and 16 km, respectively, from the shores.

These wind-stress episodes were also accompanied on the north and south shores by appropriate level changes. Putting $l = 10$ km for the range of the coastal current and using typical velocity values of 0.1–0.3 m sec^{-1}, Eq. (11) yields 1–3 cm for the coastal level disturbance, pretty much as observed.

The cross-shore adjustment drift in these episodes is not evident in the current meter records (quoted both by Simons, 1975, and Blanton, 1975), presumably because of the dominance of inertial oscillations, with velocity amplitudes of up to 0.3 m sec^{-1}. The expected adjustment drift in this episode is $u = 0.01$–0.03 m sec^{-1}, which is within the noise level of the observations. However, the change in the position of the constant-temperature surfaces from Aug. 8 to Aug. 10 allows one to deduce onshore water particle displacements of about the correct magnitude to explain longshore current generation by adjustment drift alone.

Indirect support for proposition (1) above comes from statistical measures of the climatology of coastal currents in the Great Lakes. Boyce (1974) quotes kinetic energy spectra due to Blanton (see Fig. 4) showing clear differences in the proportion of energy within low-frequency current fluctuations with distance from shore. Blanton (1974) showed this proportion as a function of distance from shore. Similar differences were noted earlier by Birchfield and Davidson (1967) and Malone (1968). Verber (1966) classified observed currents into more or less straight-line or oscillatory, according to the appearance of progressive vector diagrams, and pointed out that straight-line flow occurs predominantly nearshore. Verber's progressive vector diagrams show striking similarity to Crépon's (1967) calculated particle movements. From data of this kind one concludes that currents within 10 km or so of the shores of the Great Lakes are predominantly shore-parallel; further away they are predominantly oscillatory. The longshore currents are more or less coherent with longshore winds, although Blanton (1975) warns that direct driving by the wind does not explain all of the record. Owing to the increased importance of friction in very shallow water, the mean square longshore current fluctuation peaks at some distance from shore, much as transport in Fig. 3. Similar results for Lake Michigan have been reported by Sato and Mortimer (1975).

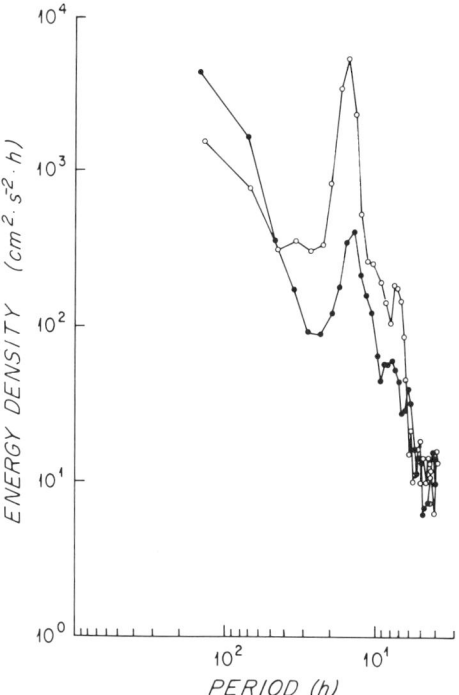

FIG. 4. Kinetic energy spectra of nearshore currents in Lake Ontario (due to Blanton, from Boyce, 1974). Low-frequency motions dominate the flow at 6 km from shore (●), while inertial motions are more prominent at 11 km (○).

The statistical evidence of fixed-point (Eulerian) current measurements is also supported by studies of the movement of Lagrangian tracers, notably fluorescent dye. Wind-driven longshore currents have been shown to carry tracers alongshore for considerable distances in the Great Lakes, if only they are released nearshore (Csanady, 1970, 1974c). Release at an offshore location results in much more erratic movement (Murthy, 1970). Figure 5, redrawn from data of Pritchard-Carpenter (1965), shows long dye plumes generated by a continuous source placed in the coastal current along the south shore of Lake Ontario.

Similar Lagrangian tracer studies also show the more or less complete disappearance of a dye plume on the reversal of the coastal current due to an opposing wind impulse (Csanady, 1974c). The adjustment drift is thus seen to perform the very important practical task of renewing the coastal water mass. For a sufficiently strong wind impulse the renewal is complete, for all practical purposes.

Early in the season, stratification is weak in the Great Lakes and the

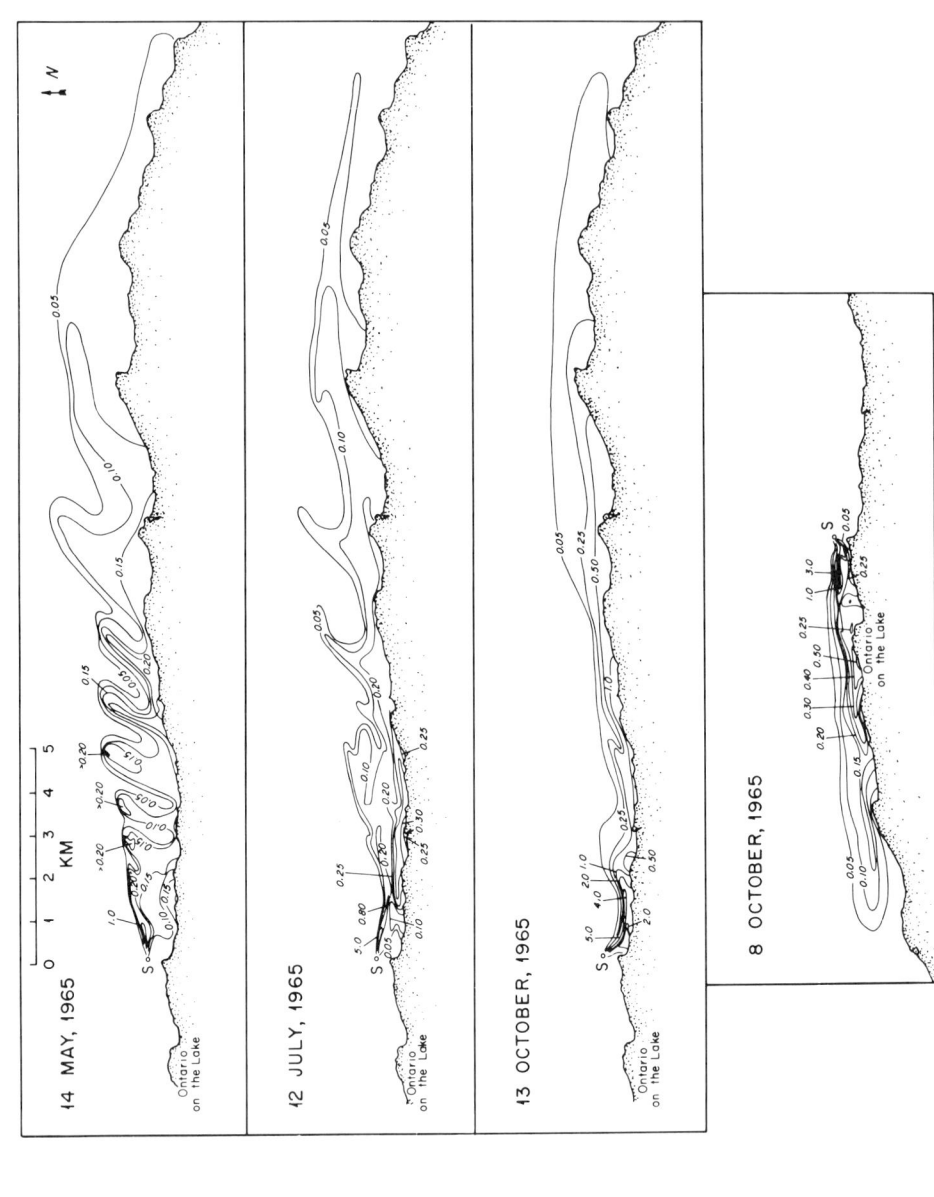

dynamic complications introduced by stratification (see below) are less important. Temperature differences are nevertheless easily observable and allow an analysis of cross-shore particle displacements. Such an analysis demonstrated particularly clearly the importance of adjustment drift in the generation of currents below the surface frictional layer (Csanady, 1974b). The considerable change of longshore momentum below the warm lens (below 10 m depth or so) was all generated by adjustment drift, as quantitative estimates clearly showed.

For the early season, Simons (1974) has shown also some particularly clear examples of water level rise or fall in the Great Lakes accompanying the establishment of coastal currents, in magnitude according to Eq. (11), with a coastal current range l of order 10 m.

2.6. Pacific Type Continental Shelves

Continental shelves of the Pacific type are narrow and slope down relatively steeply to the deep ocean. Tidal currents over them are consequently weak and bottom friction does not develop very rapidly, especially under stratified conditions. In this respect, there is a degree of similarity between nontidal basins and Pacific type shelves. On the other hand, the narrowness of these shelves makes them more directly subject to deep ocean influences than is usual over Atlantic type shelves. Whatever complications this gives rise to, transient longshore wind impulses should generate the same effects found to be prominent in nontidal basins, although in a more complex environment they may not be as easy to detect.

The continental shelf off Oregon has been the subject of intensive observational studies now for almost two decades (Collins *et al.*, 1968; Collins and Patullo, 1970; Huyer and Patullo, 1972; Halpern, 1974; Cutchin and Smith, 1973; Huyer *et al.*, 1974, 1975, 1978, 1979; Smith, 1974; Halpern, 1976; Mooers *et al.*, 1976a; Kundu and Allen, 1976; further references are given in these papers). Much of this work has been oriented toward the understanding of the seasonal upwelling cycle and its biological implications, but a considerable amount of evidence was also accumulated on the dynamics of wind-driven transient currents. Smith (1974) pointed out that longshore wind impulses were associated with longshore

---

FIG. 5. Tracer concentration distribution (ppb) in rhodamine B dye plumes generated by continuous source in coastal waters of Lake Ontario near Rochester, New York. (From Pritchard-Carpenter, 1965.)

current fluctuations distributed more or less evenly over the water column (below the surface layer). The coastal sea level rose and fell in step with such fluctuations (Fig. 6), according to Eq. (11), with a typical range l of the coastal current of order 60 km. The presence of an adjustment drift could again be inferred from the movement of the constant-property surfaces (see, e.g., Halpern, 1976).

Huyer *et al.* (1978) and Hickey and Hamilton (1980) further analyzed the dynamics of transient currents on the Oregon shelf. They confirmed Smith's (1974) results and specifically demonstrated that a model forced locally by longshore wind stress accounts for most current observations at midshelf.

In a particularly revealing study of the onset of the summer upwelling regime off Oregon, Huyer *et al.* (1979) analyze the details of the water column's response to a major southward wind-stress event. They find that the coastal sea level responds to the wind with a time lag of about 2 hr, while the longshore current along the 100 m isobath develops another 7 hr later. These values are very much as one would expect on the basis of geostrophic adjustment models. The coastal sea level also drops at a rate consistent with Eq. (11). The importance of the adjustment drift in this development can again be inferred from the movement of the constant-property surfaces.

A detailed analysis of the longshore and cross-shore momentum balances of Oregon shelf currents has been carried out by Allen and Kundu (1978). Figure 7, taken from this analysis of currents along the 100 m isobath, i.e., about 12 km from the coast, at a depth of 80 m, shows a direct comparison of the longshore Coriolis force due to adjustment drift, fu, with the observed acceleration $v_t = dv/dt$. If adjustment drift alone had been responsible for longshore acceleration, one would have $Y_t = v_t + fu = 0$. In fact this quantity was found to be different from zero. Some of the fluctuations of Y_t in Fig. 7 may be noise, but a smoothed version presumably represents a longshore pressure gradient force. There is a weak inverse relationship between Y_t and longshore wind stress, suggesting a phenomenon similar to setup in a closed basin, with the pressure gradient force opposing and partially neutralizing the wind stress. Very crudely, the split seems to be 50–50 between adjustment drift-related Coriolis force and pressure gradient force. This suggests that a wind stress of 0.2 Pa tends to generate an opposing sea level gradient of $\partial \zeta/\partial y = 10^{-7}$, as it would in a closed basin of 200 m depth.

As in the Great Lakes, the dynamical response of the Oregon coastal waters to wind stress is complicated by the surface outcropping of constant density surfaces—a topic considered next.

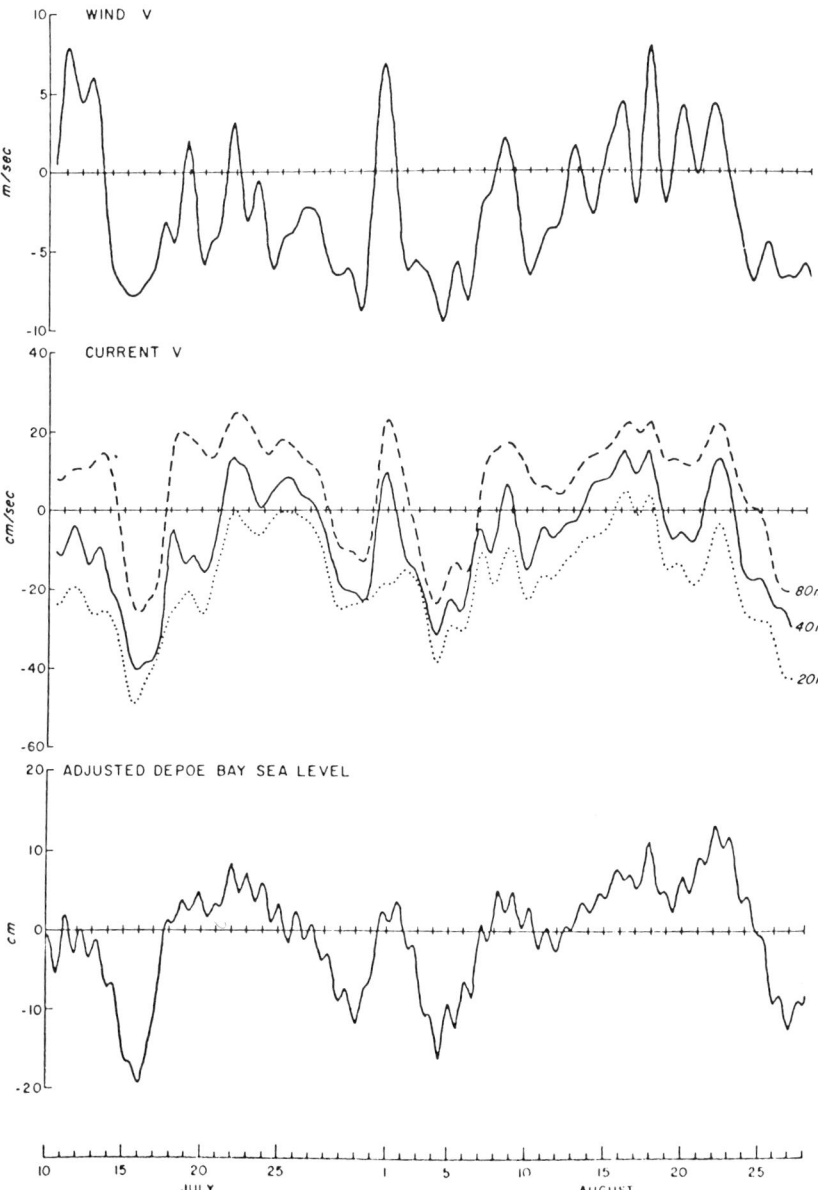

FIG. 6. Low-pass filtered wind velocity, currents, and adjusted sea level off the Oregon coast. (From Smith, 1974.) Correlation of all records is visually obvious, northward (positive) flow being correlated with elevated coastal sea level.

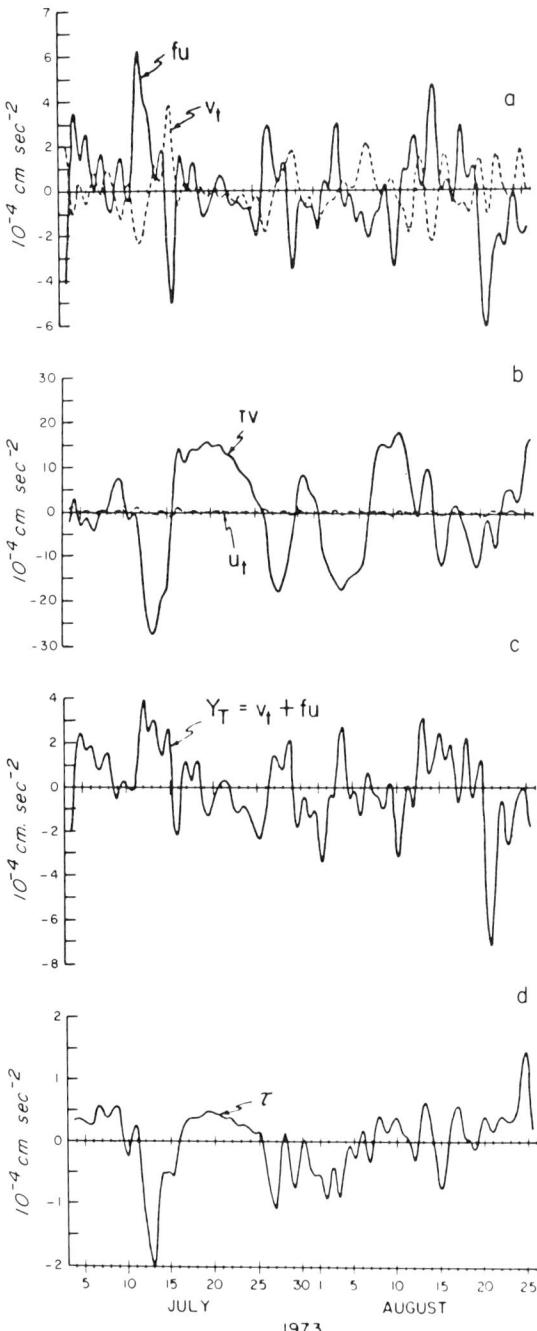

FIG. 7. Longshore momentum balance of Oregon shelf currents. (From Allen and Kundu, 1978.) (a) Longshore acceleration due to adjustment drift fu and observed acceleration v_t; (b) cross-shore acceleration; (c) inferred longshore pressure gradient; (d) wind stress.

3. Upwelling, Downwelling, and Coastal Jets

The distribution of water properties, salinity, temperature, nitrate and phosphate concentration, etc. is particularly sensitive to the circulation in a cross-shore transect because the streamlines of such circulation often cross sharp gradients. As already discussed, cross-shore circulations appear in consequence of earth rotation even when the wind blows parallel to the shore. A cross-shore wind acting over moderately deep water causes similar circulations in a more elementary way, without Earth rotation effects, as is easily demonstrated in a laboratory tank. The precise structure of cross-shore circulation cells is significantly influenced by buoyancy forces, if the temperature and salinity differences between top and bottom layers are large enough to produce density differences of the order of 1 part in 1000. As may be surmised at once, changes in the cross-shore circulation due to the hydrostatic stability associated with density stratification bring about modifications of the structure of longshore currents as well.

3.1. Response of Stratified Water Column to Wind

In a stratified water column in static equilibrium, surfaces of constant temperature and salinity are horizontal. Cross-shore particle displacements associated with transient winds distort these surfaces in a characteristic way, depending on whether the cross-shore circulation is "upwelling" or "downwelling." These terms refer to the upward motion of bottom water or the downward motion of surface water, respectively. Upwelling may cause those surfaces of constant temperature and salinity that in static equilibrium form a pycnocline, or relatively sharp density interface, to intersect the free surface. Conversely, downwelling may lead to the same surfaces intersecting the bottom at a depth several times their equilibrium depth.

As surfaces of constant density depart from their horizontal equilibrium position, horizontal pressure gradients arise in the fluid and affect the adjustment process to geostrophic balance and any resulting steady state of motion. A simple and realistic theoretical model consists of two layers of constant density, separated by a frictionless interface. Charney's (1955) analysis dealt also with this model, resulting in a quasi-geostrophic solution for an infinite straight coast, constant depth, and suddenly imposed longshore wind. The principal difference compared with the homogeneous fluid case is that within a nearshore band only the top-layer fluid responds to the wind by longshore acceleration, the bottom layer remaining quiescent. Consequently, higher longshore velocities arise in the top layer. At the same time the interface begins to rise or sink (depending on the direction of the wind) in such a way as to compensate for the surface level rise and to hold bottom pressure (nearly) constant. The strong surface layer current is then in geostrophic equilibrium with the horizontal pressure gradient associated with the inclination of the density interface and is legitimately called a coastal jet (in analogy with the atmospheric jet stream, which has a similar dynamical structure). The characteristics of a rising coastal just are illustrated in Fig. 8.

According to this simple theoretical model, the interface rises or sinks in a nearshore band of e-folding width

$$R_i = f^{-1}[\epsilon g h h'/(h + h')]^{1/2} \tag{12}$$

which is known as the internal radius of deformation (h and h' are the depths of top and bottom layers; $\epsilon = (\rho' - \rho)/\rho'$, proportionate density defect of top layer). In typical coastal oceanic cases R_i is of the order of 5–10 km. Far outside a band of scale width R_i the bottom layer moves bodily shoreward, or seaward, while the top layer must accommodate the Ekman drift in the opposite direction. Consideration of interior velocities then reveals a pattern exactly as if without any density gradients, which was illustrated for a sloping beach model in Fig. 1. Very close to the coast, however (at a distance small compared to R_i), there is no motion in the bottom layer. In the top layer, in the usual case when top-layer depth

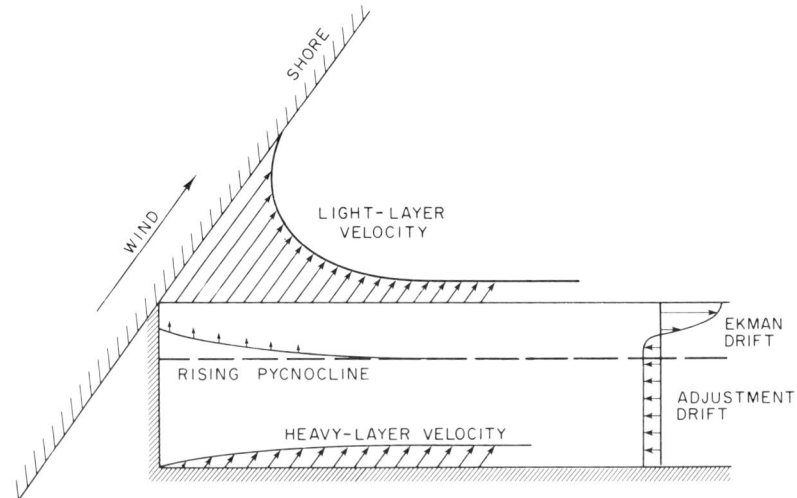

FIG. 8. Rising coastal jet in linear model of constant-depth, two-layer fluid.

is about equal to Ekman-layer depth, cross-shore motions are negligible, the force of the wind is evenly distributed over the top layer and causes uniform longshore acceleration. Where the top-layer depth is considerably greater than the Ekman-layer depth, adjustment drift occurs below the Ekman layer, but in the top layer only.

The constant-depth model applied so close to shore is overidealized, because the intersection of the equilibrium density interface with the bottom is located at a distance h/s from shore (s is the slope of the beach), which is typically 3 km or comparable to R_i. Calculations made for a coastal zone model of constant slope show, however, very similar behavior. Longshore wind stress again causes the interface to rise or sink in a coastal band of scale with $R_i^* = f^{-1}(\epsilon g h)^{1/2}$, which is not very different from R_i of Eq. (12) (Csanady, 1977a). Simultaneously, the interface–bottom intersection moves shoreward or seaward (a kinematic necessity over a sloping bottom), so that cross-shore particle movements occur also in the bottom layer adjacent to the interface–bottom intersection. These particle movements are not geostrophically balanced, so that the Coriolis force associated with them generates a longshore current in the bottom layer. Over a slope of $s = 10^{-3}$ (which is rather flat, but not atypical), a 1-m rise in the interface implies a 1-km shoreward displacement of the bottom water at the interface–bottom intersection. The integrated Coriolis force associated with this displacement generates a longshore velocity of 0.1 m sec^{-1}. The net effect is that the longshore velocity difference between top and bottom layers is reduced, requiring also a reduced slope of the isopycnals for geostrophic balance.

3.2. Coastal Jet Generation

The quantitative relationships applying to coastal jet generation are as follows. The wind-stress impulse generates a current in the top layer of total transport

$$(13) \qquad V = I$$

where now $V = \int_{-h}^{0} v \, dz$, h being the depth of the top layer. Geostrophic balance, with the bottom layer supposed stagnant, requires a pycnocline slope

$$(14) \qquad \frac{\partial \zeta'}{\partial x} = -\frac{fV}{\epsilon g h}$$

The cross-shore scale of the pycnocline deformation is R_i, so that the pycnocline slope is also

$$\text{(15)} \qquad \frac{\partial \zeta'}{\partial x} = -\frac{\zeta'_0}{R_i}$$

where ζ'_0 is the pycnocline displacement at the shore. These relationships also imply

$$\text{(16)} \qquad \zeta'_0 = I/c_i$$

where $c_i = fR_i$ is the internal wave propagation velocity or "densimetric velocity."

Typical values are $R_i = 5$ km and $c_i = 0.5$ m sec^{-1}. A very moderate wind impulse (7 m sec^{-1} wind for 10 hr) is $I = 3$ m^2 sec^{-1}, giving $\zeta'_0 = 6$ m, which is quite large, already comparable to the typical top layer depth of 10–30 m.

The theoretical models of two-layer geostrophic adjustment presented so far were based on a linearized approach, in which it is assumed that the displacement of the isopycnals from equilibrium remains small compared to their equilibrium depth. It is clear that this fundamental assumption of linearized theory is violated even for a very moderate wind-stress impulse.

The linearized theory may be extended to the case of continuous stratification, in order to explore to what extent the two-layer simplification is unrealistic. Such calculations may be made on the basis of two different idealizations: (1) supposing that the impulse of the wind is distributed over a layer depth equal to or less than the surface mixed-layer depth and ignoring friction, mixing, or surface heating in the adjustment process (Csanady, 1972); (2) modeling the offshore displacement of boundary-layer fluid at the surface by a source placed at the shore and calculating again the frictionless interior response (Allen, 1973). In the later idealization, the boundary-layer fluid must be supposed to be approriately heated or freshened in order to stay at the surface as an Ekman layer. Both approaches lead to much the same conclusions and effectively confirm the two-layer results, with no significant new insights emerging.

3.3. Large Pycnocline Movements

The most impressive results of upwelling and downwelling are the surfacing of isopycnals some distance from shore, or their sinking to a depth several times their equilibrium depth. The linear theory models certainly do not apply to these cases and they cannot answer such important questions as, how strong an impulse is required to bring a pycnocline to the surface, or how far from shore the pycnocline-free surface or pycnocline–bottom intersection will be found after a given stronger impulse. A strong longshore wind impulse ($I = 10$ m^2 sec^{-1} or more) is usually the cause of such dramatic upwelling or downwelling events, with the large pycnocline displacements developing quite rapidly, often within hours or at most a day. It is reasonable to idealize these events by supposing that the wind impulse is evenly distributed over the top layer by vigorous turbulence. One may ask then, how the two-layer fluid adjusts to geostrophic equilibrium following such an impulse, with interface and bottom friction neglected, and the density of each layer separately conserved. To neglect mixing altogether in the adjustment process is something of an overidealization—not to be forgotten in the application of the theory.

Calculations based on this idealization are relatively straightforward, making use of the principle of potential vorticity conservation (Csanady, 1977b). Some results are illustrated in Fig. 9. Quantitatively, the principal new result is that the velocity of the coastal jet is limited to a value of about $v = c_i \cong (\epsilon g h)^{1/2}$, analogously to certain "critical flow" problems in hydraulics. Thus, instead of $V = I$, one has

(17) $$V = V_{\max} = c_i h$$

The longshore momentum balance is completed by Coriolis force associated with the bodily displacement (adjustment drift) of the entire top layer from shore to a distance (for upwelling circulation, in a constant-depth, two-layer model):

(18) $$x_u = (Ih'/fh(h + h')) - R_i$$

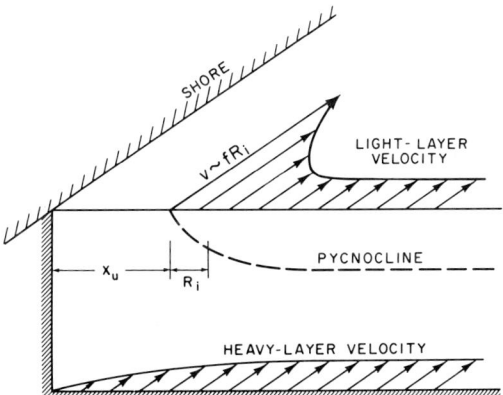

FIG. 9. Coastal jet development after surfacing of the pycnocline, in two-layer, constant-depth model.

Thus the minimum impulse needed to bring the density interface to the surface at the shore ($x_u = 0$) is

(19) $$I_u = fh(h + h')R_i/h'$$

Given the typical values $h = 20$ m and $R_i = 5$ km, I_u is of the order of 10 m^2 sec^{-1}, which is produced by a moderate longshore wind (7 m sec^{-1}) blowing for 28 hr. An impulse twice as great moves the pycnocline–surface intersection to $x_u = R_i = 5$ km, an impulse three times as great to $2R_i$, etc. Once surfaced, the pycnocline slope remains the same $\partial \zeta'/\partial x|_u = h/R_i = $ order 4.10^{-3}, the entire pattern moving bodily further offshore, causing a compensating onshore displacement in the bottom layer. Accordingly, adjustment drift causes the development of longshore flow in the bottom layer.

It should be pointed out here that an offshore wind also causes upwelling, and if strong enough, may bring the interface to the surface. However, the flow pattern so generated is not in equilibrium without the wind acting and the interface relaxes to a horizontal position on the cessation of the wind. Thus in theory the upwelling caused by a longshore wind is long-lived, that due to an offshore wind ephemeral. In practice, of course, dissipative processes cause the inclined interface in geostrophic equilibrium with a coastal jet also to relax toward static equilibrium, but this is usually a slow process, with a typical time scale of 5 days or so. For a more detailed discussion of upwelling caused by cross-shore wind see Csanady (1977b).

Another point that should perhaps be emphasized is that outside the range of pycnocline adjustment (i.e., more than $2R_i$ or so from the pycnocline–free surface intersection) the fluid behaves as if it were not stratified, i.e., the results of the earlier discussion hold. This *a posteriori* justifies the application of the simple homogeneous fluid theory to some observations made in stratified seas.

The above discussion involved principally upwelling events, at least in the examples cited. The analogous downwelling events can be treated similarly, and give rise to somewhat less spectacular consequences. This is partly because, as already pointed out, bottom layer velocities are generated over a sloping bottom due to the kinematically constrained cross-shore excursion of the bottom–interface intersection, reducing the velocity contrast across the inclined pycnocline.

3.4. Comparison with Observation

Intense upwelling events are known to occur in a number of coastal locations, notably in the Great Lakes and along the Oregon coast. Early reports described the hydrography of upwelling (Church, 1945; Ayers *et al.*,

1958; Smith *et al.*, 1966), while later systematic studies in the course of large-scale cooperative experiments provided detailed information also on longshore and cross-shore currents. In the course of these investigations some clear-cut upwelling events have been documented, produced by a local longshore wind impulse, which may be compared with the above simple conceptual picture.

Huyer *et al.* (1979) have shown that the summer upwelling regime of the Oregon shelf is initiated by a sufficiently large southward wind-stress impulse. In 1975, for example, such a large impulse occurred between March 25 and April 1. The total longshore impulse was about $I = 50$ m^2 sec^{-1}, but the pycnocline became strongly tilted at an earlier stage of the storm, when the impulse only reached about 20 m^2 sec^{-1}. Huyer *et al.* do not show the distribution of the isopycnals before and after the storm, but Halpern (1976) gives such a sequence for a similar isolated wind-stress impulse that occurred in July, 1973. The longshore wind-stress impulse in this latter storm occurred between July 10–15 and reached a final value of about 60 m^2 sec^{-1}. It is difficult to select a good two-layer, constant-depth representation of the coastal water mass for this case, but if one puts $\epsilon = 3 \times 10^{-3}$, $h = 30$ m, $h' = 70$ m, $f = 10^{-4}$ sec^{-1}, an internal radius R_i of 8 km is calculated. Equation (18), with $I = 60$ m^2 sec^{-1}, gives $x_u = 6$ km, which is not a too unrealistic representation of the hydrographic survey toward the end of the storm.

The coastal jet in geostrophic equilibrium with the inclined isopycnals accompanying upwelling has been well documented. In the study just cited, Halpern (1976) quotes a near-surface velocity of 0.8 m sec^{-1} on July 13, about equal to the expected maximum velocity $(\epsilon g h)^{1/2}$ in the simple two-layer model representation. However, the offshore e-folding scale of the region within which the isopycnals were rising was about 15 km, or almost twice as great as the simple model value R_i. Halpern calculated this scale from the displacement of the $\sigma_t = 26.0$ isopycnal, which has an apparent offshore equilibrium depth of about 70 m. The apparent e-folding scale of the $\sigma_t = 25.0$ isopycnal (which lies at about 30 m depth far offshore, assumed above for h) is closer to 8 km.

Upwelling events in Lake Ontario could be observed with greater spatial resolution. Figure 10 shows a well-documented event that occurred in October 1972. The wind-stress impulse on this occasion was 27 m^2 sec^{-1}, which should have produced an offshore displacement, by Eq. (18), of 2.3 km, more or less as observed. The structure and intensity of the coastal jet, as well as the isotherm (constant-density surface) distribution was very much as expected from the simple theoretical model.

Examples of downwelling and associated coastal jets have also been documented in the Great Lakes. An example is shown in Fig. 11. This also conforms in all essential aspects to the quasi-geostrophic model.

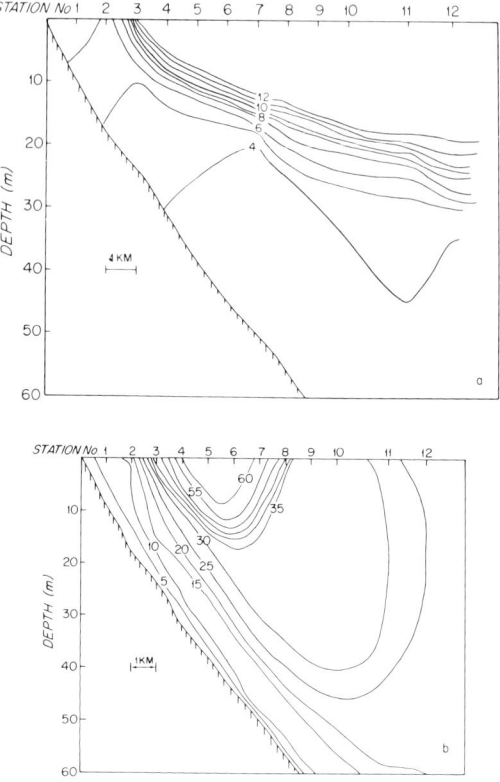

FIG. 10. Observed instance of coastal jet accompanying surfaced thermocline in Lake Ontario, Oct. 6, 1972. (From Csanady, 1976a.) (a) Isotherm contours in transect perpendicular to shore (°C); (b) longshore velocity component in same transect (cm sec^{-1}).

4. Trapped Waves and Propagating Fronts

Some flow phenomena observed in the coastal ocean are clearly unrelated to the wind. One of the most spectacular is the apparently spontaneous reversal of a well-established coastal current, set up previously by wind. Such reversals have been documented in detail in Lake Ontario (Csanady and Scott, 1974), but have also been noted in other locations (e.g., Kundu *et al.*, 1975). Current reversals, and other less drastic changes in the flow, are often found to propagate alongshore, in one specific direction only—the negative y direction in the coordinate system of Fig. 1. Thus in the Great Lakes such "signals" propagate cyclonically or counterclockwise around a basin, along the east coast of North America

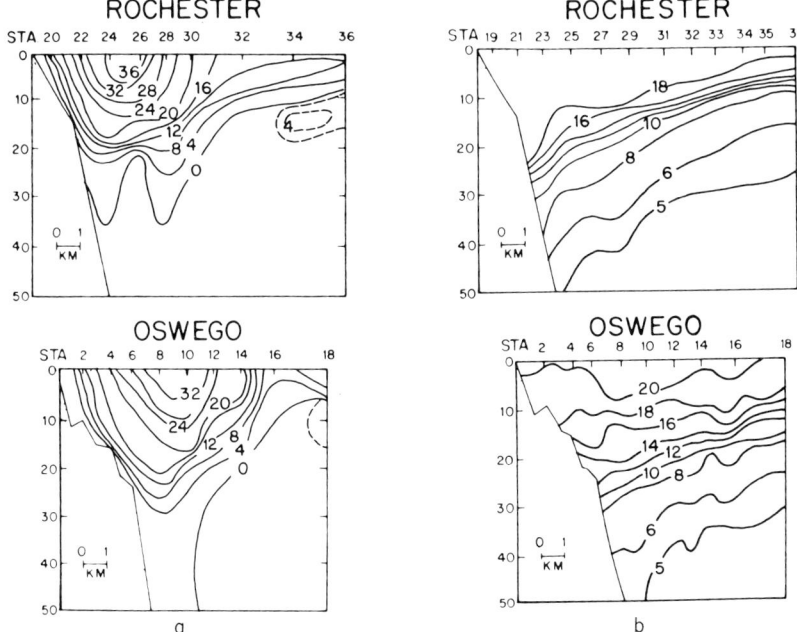

FIG. 11. Coastal jet associated with downwelling along south shore of Lake Ontario, at two locations, observed July 17, 1972. (a) Contours of longshore velocity (cm sec^{-1}); (b) isotherms (°C). (From Csanady and Scott, 1974.)

southwestward, along the west coast northward. For brevity, this direction will be called "cyclonic" in all applications. Linear theory models of the phenomenon are waves the amplitude of which is significant only within the coastal zone, i.e., which are "trapped" along the coast.

4.1. Linear Theory Models

The basic dynamic principles that govern the propagation of the relevant types of trapped waves are as follows (Csanady, 1976a). The longshore component of the wind stress, which primarily generates longshore currents, is nonuniform over the scale of basin topography or of weather systems. In the case of a closed basin, a uniform wind drives the coastal current in a cyclonic sense along some portion of the coast, in an anticyclonic sense along another portion. Similar, though less drastic, variations occur along continental-oceanic coasts, owing to changes in coastline orientation. In addition, the wind-stress field varies on a scale of sometimes only a few hundred kilometers, and becomes an important factor in causing longshore current variations along a more or less straight coastline.

Where the wind stress, and the longshore current it generates, is directed anticyclonically, a sea level depression develops as the longshore current adjusts to geostrophic equilibrium. A cyclonically directed coastal current is associated with a rise in coastal sea level. The nonuniformity of wind stress along the coast thus causes a longshore sea level gradient to come into existence along the coast, associated with a system of longshore currents. If the coastal current has a typical velocity v_0 and scale width l, then the coastal elevation/depression ζ_0 must be of the magnitude required by geostrophic balance: $\zeta_0 = f v_0 l / g$ [see Eq. (11)].

In a closed basin of perimeter length L the main variation of the longshore wind stress has a wavelength L and wavenumber $k = 2\pi/L$, and the longshore elevation gradients are of order $k\zeta_0$, or

$$\text{(20)} \qquad \frac{\partial \zeta}{\partial y} = O\left(\frac{kfvl}{g}\right)$$

where Eq. (11) was used to substitute for ζ_0. In a closed basin where a wind impulse had established coastal currents along opposite shores (Fig. 12), this longshore sea level gradient accelerates the water in a "down-

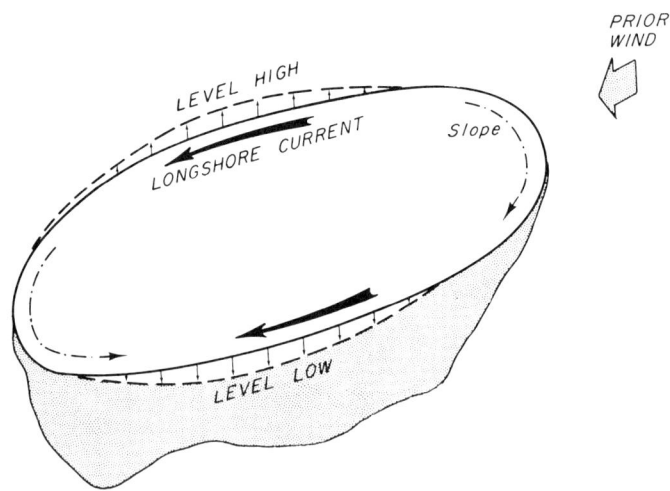

FIG. 12. Propagation of "trapped" long wave, associated with a system of coastal currents generated by prior wind impulse. Geostrophic balance of longshore currents requires coastal elevation/depression of levels, which results in longshore accelerations (dash–dot line marked "slope").

hill" direction, as indicated by dash–dot lines. Thus eventually the originally downwind end acquires a cyclonic coastal current, the upwind end an anticyclonic one. The two stagnation points between cyclonic and anticyclonic coastal currents shift, as a result, in a cyclonic sense. The longshore acceleration is of magnitude $g(\partial \zeta/\partial y)$, which causes a buildup (or reduction) of the longshore velocity comparable to the initial value v_0 in a period of order initial velocity divided by acceleration, or $(kfl)^{-1}$, using Eq. (20). The typical frequency of the change, σ, is the reciprocal of this or

(21) $$\sigma = O(kfl)$$

The pattern also propagates around the perimeter at the speed $c = \sigma/k$, or

(22) $$c = O(fl)$$

More detailed shelf models give $c = \lambda fl$, with λ usually between 0.3 and 0.6 (Mysak, 1980). In homogeneous water the scale width of the coastal current is primarily determined by the depth distribution and l is typically of the order of shelf width, or, in a closed basin, the distance to the locus of average depth. In the Great Lakes l is of order 10 km, which makes c by Eq. (22) typically 0.5 m sec^{-1}. With stratification present, the scale of the fast coastal current is R_i and $c = fR_i$, where R_i is the internal radius of deformation. This is often of a magnitude comparable to l, so that propagation speeds of the coastally trapped waves do not change much with stratification. However, the nearshore isopycnal tilts associated with cyclonic and anticyclonic coastal currents in stratified water are opposite, and the change from one to the other involves the replacement of a considerable coastal water mass by lighter or denser water drawn from offshore—a practically very important difference.

Linear theory models are the internal Kelvin wave and the shelf wave or topographic wave. The internal Kelvin wave is a classical constant-depth, two-layer fluid model, discussed, e.g., by Proudman (1953). Trapped waves also arise in a homogeneous fluid on account of sloping topography, and the low-frequency, Earth-rotation-dependent modes of these have properties much as discussed above, i.e., nearly geostrophic longshore flow, low frequency, and modest propagation velocity. Topographic waves have been identified by Reid (1958) and used by Robinson (1964) in modeling the propagation of sea level disturbances along the east coast of Australia. A considerable literature has arisen on this topic since then, recent reviews of which have been given by Leblond and Mysak (1977) and Mysak (1980). Figure 13 illustrates the structure of flow in a simple topographic wave model. In shallow seas both bottom slope and

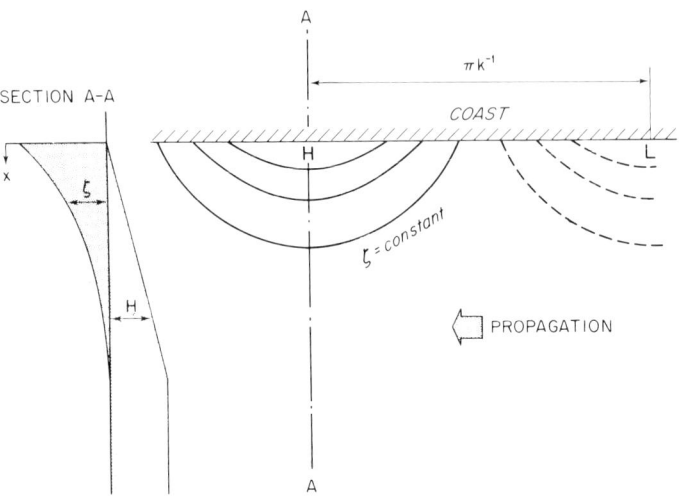

FIG. 13. Topographic wave in simple coastal zone model with inclined plane beach to a limiting-depth, homogeneous fluid, showing elevation contours. Longshore velocities are geostrophic.

stratification are important in determining the structure of trapped waves, and the most important modes are hybrids between the internal Kelvin wave and the topographic wave in a homogeneous fluid (Wang and Mooers, 1976; Huthnance, 1978). Figure 14 illustrates the modal structure of such a hybrid wave.

4.2. Surface Fronts

Linear theory models again suppose small displacements of the isopycnals from equilibrium and do not relate directly to situations involving, e.g., the surface outcropping of the pycnocline. At present there is not much theoretical guidance to understand the effects of such complications. Consider, for example, the case illustrated in Fig. 15, which occurs frequently in the Great Lakes. The pycnocline intersects the free surface some distance offshore, along a finite portion of the coast only. The pycnocline–surface intersection meets the coast in two locations, more or less at a right angle. Over a substantial portion of the coast the surfaced pycnocline can move in a cross-shore direction without affecting coastal sea levels, so that presumably it is not involved in the propagation of sea level disturbances. One would expect that in such locations disturbances would propagate more or less as a "pure" topographic wave in a

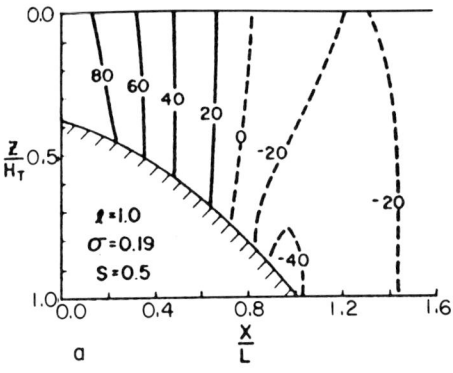
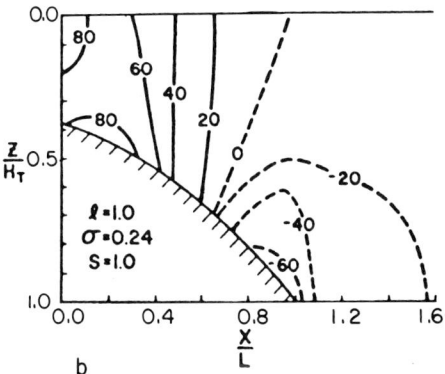
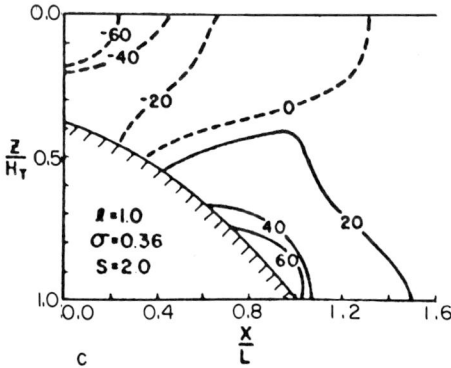

Fig. 14. Distribution of velocity amplitude (% maximum) in hybrid topographic–internal Kelvin wave, in weakly stratified fluid (a) contrasted with intermediate (b) and strong stratification (c). (From Wang and Mooers, 1976.)

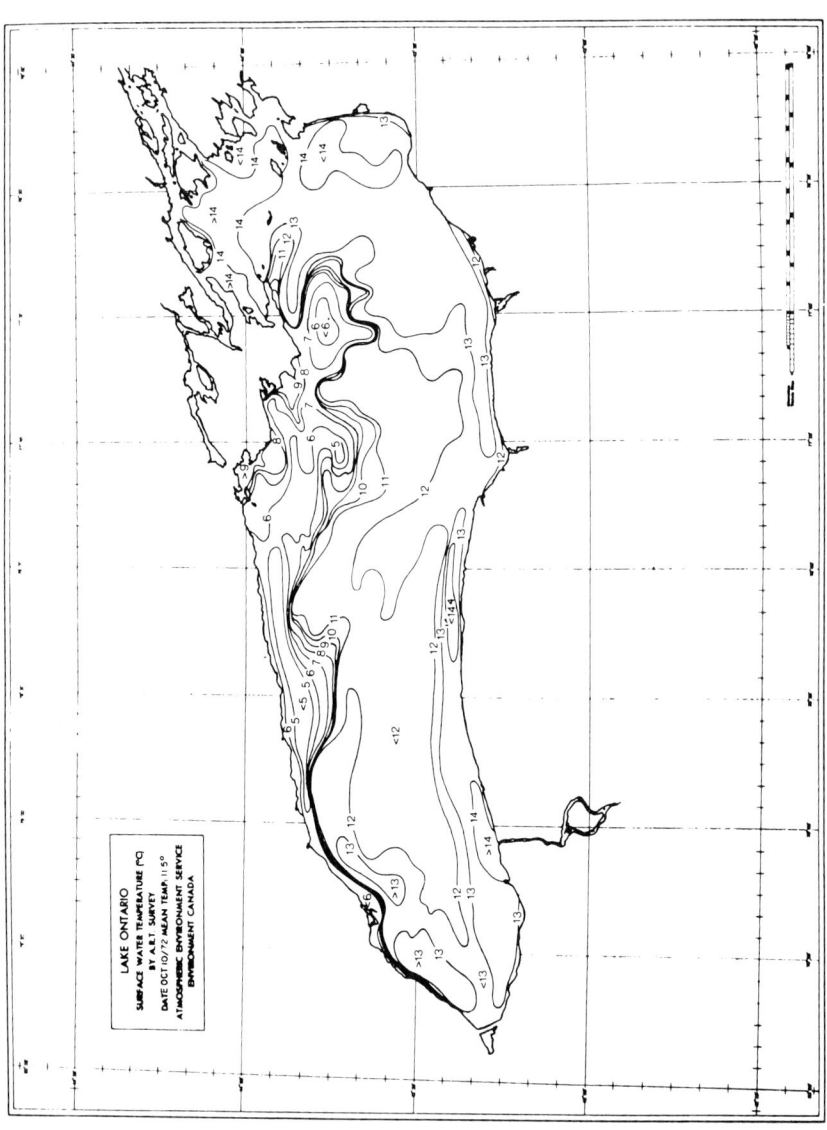

homogeneous fluid, without interference with the coastal jet flow structure that accompanies the surface pycnocline.

Where the surfaced pycnocline runs into the coast, however, geostrophic flow along the isopycnals is prevented by the coast, and the sharp pressure gradients in the longshore direction (present in the surface layer, associated with isopycnal slope) generate fluid particle velocities affecting the propagation of the front. Yamagata (1980) has recently proposed a simple theoretical model of the propagation of "warm" fronts of this kind. As in the linear theory, such a front, i.e., the eastern end of the cold pool in Fig. 15, propagates in a cyclonic direction around the basin. However, the propagation is slowed down by self-advection, particle velocities opposing the direction of wave propagation. Similar results have been found by Bennett (1973) for a quasi-linear model of an internal Kelvin wave. For a fully surfaced pycnocline Yamagata's results show that the propagation speed is changed by 50% (of the internal Kelvin wave speed).

The excitation of trapped waves is modeled in linear theory either by a superposition of a static setup solution and quasi-geostrophic wavelike modes (Csanady, 1968b; Csanady and Scott, 1974; Birchfield and Hickie, 1977), or by integrating the forcing due to wind back along wave characteristics (Gill and Schumann, 1974; Gill and Clarke, 1974; Clarke, 1977). The latter approach amounts to calculating the wind impulse integral over "retarded time," i.e., at locations where the wave had passed in its history. This is particularly convenient along a long, more or less straight coastline, for periods long compared to the typical period of a wind-stress impulse, provided that bottom friction is suitably low. The combination of static and wavelike modes is more illuminating for closed basins, especially if the period of interest includes the duration of the wind-stress impulse. The results are entirely equivalent, however, and conform to the dynamic principles described before. As pointed out already, linear theory models have limited applicability.

4.3. Observational Evidence on Wave and Front Propagation

Whether one regards a propagating, coastally trapped flow feature as a "wave" or as a "front," it should be accompanied by a pressure field ex-

---

FIG. 15. Lake Ontario surface water temperature (°C) on Oct. 10, 1972, determined by airborne radiation thermometer. Much of the north shore is occupied by upwelled cold water, separated from the rest of the lake by a clear surface front. The longshore extension of the cold cell is about 150 km. (From Irbe and Mills, 1976.)

tending over a coastal region of limited "trapping" width l, propagating in a cyclonic direction along the coasts of oceanic basins at a speed of λfl with $\lambda = 0.3-1.0$, and its principal signature should be strong longshore flow. As for directly wind-driven longshore currents, geostrophic balance [Eq. (11)] implies a relationship between the coastal sea level signal and the longshore velocity amplitude, $\zeta_0 = lfv_0/g$. At midlatitudes, this means that, given a narrow trapped field, $l = 10$ km, a velocity amplitude of $v_0 = 0.1$ m sec^{-1} corresponds to a sea level signal of $\zeta_0 = 1$ cm. A sea level signal of this amplitude is small compared to short-term level fluctuations associated with tides, seiches, etc. and its detection is possible only through statistical time series analysis. By contrast, a velocity signal of 0.1 m sec^{-1} is of the same order of magnitude as fluctuations due to other causes and is more immediately apparent in any record. Where the trapped wave is of the internal Kelvin wave type, isopycnal movements accompanying it are also large and conspicuous.

A statistical analysis of sea level records from eastern Australia by Hamon (1962) and their interpretation by Robinson (1964) in terms of continental shelf waves originated current interest in coastally trapped waves. Similar analyses of sea level records from the west coast of North America (Mooers and Smith, 1968; Cutchin and Smith, 1973; Smith, 1974) yielded some evidence to suggest northward (cyclonic) propagation of certain low-frequency signals, but also southward propagation of other sea level disturbances, apparently in phase with weather systems. Although it is possible to think of such a southward-traveling disturbance in terms of linear theory as a "forced" wave, it is more straightforward to view it as directly forced wind-driven transient flow, discussed at some length earlier in this article. Given the large spatial extent of the weather systems involved (order 2000 km), the distinction between local and nonlocal forcing becomes more or less academic. As already pointed out, most of the time west coast currents and sea level do behave as if locally forced.

Current meter observations on the Oregon–Washington shelf in the early 1970s provided more direct and convincing evidence for the northward propagation of flow episodes unrelated to the wind. Figure 16 from Kundu *et al.* (1975) shows a comparison of winds and currents observed at the central transect of the 1973 CUE-2 experiment. One flow episode centered at July 31 is clearly not wind-driven. Figure 17 from Kundu and Allen (1976) shows the northward propagation of the same event over a distance of about 80 km, at a speed of 100–150 km day^{-1}. Along the 100 m isobath, the current fluctuations involved the entire water column and were strong enough to reverse the direction of the flow for about a 2-day period. During the same period the temperature at 40 m depth was higher

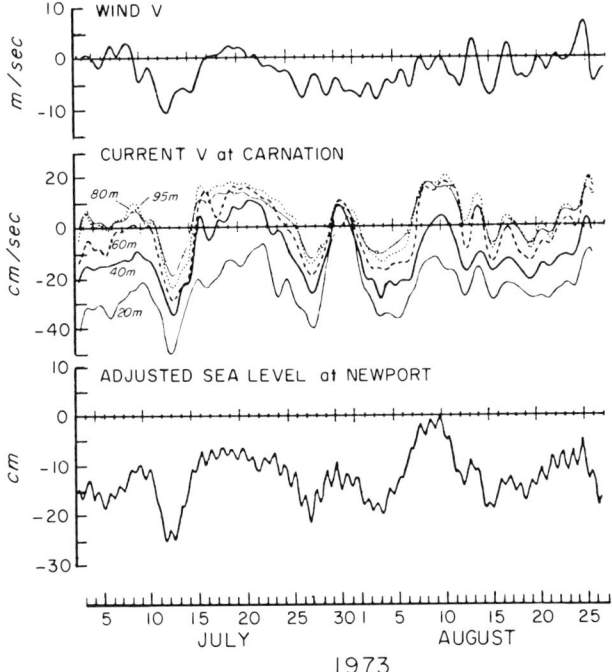

FIG. 16. Longshore component of wind velocity, current velocity along 100 m isobath, and adjusted sea level at shore over Oregon shelf during summer of 1973. Note "event" at end of July unrelated to wind. (From Kundu et al., 1975.)

by about 0.5°C than before or after, showing that substantial isopycnal adjustment accompanied the episode. Although the precise modal structure of the event is not clear from the data, it seems reasonable to regard it as a hybrid Kelvin–topographic wave with an offshore trapping scale of the order of 20 km. A detailed analysis of the velocity structure by Kundu et al. (1975) supports this notion.

Faster waves (300–500 km day^{-1}) have also been inferred to travel along the same shelf at other times (Huyer et al., 1975; Kundu et al., 1975). The trapping width of these is presumably somewhat greater and they may be closer to a homogeneous fluid or "pure" topographic wave in modal structure. The occurrence of different wavelike modes, and their relative rarity, given the dominance of direct quasi-local forcing by weather systems or large horizontal extent, explains why the statistical time-series analysis of sea level fluctuations, or of longshore velocities, tends to be confusing. One must also remember that isopycnals generally intersect the free surface off the Oregon shelf during the summer, ren-

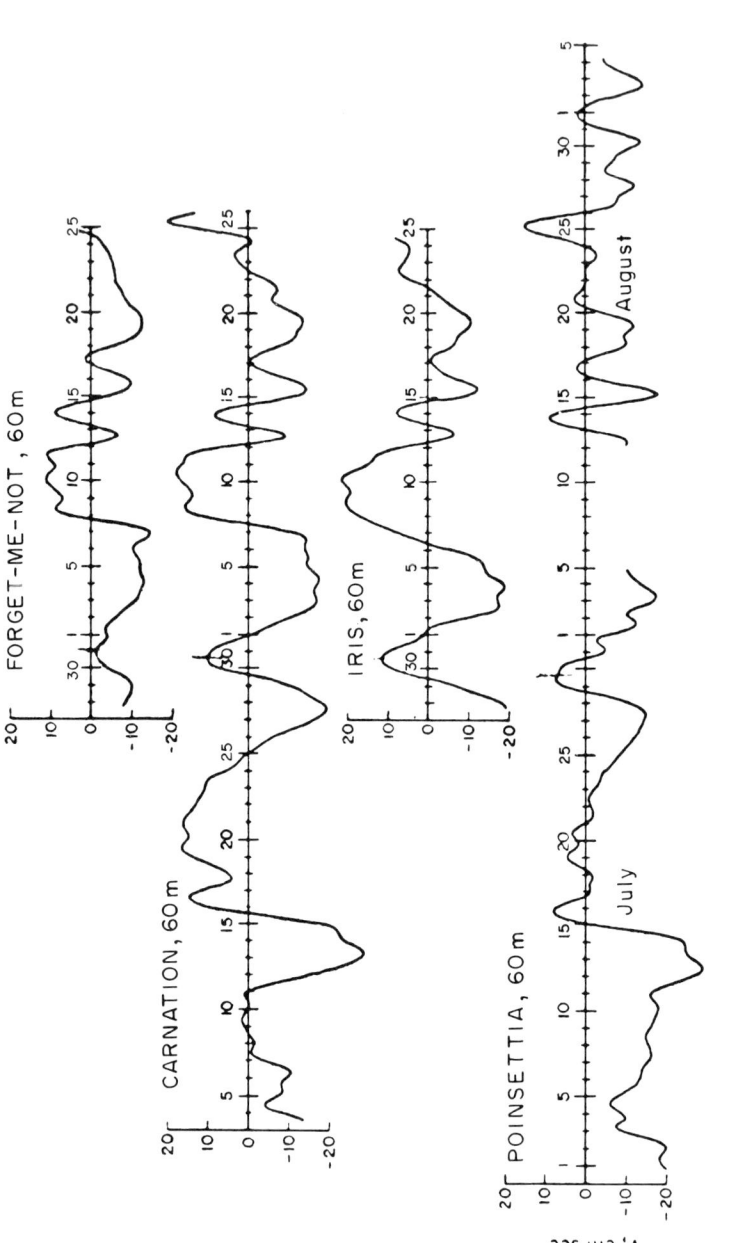

FIG. 17. Time series of longshore current at various stations separated alongshore. Non-wind-related event at end of July propagates northward at about a speed of 120 km/day. (From Kundu and Allen, 1976.) Total distance from Poinsettia to Forget-Me-Not stations was about 80 km.

dering linear models of Kelvin type waves, strictly speaking, inappropriate. With these qualifications, it is fair to assert that the Oregon shelf data provide conclusive evidence for the presence of coastally trapped waves, although their importance to coastal current climatology is on that shelf limited.

Along the east coast of North America, the evidence for propagating waves or fronts is not nearly as comprehensive. Wang (1979) and Brooks (1979) demonstrate by statistical time-series analysis of sea level records the southward propagation of signals in the southern half of the Mid-Atlantic Bight, at a speed of about 7 m sec^{-1}, corresponding to a trapping width of 100 km, which is comparable to shelf width. Event propagation, especially propagation of current reversals, does not seem to have been observed and documented, however, nor is there observational evidence on the modal structure of topographic or hybrid topographic–Kelvin waves.

The evidence for cyclonically propagating sea level disturbances, warm and cold fronts, and current reversals is strongest and most detailed in the Great Lakes. The earliest clear demonstration of internal Kelvin wavelike propagation of a warm front around the southern end of Lake Michigan was given by Mortimer (1963) (see Fig. 18). This event followed a southward wind-stress impulse that caused downwelling of warm water on the western shore. Subsequently, the warm front propagated eastward along the southern end of Lake Michigan at a speed of approximately 0.5 m sec^{-1}, corresponding to an internal Kelvin wave with a trapping width of $l = 5$ km.

More detailed evidence on the propagation of internal Kelvin waves was obtained in Lake Ontario during IFYGL. At the end of July 1972 a series of eastward wind-stress impulses generated a system of coastal jets, associated with appropriate uptilts and downtilts of the thermocline (Csanady and Scott, 1974). By the end of the wind-stress episode the thermocline tilt and current direction were reversed at some coastal transects. At the end of an additional 4 days of calm weather the reversal propagated virtually around the entire lake. The offshore trapping width was clearly seen in the experimental data to be of order 5 km, as suggested by a two-layer, linear model, and the propagation speed was a corresponding 0.5 m sec^{-1}. The modal structure of the wave was closer to a hybrid topographic–internal Kelvin wave (see Fig. 14) than to a constant-depth model internal Kelvin wave, with which these observations have originally been compared.

Late in the season (October 1972) a wave propagation event was documented on the north shore of Lake Ontario (Csanady, 1976a) that differed from the July event in that it did not involve isopycnal movements. The

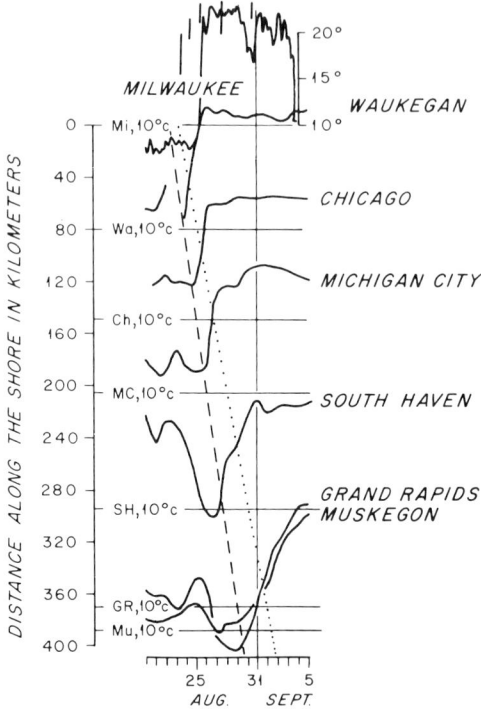

FIG. 18. Alongshore propagation of warm front in Lake Michigan, demonstrated by temperature records at water intakes. (From Mortimer, 1963.)

thermocline intersected the surface some distance offshore and was apparently unaffected by the longshore pressure gradient that more or less uniformly accelerated the entire water column. The modal structure of this event was thus similar to a "pure" topographic wave of the linear theory (see Fig. 13). The propagation speed was close to 0.5 m sec^{-1}, or about as expected for the observed offshore trapping scale of 10 km.

At a speed of 0.5 m sec^{-1}, a wave travels around the perimeter of Lake Ontario in about 15 days. Blanton (1975) examined current meter spectra from Lake Ontario and found pronounced periodicities at 12–14 days. Bennett and Saylor (1975) have also noted periodicities of 8 days, which correspond to a propagation velocity of about 1 m sec^{-1}. Marmorino (1979) analyzed current meter records taken during the winter (unstratified) season and found a signal propagation velocity of about 0.5 m sec^{-1}.

The October 1972 upwelling event on the north shore of Lake Ontario was also documented using airborne radiation thermometry by Irbe and Mills (1976). The sequence of surface isotherm maps for the period following the development of the upwelling clearly shows the slow propagation of the warm front, which marks the eastern end of the upwelling zone, toward the west, and a similar movement of the cold front at its western end toward the east (Fig. 19). The theory of Yamagata (1980) should be applicable to the propagation of the warm front, which should propagate therefore at a speed noticeably below the linear Kelvin wave celerity. There is a suggestion in the data that this was so, but the estimates are too coarse to establish such an effect conclusively.

The climatology of coastal currents in the Great Lakes has already been commented upon: it differs from offshore current climatology in that it is dominated by relatively strong longshore flow. Coastally trapped pressure fields, propagating cyclonically around the lakes, contribute a portion of the observed longshore currents. Their greatest practical importance, however, is that they cause reversals, and associated pycnocline adjustment, on the passage of a warm or cold front. The accompanying massive water exchange presumably limits coastal pollution levels and is important in other phenomena depending on large-scale mixing (e.g., in the supply of nutrients to the coastal zone). Without cyclonic propagation of the fronts, current reversals along at least some sections of the coasts would be less frequent (if they were caused only by the prevailing winds). Particle displacements associated with propagating, coastally trapped long waves typically consist of longshore excursions of several tens of kilometers, followed by offshore displacement through several kilometers. These are clearly important components of "circulation," in the sense of this term defined in the Introduction.

5. Steady Currents

In all of the discussion so far, an important part of the problem was the acceleration of the fluid, accompanying the development of the flow, geostrophic adjustment, or wave propagation. In the following, the focus will be on steady-state equilibrium flow patterns in which bottom friction plays a dominant role. Over continental shelves of Atlantic type, for example, tidal currents are generally strong, with the result that the adjustment time to frictional equilibrium is short. Under such circumstances circulation is dominated by steady-flow episodes in which forcing by the wind is generally opposed by bottom stress, although not always or everywhere, the pattern of flow being also strongly dependent on topography.

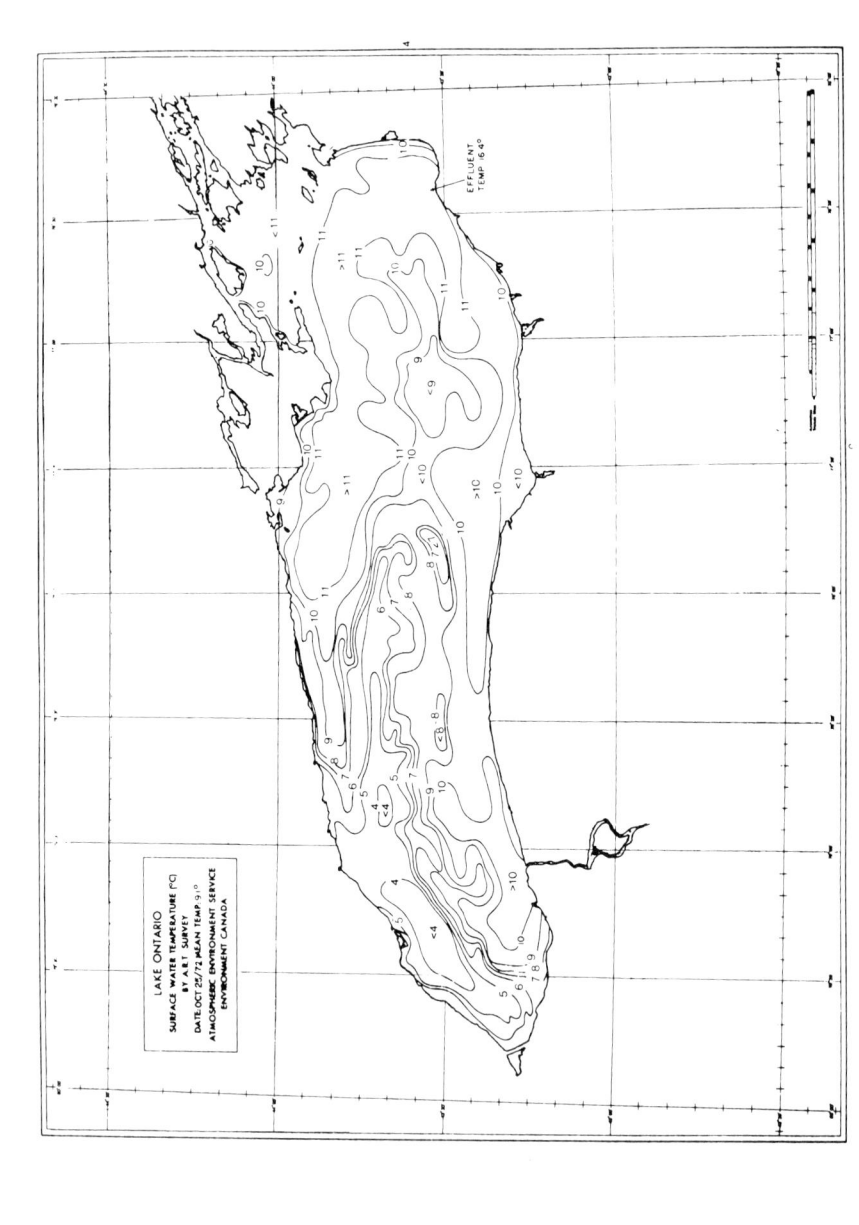

5.1. Frictional Adjustment

The simple model of a developing longshore current that resulted in Eq. (1) did not take into account bottom friction. This is clearly unrealistic in very shallow water, where a substantial longshore transport V could only develop if the velocity became very high. As the longshore velocity increases, so does bottom stress, until it balances the applied wind stress, or the longshore pressure gradient if the latter is the driving force. The longshore current, and with it the transport, is thus limited in intensity by bottom friction. Important questions are the following: How large do the limiting velocity and transport become? How long does it take for the flow to adjust to frictional equilibrium? How does the frictional adjustment time vary with depth?

The problems of estimating bottom stress and parameterizing internal friction deserve a review article on their own. Here a standard turbulent flow model will be assumed, with a thin (order 1 m), highly sheared "wall layer" above the sea floor, at the top of which the velocity has components u_b and v_b. These are related to the bottom shear stress by a quadratic law:

$$\tau_{bx} = c_d \rho u_b (u_b^2 + v_b^2)^{1/2}$$
$$\tau_{by} = c_d \rho v_b (u_b^2 + v_b^2)^{1/2} \tag{23}$$

where c_d is a bottom drag coefficient. The magnitude of this coefficient has not been satisfactorily explored and a standard value of 2×10^{-3} is often used (Jeffreys, 1923; Bowden, 1970). Recent observations of Weatherly and Van Leer (1977) and of Kundu (1977) suggest a value in the neighborhood of $c_d = 1.5 \times 10^{-3}$. Similar values of c_d are presumably appropriate for steady flow over average bottom roughness, but they do not take into account such potentially important influences as the interaction of wave-orbital motions with the flow (Smith, 1977; Grant and Madsen, 1979), which is likely to increase the "effective" value of c_d.

The simple impulse–momentum balance of Eq. (1) may be amended in a straightforward way to take into account bottom friction by supposing $v_b \cong V/H$, $u_b \cong 0$, and using Eq. (23) to describe bottom stress. The simplification is realistic in a well-stirred water column close to shore and gives a first-order answer to the questions raised earlier. With bottom

---

FIG. 19. Lake Ontario surface water temperature (°C) on Oct. 25, 1972, 15 days after the survey shown in Fig. 15. Warm front at eastern end of cold cell has moved westward by about 150 km; cold front forming western boundary moved somewhat less (approximately 100 km) to just west of Niagara River. (From Irbe and Mills, 1976.)

stress included in the longshore momentum balance, and the coastal constraint (2) supposed valid, one finds that the longshore transport develops according to

$$V = \frac{u_* H}{\sqrt{c_d}} \left(\frac{1 - \exp(-2u_* t \sqrt{c_d}/H)}{1 + \exp(-2u_* t \sqrt{c_d}/H)} \right) \tag{24}$$

where $u_*^2 = \tau_y/\rho$ is kinematic longshore wind stress, supposed constant. The longshore pressure gradient was here assumed to vanish, but it is easily taken into account in a similar calculation (Csanady, 1974a). At short times Eq. (24) reduces to Eq. (1), for constant wind, i.e., $V = I = u_*^2 t$. At long times the transport tends asymptotically to $u_* H/\sqrt{c_d}$, or the depth-average velocity to a constant $u_*/\sqrt{c_d}$, which is of course the velocity required for a bottom stress balancing the wind stress. The e-folding time scale of the frictional adjustment is seen to be

$$t_f = H/2u_* \sqrt{c_d} \tag{25}$$

For the "typical" value of the wind stress of 0.1 Pa, $u_* = 0.01$ m sec^{-1}, and supposing $c_d = 2 \times 10^{-3}$, one finds t_f to be slightly more than 3 hr in 10 m and more than 30 hr in 100 m. The frictional adjustment time varies directly with depth, and inversely with the square root of the wind stress and of the bottom drag coefficient. In a hurricane, for example, both u_* and c_d are high and frictional adjustment time is short.

Equation (25) may also be written in terms of the surface Ekman layer depth D_E [from Eq. (5)]:

$$t_f = f^{-1} H / 10 \sqrt{c_d} D_E \tag{25a}$$

which is typically $2.2(H/D_E)f^{-1}$. Thus for H/D_E large, ft_f is also large so that frictional equilibrium flow is confined to depths of the order of D_E or less, for typical storm durations of a few times f^{-1}.

These order of magnitude estimates may be checked against the observations shown in Fig. 3, which was cited earlier in support of the frictionless quasi-geostrophic theory. The agreement with that theory is in fact confined to water deeper than about 30 m. The vanishing transport at the shore and its slow increase with distance shows a behavior consistent with Eq. (24), the asymptotic solution $V \sim H$ being a more or less realistic description of the observations in shallow water.

The conclusions drawn here from order of magnitude estimates are valid if the velocity of a wind-driven developing current is the total velocity. When a large tidal oscillation, or some other short-term velocity fluctuation of large amplitude, is superimposed, however, the instantaneous stress is determined by the fluctuating velocity. Referred to the "circulation" component of the motion, i.e., to the average velocity over a full tidal cycle, the quadratic drag law reduces approximately to a linear bottom stress formula:

(26) $$\tau_{by} = \rho r v_b$$

where r is a resistance coefficient of the dimension of velocity. Such a linearized drag law appears to give good results in describing the mean circulation component in the Mid-Atlantic Bight, as well as frictional equilibrium storm currents in the same location, or hurricane-driven currents in the Gulf of Mexico (Scott and Csanady, 1976; Bennett and Magnell, 1979; Forristal et al., 1977). The appropriate value of r in all the cited cases was empirically found to be about 10^{-3} m sec^{-1}. Such a value of r may be interpreted physically as an average bottom velocity magnitude of order 0.3 m sec^{-1} multiplied by a drag coefficient c_d close to 3×10^{-3}.

A developing flow model with bottom friction parameterized according to Eq. (26) yields an expression similar to Eq. (24) for longshore transport, with the asymptotic value of the transport being $u_*^2 H/r$ and the frictional adjustment time

(25b) $$t_f = H/r$$

Given $r = 10^{-3}$ m sec^{-1}, this yields estimates very similar to Eq. (25), for "typical" u_*. The period of the semidiurnal tide is the minimum period for which a "circulation" component of the flow can be usefully defined or observed in such an environment. The value of the frictional adjustment time t_f equals the semidiurnal period in water about 60 m deep, given the "typical" values of wind stress etc. assumed earlier. Thus even under average conditions a typical Atlantic type shelf is likely to respond in the frictional mode to forces affecting its circulation. This is even more likely to be the case in stronger winds.

5.2. Vorticity Tendencies

In the discussion of coastal zone physics, so far the concept of vorticity has been intentionally avoided for maximum simplicity. Although the intellectually most satisfying explanation of oceanographic phenomena is often formulated in terms of vorticity tendencies, the elegance of this is lost on colleagues not intuitively comfortable with the notion of vorticity. It seems, however, that models of frictional equilibrium flow over a sloping shelf remain altogether too puzzling without an explanation of their balance of vorticity tendencies. Since it will be necessary to invoke the notion of vorticity repeatedly in what follows, advantage will be taken here of the opportunity to explain the balance of vorticity tendencies for some flow models discussed earlier and thereby tie up some loose ends.

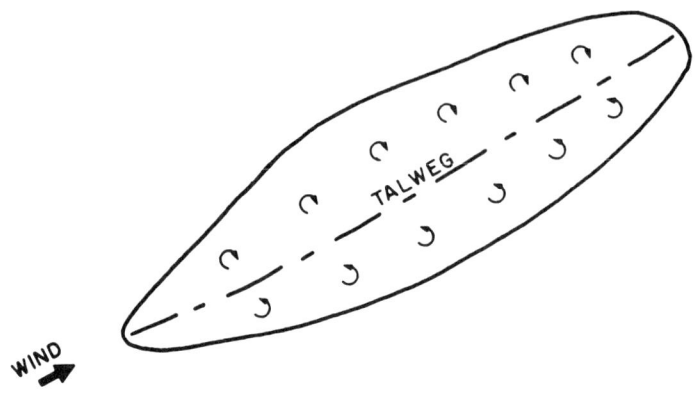
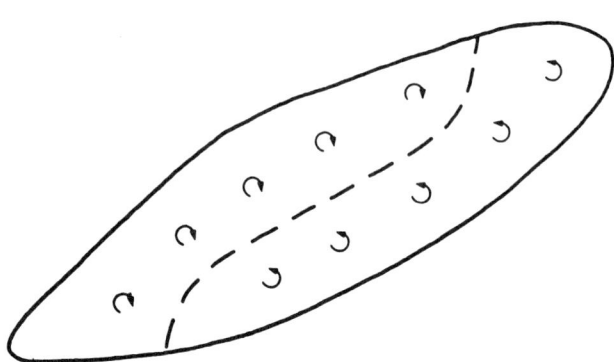

FIG. 20. Vorticity generation and propagation in closed basin. Wind impulse generates vorticity of opposite sign on either side of the locus of maximum depth ("Talweg"). Currents flowing into deeper water stretch the water columns and generate positive vorticity at downwind end. The opposite effect takes place at the upwind end. Net result is that the line separating positive and negative vorticity shifts in a cyclonic direction. (From Csanady, 1974d.)

Consider first the quasi-geostrophic model of current generation in a coastal zone of constant depth without bottom friction [Eqs. (1)–(3)]. The curl of the linearized depth-integrated momentum equations yields the following relationship for spatially uniform wind:

(27) $$\frac{\partial}{\partial t}\left(\frac{\partial V}{\partial x} - \frac{\partial U}{\partial y}\right) = f\frac{\partial \zeta}{\partial t}$$

The left-hand side is the rate of change (tendency) of depth-integrated vorticity, while the right-hand side expresses the physical effect of vortex line stretching on a rotating earth, associated with a rise of surface level. For a quasi-geostrophic longshore current Eq. (9) may be substituted for V and Eq. (27) integrated with respect to time to arrive at

$$(28) \qquad \frac{gH}{f} \frac{\partial^2 \zeta}{\partial x^2} = f\zeta$$

This now has the coastally trapped solution

$$(29) \qquad \zeta = \zeta_0 \exp(-x/R)$$

where R is the radius of deformation given in Eq. (3). Thus the physical explanation of why the trapping width of such a current is R may be said to lie in the balance of vorticity tendencies.

When some factors affecting the principal vorticity-tendency balance change, so does the trapping width. Consider the case of a sloping bottom $H = H(x)$, but otherwise the same problem as above. Instead of Eq. (27), one now finds

$$(30) \qquad \frac{\partial}{\partial t}\left(\frac{\partial V}{\partial x} - \frac{\partial U}{\partial y}\right) = f\frac{\partial \zeta}{\partial t} - g\frac{dH}{dx}\frac{\partial \zeta}{\partial y}$$

The second term on the right is a new vorticity source term and is most easily interpreted in terms of the cross-shore geostrophic velocity:

$$(31) \qquad u_g = -\frac{g}{f}\frac{\partial \zeta}{\partial y}$$

In terms of this velocity the new source term is $f(dH/dx)u_g$, representing vortex line stretching that results when the fluid columns move across depth contours and increase their depth.

In the absence of a longshore pressure gradient the new source term vanishes. Over a beach of constant slope, $H = sx$, the time-integrated Eq. (27) now has a trapped solution slightly more complex than Eq. (29):

$$(32) \qquad \zeta = \zeta_0 K_0(4x/L)^{1/2}$$

where K_0 is a Bessel function and L is as defined in Eq. (4). This new measure of the trapping width arises from the balance of the same terms in the vorticity equation as Eq. (29), the result being modified only by the different geometry of the coastal zone.

With an opposing pressure gradient present, the most interesting case is when the new source term in Eq. (30) dominates. Neglecting $\partial U/\partial y$ and the first term on the right of Eq. (30), and substituting the geostrophic relationship for V, one arrives at

(33) $$\frac{\partial^2}{\partial x\,\partial t}\left(\frac{H}{f}\frac{\partial \zeta}{\partial x}\right) + \frac{dH}{dx}\frac{\partial \zeta}{\partial y} = 0$$

This is the equation describing the structure and propagation of topographic waves, discussed in detail by Gill and Schumann (1974). Setting

(34) $$\zeta = Z(x)\,\exp(iky + ict)$$

an eigenvalue problem for $Z(x)$ arises that determines the trapping width l of the wave and its propagation speed $c = \lambda f l$. As is clear from the form of Eq. (33), these quantities depend on the precise geometry of the coastal zone.

In terms of the vorticity concept, the generation and propagation of topographic waves may be explained as follows. An initial longshore wind-stress impulse generates a strong coastal current over a portion of the coast. This is associated with a distribution of vorticity along a finite portion Y of the coastline (Fig. 20, p. 144). Forward of this distribution (in the cyclonic sense) a current of the direction illustrated turns slightly seaward: one may think of the vortices "inducing" cross-isobath flow. This,

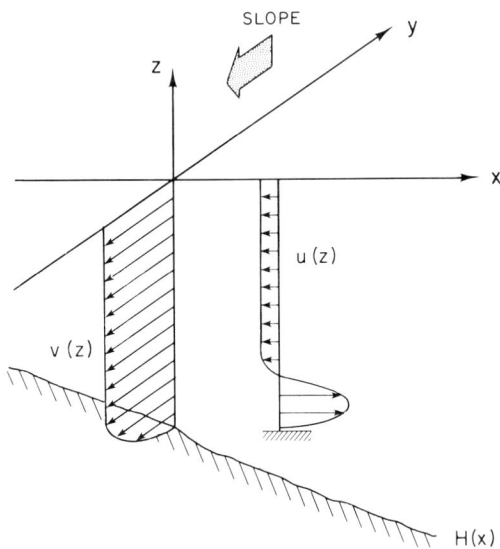

FIG. 21. "Parallel flow" driven by longshore pressure gradient ("Slope"). Cross-shore transport vanishes by hypothesis, so that onshore flow above bottom boundary layer must be balanced by offshore bottom Ekman transport. For the same longshore gradient, this requires greater bottom stress in deeper water, hence longshore velocity increasing with depth.

however, generates positive vorticity by the stretching of vortex lines (increasing depth of fluid columns) and thus extends the original region Y of vorticity distribution into the forward direction, causing effectively a cyclonic propagation of the whole coastal current pattern. The balance of the two terms in Eq. (33) expresses exactly this vorticity growth due to vortex line stretching. The argument here is the same as given by Longuet-Higgins (1965) in his discussion of Rossby waves.

The development of frictional equilibrium flow in shallow water means that the velocity becomes (nearly) constant with distance from shore, to some limiting depth. This corresponds to negligible vorticity very close to shore, apparently exempting this region from the inertial response envisaged in the above discussion of topographic wave generation and propagation. The importance of cross-isobath flow in the vorticity balance remains, however, and leads to some interesting properties of frictional equilibrium circulation over a sloping coastal region.

5.3. Steady Parallel Flow over a Straight Continental Shelf

As in the aforementioned simple transient flow models, a realistic idealization of some continental shelves is to suppose isobaths straight and parallel to the y axis and to the coast. Also as in the simplest transient cases, the cross-isobath transport may be supposed to vanish within some range of the coast, i.e., Eq. (2) is postulated to be valid for $x \leq l$.

For steady flow, the continuity equation now implies $V = V(x)$—the longshore transport is a function of distance from shore only, i.e., it remains constant along isobaths between upstream and downstream transects. This idealization can clearly be valid for certain portions of the continental shelf only and then to a certain degree of approximation, so that its limitations will have to be explored later. The fluid density will also be supposed constant at first, and the effects of density variations will be discussed subsequently.

A further simplification that can safely be made is to suppose the cross-isobath component of the bottom stress small and neglect it in the depth-integrated momentum balance, as well as in the vorticity equation. In the momentum balance the cross-isobath bottom stress competes with the Coriolis force of longshore flow, compared to which it is typically an order 1% perturbation. In the vorticity tendency balance the longshore derivative of the cross-shore bottom stress appears along with the cross-shore derivative of the longshore stress, these being small on two counts: $v_b \gg u_b$ and $L_y \gg L_x$, where L_x and L_y are typical scales of variation cross-shore and alongshore.

The balance equations for depth-integrated longshore momentum and vorticity tendency are, with the above simplifications:

$$gH\frac{\partial \zeta}{\partial y} = \frac{\tau_y}{\rho} - rv_b$$

(35)

$$g\frac{dH}{dx}\frac{\partial \zeta}{\partial y} = W - \frac{\partial}{\partial x}(rv_b)$$

where $W = (1/\rho)(\partial \tau_y/\partial x - \partial \tau_x/\partial y)$ is the wind-stress curl. When this vanishes, Eqs. (35) are consistent only provided that

(36) $\qquad\qquad g(\partial \zeta/\partial y) = \text{constant}$

The simple idealization of steady along-isobath flow over the contours of a sloping coastal region is thus seen to have some surprisingly far-reaching consequences (Csanady, 1976b). Equation (2) in these circumstances implies Eq. (36)—the longshore pressure gradient is constant with distance from shore. The magnitude of the longshore gradient is arbitrary, and is the only free parameter available to represent the local effect of the basin-wide (global) circulation on a limited coastal region, or in other words, to match the along-isobath flow supposed to occupy a given coastal region to the flow outside. Both the coastal constraint [Eq. (2)] and its consequence [Eq. (36)]—longshore pressure gradient constant with distance from shore—are clearly limited in validity to some distance range from the coast. The question of what this distance range is will be pursued further below.

The depth-integrated force balance between a longshore pressure gradient constant with distance from shore, wind stress, and bottom stress, for a variable-depth water column is essentially the same as was discussed above for transient flow in a closed basin, with "setup" opposing the wind stress. Along an open coast, however, the pressure gradient need not oppose the local wind. Where the pressure gradient supports the wind in generating longshore flow in a given direction, bottom stress must be strong enough to balance both. If the two are in opposition, the bottom stress must make up the difference. In shallow water, the integrated pressure gradient force $gH(\partial \zeta/\partial y)$ is small and bottom stress must always oppose wind stress. Where the depth is large enough for $gH(\partial \zeta/\partial y)$ to overwhelm the wind stress, bottom stress opposes the pressure gradient force. Thus, if wind stress and longshore pressure gradient are in opposition, bottom stress vanishes along a critical isobath (where $gH(\partial \zeta/\partial y) = \tau_y/\rho$) and changes sign on crossing this isobath. This is analogous to the change of acceleration in the transient, developing current along the same isobath.

One physical interpretation of the above results is to suppose the water column perfectly stirred to distribute the wind force and bottom drag evenly in the vertical. The velocity is then also equal to the depth-average value at all levels, i.e., $v = V/H$, $u = 0$—all fluid particles move along depth contours. There is under these circumstances no vortex line stretching, and the curl of the net frictional force, wind stress less bottom stress distributed in the vertical, must vanish, i.e.,

$$(37) \qquad \frac{\partial}{\partial x}\left[\frac{\tau_y/\rho - rv_b}{H}\right] = 0$$

The first of Eqs. (35) shows that this result implies Eq. (36).

The interpretation in terms of a perfectly stirred water column may be thought to apply in very shallow water. Equations (35) hold, however, independently of the details of interior friction. In deeper water, it is instructive to state the depth-integrated longshore momentum balance in terms of Ekman transports, simply dividing the first of Eqs. (35) by the Coriolis parameter:

$$(35a) \qquad \frac{\tau_y}{\rho f} - \frac{rv_b}{f} = \frac{gH}{f}\frac{\partial \zeta}{\partial y}$$

The left-hand side is Ekman transport in the surface layer plus Ekman transport in the bottom layer (due note being taken of signs), while the right-hand side is geostrophic cross-shore transport associated with the longshore pressure gradient. In the interior of the water column frictional influences are negligible, and only the geostrophic cross-shore flow is in evidence. As the depth varies, so does the geostrophic cross-shore transport. For longshore wind stress constant with distance from shore, the divergence of the geostrophic cross-shore transport is absorbed by a bottom Ekman layer. In the simple case when the wind stress vanishes, $\tau_y = 0$, v_b points along negative y (for positive $\partial \zeta/\partial y$) and varies as H. The pressure gradient-driven flow becomes stronger in deeper water, and generates higher bottom stress and a thicker bottom Ekman layer, which is able to conduct away the excess geostrophic cross-shore transport (Fig. 21, p. 146). With τ_y finite and in opposition to pressure gradient, bottom-layer Ekman transport changes sign at the critical isobath, which thus becomes a line of divergence or convergence.

5.4. Shelf Circulation as a Boundary-Layer Problem

The above simple shelf circulation model is clearly restrictive: the coastal constraint, Eq. (2), cannot be valid everywhere, and the local longshore pressure gradient is not a "deus ex machina." To remove the restrictions of the model, it is necessary to consider the "global"

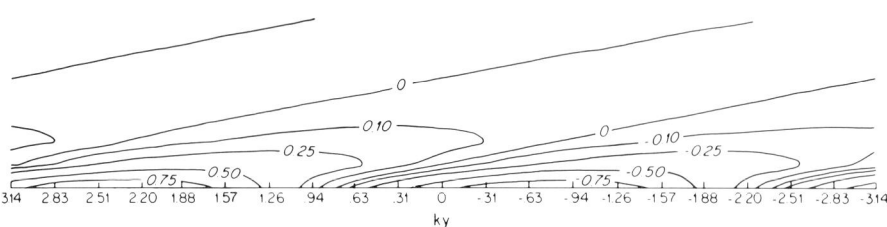

FIG. 22. Coastally trapped pressure field evoked by sinusoidally varying longshore wind over a sloping beach. Maximum positive wind stress at $ky = 0$ is opposed by a longshore gradient near shore of a magnitude close to $u_*^2/gs(\kappa k^{-1}/2)^{1/2}$, which is typically 2×10^{-7}. (From Csanady, 1978c.)

problem: What is the distribution of transports and of sea level over an extended region of the coastal ocean? If this is too difficult to answer in general, what is the likely range of validity of the coastal constraint, and by what physical mechanism are longshore pressure gradients generated along an open coastline?

A reasonably simple approach to this problem emerges if one neglects the small-scale variations of topography or of forcing in the longshore direction, and focuses on such physical influences as major changes in coastline orientation or wind-stress variations on the scale of weather systems. It is then still realistic to postulate $L_y \gg L_x$, as in the simple parallel flow model above, i.e., to regard the coastal zone of interest as a narrow boundary region of a larger scale circulation pattern. Under these circumstances the contribution of the cross-isobath component of bottom stress to the vorticity equation is negligible, and the second of Eqs. (35) remains valid. The near-bottom velocity may be approximated by its geostrophic value, $v_b = (g/f) \, \partial\zeta/\partial x$, to arrive at the following equation:

$$(38) \qquad \frac{\partial^2 \zeta}{\partial x^2} + \kappa^{-1} \frac{\partial \zeta}{\partial y} = W$$

where $\kappa^{-1} = (f/r) \, dH/dx$. Equation (38) is a parabolic equation for the sea level ζ analogous to the one-dimensional heat conduction equation, with conductivity κ and negative y playing the role of time (for positive f). In the coordinate system used here (the same as in Fig. 1 and later figures), negative y is the direction of propagation of Kelvin or topographic waves, called the "cyclonic" direction before. In the absence of any forcing, an arbitrary $\zeta(x)$ distribution at a chosen transect ($y = 0$, say) spreads out along x, proceeding in the negative y direction, much as a hot spot spreads out in a conducting rod. A concentrated high spot spreads out in the cross-shore direction over a distance of order $(2\kappa|y|)^{1/2}$. The effective "conductivity" has the physical dimension of distance and is proportional to the bottom resistance coefficient r. Typical values are $r = 10^{-3}$ m sec^{-1}, $f = 10^{-4}$ sec^{-1}, $dH/dx = 3 \times 10^{-3}$, which result in $\kappa = 3$ km. Thus a concentrated high spot spreads out to occupy the entire shelf width l in a longshore distance Y of order $l^2/2\kappa$. If l is 100 km, Y is 1700 km, or comparable to typical weather systems and continental dimensions. A shelf "remembers" a disturbance for this long, or a disturbance affects a forward ($y < 0$) portion of the shelf of this order of length.

In order to solve Eq. (38), it is necessary to specify boundary conditions at the coast ($x = 0$) and at the shelf edge ($x = l$) or asymptotically as $x \to \infty$, as well as an "initial" distribution at some transect, which may without loss of generality be taken as $y = 0$. The coastal boundary condition is straightforward, $U = 0$, which, expressed in terms of ζ, yields the equivalent of a heat flux condition at the end of a conducting rod. However, the necessity for imposing conditions at the other two boundaries reflects weaknesses of the boundary layer model. A prescription of $\zeta(y)$ along $x = l$ amounts to parameterizing the influence of the deep ocean on the shelf region, while $\zeta(x)$ at $y = 0$ effectively defines the inflow from a backward ($y > 0$) shelf region. In comparison with the parallel flow model, the only gain is that no assumptions need be made regarding some "forward" boundary transect at $y = y_2 < 0$.

With all its weaknesses, however, the boundary layer model allows considerable insight into the mechanism of longshore pressure gradient generation by such local influences on a shelf region as longshore or cross-shore wind stress, distributed in an arbitrary way, subject only to the underlying hypothesis of long longshore scales. Equation (38) seems to have been first derived and discussed by Birchfield (1972), who has clearly pointed out the important underlying vorticity tendency balance between vortex line stretching and bottom-stress curl. Birchfield's derivation of this equation was based on a boundary layer expansion of the viscous flow problem in terms of powers of the Ekman number, along principles developed in detail by Greenspan (1968). Using the same approach, Pedlosky (1974b) and Hsueh et al. (1976) also arrived at the same equation and noted some of its general consequences. The Ekman number expansion approach seems to imply certain assumptions about interior stresses. The derivation given here (the same as in Csanady, 1978c) shows that these assumptions only involve the parameterization of bottom stress and do not exclude a nearshore region where the depth is comparable to Ekman depth. The general "boundary layer" assumption of $L_x \ll L_y$ also underlies Gill and Schumann's (1974) analysis of topographic waves, an analogy which suggested the term "arrested topographic wave" for solutions of Eq. (38) (Csanady, 1978c). It seems best to think of this equation, however, as describing the behavior of a frictional boundary layer at the coast that exists only over sloping bottom.

Figures 22 (p. 150) and 23 illustrate the effect of a sinusoidally varying longshore wind on a long, straight shelf of constant slope, with the deep ocean supposed "inert," i.e., $\zeta = 0$ as $x \to \infty$. Figure 22 (p. 150) shows that a varying longshore wind, of longshore wavelength $2\pi k^{-1}$, sets up a coastally trapped steady pressure field, with a trapping scale of $L = (2\kappa k^{-1})^{1/2}$. If the longshore wavelength of the forcing is 1000 km, a typical value of $\kappa = 3$ km results in $L = 50$ km, i.e., trapping in a moderately narrow nearshore band. The dynamic role of the peculiar pressure field is elucidated somewhat by the transport streamline pattern of Fig. 23, which shows the transition between the region where the costal constraint is valid ($U = 0$—longshore force balance between wind stress, bottom stress, and longshore pressure gradient, as in the parallel flow model) to an "outer shelf" region where the total transport is only what flows onshore or offshore in a surface Ekman layer, so that most of the water column is quiescent (in this simple case when the surface Ekman layer is nondivergent). The outer shelf region is at a distance of $x > 3L$ from shore, i.e., typically beyond 150 km, where the water is 450 m deep if the slope is 3×10^{-3}, as supposed above for the "typical" quantities. Forcing of shorter wavelength results in trapping closer to shore, however.

This model suggests that a trapped pressure field, with an associated open circulation cell, may be generated on the inner shelf on account of

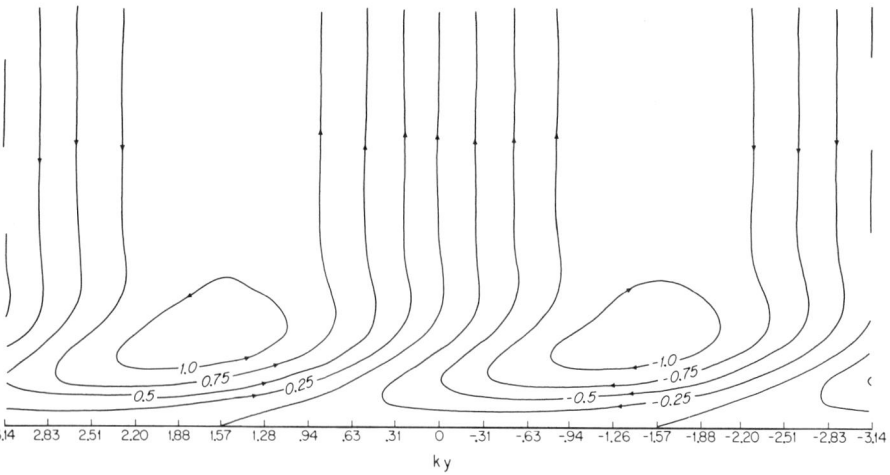

FIG. 23. Pattern of transport streamlines corresponding to the pressure field of Fig. 22. Far from shore the longshore wind stress is balanced by the Coriolis force of cross-shore Ekman transport. Right at the shore, in zero depth the wind stress and bottom stress balance, hence transport is exactly in phase with the wind.

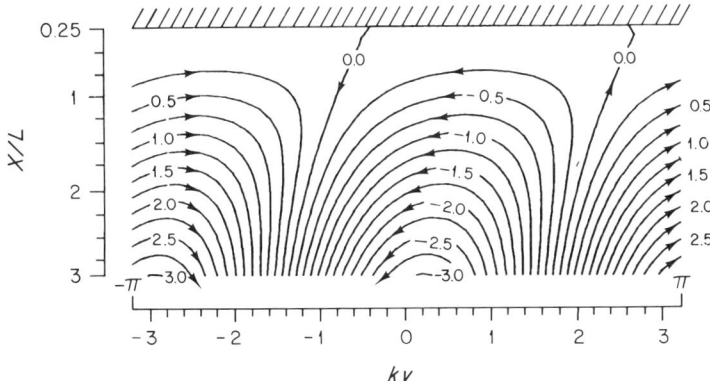

FIG. 24. Pattern of transport streamlines generated by sinusoidally varying cross-shore wind over a sloping beach. Far from shore the transport is surface Ekman drift, the convergence/divergence of which stretches vortex lines and generates massive circulation cells. Friction nearshore distorts the streamlines.

the longshore variations of forcing alone. Physically, the key to this possibility is the variability of bottom stress in shallow water. Cross-isobath transport below the surface Ekman layer implies vorticity generation by stretching or squashing fluid columns, which must be balanced by the curl of the bottom stress. Equation (38), with $W = 0$, expresses just this balance of vorticity tendencies. A specific distribution of surface levels is required to bring about the required balance of vorticity tendencies and this is associated with a peculiar distribution of interior velocities over the inner shelf. Certainly physical intuition would not lead one to expect such a distribution. Of particular practical importance is the model prediction that the cross-shore transport U can be of significant magnitude as close to shore as 10 km, depending on topography, forcing, etc. The inner shelf longshore current transports varying amounts of fluid, accepting the inflow from a surface Ekman layer further offshore where the longshore wind stress drives it shoreward, and supplying the outflow where the longshore stress is oppositely directed.

In a qualitative application of this model to realistic coastline geometry, one may think of longshore variations in forcing as being due to changes of coastline orientation. The calculated results suggest that the flow accommodates itself to such changes within an inner shelf boundary layer, with the outer shelf not being affected. The open circulation cells associated with the flow adjustment should have considerable practical importance as a mass-exchange mechanism.

The solution just discussed was particularly simple because a longshore wind stress, varying in the longshore direction only, has zero curl and contributes nothing to the distributed forcing term W. The wind stress enters the problem through the boundary condition at the shore, which requires balance between longshore wind stress and bottom stress (the pressure gradient force vanishing with depth), which results in a condition specifying the offshore surface level gradient, $\partial \zeta/\partial x$.

More complex wind fields lead to more complex solutions. Whatever the details, however, the parabolic nature of Eq. (38) and the intrinsic length scale κ, related to bottom friction, govern the character of the solutions. In view of the linearity of Eq. (38), different driving forces may be thought of as driving different components of shelf circulation, which are simply additive.

A cross-shore wind varying in the longshore direction creates a pressure and flow field that also must satisfy the boundary condition at the shore, effectively a condition on $\partial \zeta/\partial x$. In addition, the wind-stress curl W is now not zero and creates a field not trapped within a nearshore band. This clearly must merge with the pattern of the deep-water gyres offshore, although the sloping bottom provides a powerful constraint likely to dominate the response of the ocean to this type of wind field for some distance offshore. Figure 24 (p. 153) illustrates a calculated flow field for an idealized case of this kind, homogeneous water over a sloping bottom, acted upon by a cross-isobath wind-stress field varying along the isobaths. The pattern far from shore is due to direct forcing by wind-stress curl over a slope, a well-known physical effect discussed, e.g., by Pedlosky and Greenspan (1967). Near the coast a trapped field performs the transition to the nearshore dynamic balances, where the coastal constraint dominates.

A potentially important forcing effect on the shelf circulation is the

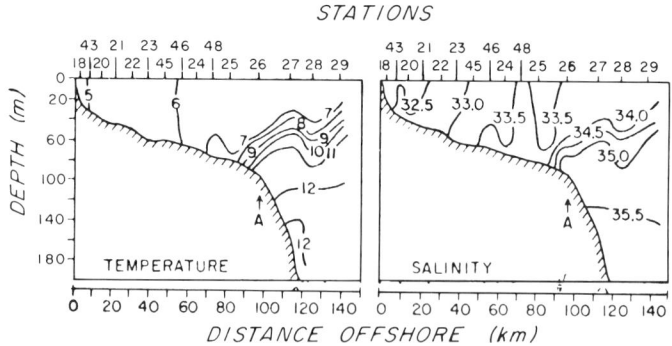

FIG. 25. Cross-shelf distribution of temperature (°C) and salinity (%o) in transect south of Long Island, in early April, 1975, under well-mixed conditions. (From Walsh *et al.*, 1978.)

pressure field impressed by the deep ocean. In the boundary layer model this appears as a boundary condition at the outer edge $x = l$, $\zeta(l, y)$. A longshore pressure gradient impressed at the edge of the shelf affects the entire shelf much as the free-stream pressure gradient affects a laboratory boundary layer, on the reasonable supposition that the longshore scale of such a gradient is comparable to oceanic dimensions. Such an "impressed" pressure gradient is thus more or less constant with distance from shore and leads to effects as discussed earlier in connection with the parallel-flow shelf circulation model.

5.5. Thermohaline Circulation

Although secondary in importance to the wind as a driving force, pressure differences within shallow seas arise also from horizontal density gradients due to freshwater influx or to rapid heating and cooling of shallow water. These may be expected to generate their own "thermohaline" circulation pattern, which combines with wind-driven circulation. The observed southwestward drift of shelf waters off the east coast of North America (north of Cape Hatteras) has repeatedly been attributed to freshwater inflow (see historical remarks by Beardsley and Boicourt, 1980). In the Great Lakes rapid early season heating of nearshore waters leads to the formation of a so-called "thermal bar" associated with a slow cyclonic circulation (Rodgers, 1965; Bennett, 1971; Huang, 1971; Csanady, 1971, 1974b).

Similar effects may be understood in the simplest terms by considering a two-dimensional infinite coast model, in which nearshore freshening or heating occurs uniformly along the coast. Although freshwater sources are concentrated in rivers, river plumes mix with shelf water within a relatively narrow coastal boundary layer so that it is not too unreasonable to idealize the freshwater inflow as uniformly distributed along the shoreline. A uniform line source of freshwater, mixing with seawater due to storm- and tide-induced turbulence, should give rise to constant-density surfaces parallel to the coastline. The typical winter salinity distribution in the Mid-Atlantic Bight, for example, is of this kind; see Fig. 25 for an illustration. Except near the shelf edge, there is little density variation at any location between top and bottom of the water column, but the horizontal density gradients are significant in generating pressure gradients in a cross-shore direction. Longshore density gradients are small over most of the shelf and may be neglected in a first approximation. Similarly, in the Great Lakes, early season heating leads to isothermal surfaces more or less parallel to the coast.

The pressure forces arising from a density distribution that is a function of x and z only accelerate the fluid in the first instance in an offshore direction at the surface, toward the shore near the bottom. A simple transient model may be formulated as follows (Csanady, 1978b). Consider an imaginary vertical membrane separating light nearshore fluid from heavier offshore fluid. When the membrane is withdrawn, cross-shore fluid motions ensue, until the Coriolis force deflects these into a longshore direction. The offshore moving surface layers develop cyclonic longshore velocity; the bottom layers, displaced shoreward, acquire anticyclonic longshore velocity. A consideration of the vorticity-tendency balance reveals that, as in the case of the surface outcropping of isopycnals in an upwelling event, the longshore velocity that develops in the course of geostrophic adjustment is limited to a value of order $(\epsilon g H/2)^{1/2}$. The Coriolis force associated with adjustment drift accelerates the fluid to the limiting longshore velocity by the time the cross-shore particle displacement grows to order $R_i = f^{-1}(\epsilon g H/2)^{1/2}$. The results are based on the somewhat overidealized supposition that the two fluid masses do not mix and that interface and bottom friction remain negligible in the adjustment process. The geostrophically balanced flow following adjustment has a velocity distribution as illustrated in Fig. 26, associated with an inclined density interface, also shown.

The time scale of geostrophic adjustment is f^{-1}, or short compared to the lifetime of a density field established seasonally by freshwater inflow or by nearshore heating. The above transient model should therefore be more or less irrelevant to the steady-state flow pattern that accompanies such a density field, unless the "steady" flow is unstable to the point where it is subject to frequent breakdown and re-establishment. When the steady-state flow is stable, it should be controlled by friction and mixing at the interface, so that the key to its realistic description is a sound representation of interior friction and mixing. At present we lack the physical understanding of turbulent flow in a stratified fluid necessary for this task. An *ad hoc* approach is to suppose constant eddy viscosity, conductivity, and diffusivity, or to postulate some relationship between these parameters and interior density gradients. There is almost no empirical background for estimating the governing quantitative parameters in such approximate theories, and only certain limiting cases can be treated at all realistically.

Steady-state thermohaline circulation models, for the case $\rho = \rho(x, z)$, have been discussed on the basis of the viscous-conducting fluid analogy by Huang (1971) and Stommel and Leetmaa (1972). The main mathematical difficulty in such models is caused by the advection of heat and salt by the cross-shore component of the mean flow. The difficulty is to some ex-

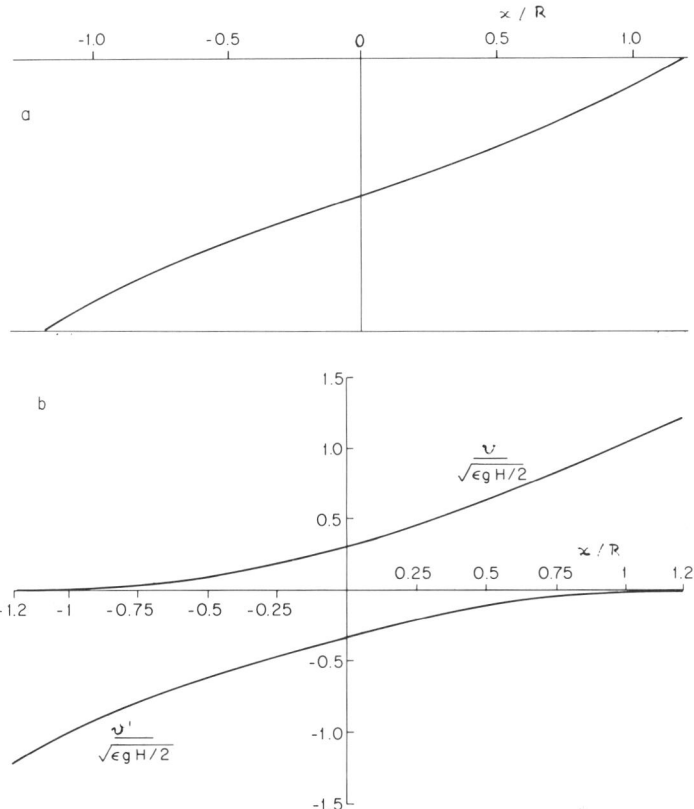

FIG. 26. (a) Shape of interface between two fluids of different density, after adjustment to geostrophic equilibrium. An imaginary membrane at $x = 0$ separated the fluids prior to adjustment, which was supposed to take place without friction. The horizontal distance scale is $R = f^{-1}(\epsilon g H/2)^{1/2}$, the internal radius of deformation. (b) Velocity distribution above and below inclined interface after adjustment.

tent an artifact of the steady-state model, because cross-shore mixing is in fact dominated by advection in transient flow episodes, especially in tidal waters. These have little relationship to the long-term thermohaline circulation and may be expected to give rise to a conventional gradient diffusion process by the mechanism known as "shear diffusion" (Csanady, 1976b; Fischer, 1970). If cross-shore advection of heat and salt by the mean flow is neglected, the density field can be calculated independently of the flow (or supposed externally impressed, or empirically determined). Under these circumstances the density distribution can be regarded as external forcing, insofar as it enters the momentum balance.

With $\rho = \rho(x, z)$ the horizontal density gradients in the cross-shore direction do not affect the momentum balance alongshore. However, the cross-shore momentum balance reveals that interior longshore velocities are affected. In the stratified interior, friction is reasonably supposed negligible, in which case the longshore velocity gradient is given by the well-known "thermal wind" equation:

$$\text{(39)} \qquad \frac{\partial v}{\partial z} = -\frac{g}{f}\frac{\partial \epsilon}{\partial x}$$

where $\epsilon = (\rho - \rho_0)/\rho_0$ is proportionate density defect, with π_0 being the density of the densest water present (so that ϵ is always negative). Because bottom friction depends on the near-bottom velocity, the vorticity-tendency equation [second of Eqs. (35)] comes to contain the depth integral of $\partial \epsilon/\partial x$. In the two-dimensional case $\rho = \rho(x, z)$, there is no reason why a longshore gradient $\partial \zeta/\partial y$ should arise, and the steady-state surface elevation field, obtained by integration of the vorticity-tendency equation, becomes simply (Csanady, 1979):

$$\text{(40)} \qquad \zeta = -\int_{-H_m}^{0} \epsilon \, dz$$

where H_m is the depth of a chosen offshore isobath at the shelf edge, and the path of integration is as illustrated in Fig. 27. It bears repetition that this model is based on a parameterization of bottom stress as proportional

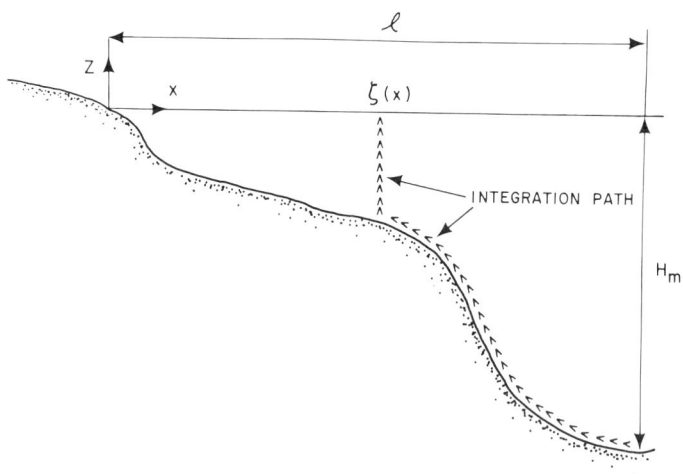

FIG. 27. Integration path in determining coastal sea level elevation from density-anomaly distribution $\epsilon(x, z)$. Method applies to equilibrium flow forced by the density distribution.

to near-bottom geostrophic velocity and the hypothesis that the density field is produced by storm and tidal stirring, independently of the mean thermohaline circulation. In typical cases the density defect near the surface is $\epsilon = -10^{-3}$, confined to a surface layer of order 100 m depth, in which case the coastal sea level stands 0.1 m higher than offshore on account of freshwater inflow or nearshore heating. If the level drop takes place over a typical horizontal distance of order 10 km, the associated surface geostrophic velocity is of order 1 m sec^{-1}.

The two-dimensional density field just discussed is a first approximation. More realistically, freshwater inflow is concentrated in major rivers and even if each of these becomes effectively distributed over a few hundred miles of coastline, there are important variations in the rate of inflow between, say, the northerly and southerly parts of continents. Such variations are responsible for longshore density gradients that may be significant producers of shelf circulation. The vorticity-tendency equation for arbitrary density distribution (impressed independently of the mean circulation) is the second of Eqs. (35) but with a forcing term added on the right, of the form:

$$(41) \qquad \phi = \kappa^{-1} \int_{-H}^{0} \frac{\partial \epsilon}{\partial y} dz - \frac{\partial}{\partial x} \int_{-H}^{0} \frac{\partial \epsilon}{\partial x} dz + \frac{fW}{rg}$$

The second of these forcing terms gives rise to the solution already given [Eq. (40)]. The first term is as significant a forcing effect as the second if L_y is of order L_x^2/κ, where L_x, L_y are the x, y scales of the density field. Taking L_x to be of the order 10 km as in the above estimate of thermohaline effects, a "typical" value $\kappa = 3$ km results in $L_y/L_x = 3$, or $L_y = 30$ km. This means that a density anomaly of 1⁰/₀₀ across a 10-km-wide band is equivalent as a thermohaline forcing effect to the same anomaly distributed over a longshore distance of 30 km. The effect of a given longshore density gradient increases with decreasing κ (low friction and steep slope). The effect of slow alongshore density variations may also be estimated simply from Eq. (40). If ϵ changes by 10^{-3} over a longshore distance of 1000 km, over a surface layer of 100 m depth, approximately constant, a longshore sea level slope of order 10^{-7} results, which is dynamically important.

5.6. *Mean Circulation of a Stratified Fluid*

Where the density gradients in the interior of the fluid are primarily determined by the adjustment of isopycnals to the dynamic balances of the mean flow, the $\partial \epsilon / \partial x$ and $\partial \epsilon / \partial y$ terms in Eq. (41) are themselves dependent upon the ζ distribution and are not, from a physical point of view, external forcing terms. Taken over to the left-hand side of the equation, they convert Eq. (35) into a similar parabolic equation for bottom pressure (Pedlosky, 1974b; Hsueh and Peng, 1978; Hendershott and Rizzoli, 1976). Interior velocities and the surface pressure field then come to depend on how the density field changes in response to external forcing.

In discussing the development of wind-driven transient flow in a stratified fluid, it was noted that the isopycnals rise or sink in a relatively narrow nearshore band [scaled by the internal radius of deformation, R_i of Eq. (12)] and undergo large vertical excursion even under modest wind-stress impulses. If a steady-state flow pattern is eventually approached in such a case, the tendency to vertical isopycnal displacement within the same nearshore band should persist and must be counteracted by dissipation processes—mixing and internal friction (Allen, 1973; Pedlosky, 1974a,b,c). A steady-state pattern of upwelling or downwelling circulation then comes into existence in which the vertical advection of temperature and salinity is balanced by mixing across (and along) isopycnals. Modeling of the flow and pressure fields associated with similar phenomena is greatly hampered by our limited understanding of dissipative processes in a stratified fluid. In a recent review article, Allen (1980) concludes that similar model studies are all more or less unrealistic for this reason. At the same time, however, these model studies clearly show that a boundary layer of scale width R_i may well accomplish mass balance closure in somewhat the same way as the frictional boundary layer over a slope (Fig. 24) accepting Ekman transport over some portions of the coast and supplying it over other portions.

Given the absence of a realistic parameterization scheme for nearshore dissipative processes in a stratified fluid, the simplest step is to ignore them and construct a steady circulation model without friction and mixing. Consider a constant-depth basin containing a two-layer fluid, acted upon by a constant wind, and suppose that the resulting interface displacements are small enough for linearized theory to apply. In the absence of mixing and friction, the fluid conserves potential vorticity and a flow pattern may be calculated without difficulty (Csanady, 1968a; Csanady and Scott, 1974). This pattern is illustrated in Fig. 28.

As may be seen from Fig. 28, the total transport equals the Ekman drift over most of the basin, so that below the Ekman layer the fluid is quiescent. A coastal boundary layer of scale width R_i accepts the Ekman drift along the right-hand shore, and transports it around the ends of the basin to the coastal boundary layer along the left-hand shore, which supplies the fluid for the interior Ekman transport. The surface remains flat, except within the coastal boundary layers. Along the shores parallel to the wind a

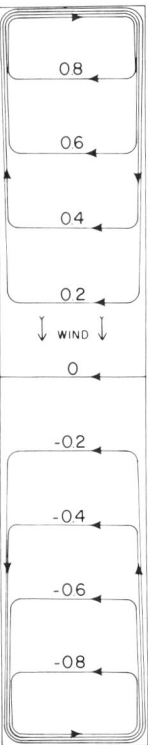

FIG. 28. Transport streamlines of top-layer circulation in two-layer rectangular basin, driven by uniform wind parallel to longer sides; frictionless case. System of coastal jets along longer sides accepts and supplies Ekman drift at basin center, returning it to other side in end-wall coastal jets.

pressure gradient opposes the wind in the upper layer, but the lower layer is entirely quiescent. The pycnocline correspondingly has a strong tilt in the longshore direction, as well as cross-shore. At the coast, there is exact balance between wind stress and pressure gradient force in the upper layers:

$$\text{(42)} \qquad \frac{\partial \zeta'}{\partial y} = -\frac{\tau_y}{\epsilon g h \rho}$$

This gradient decays with distance from the coast as e^{-x/R_i}. The longshore transport in the coastal boundary layer is geostrophic, i.e.,

$$\text{(43)} \qquad v = -\frac{\epsilon g}{f} \frac{\partial \zeta'}{\partial x}$$

At the coast, $v_0 = \epsilon g \zeta_0'/fR_i$, the subscript 0 identifying values at $x = 0$. The surface level changes in opposition to the pycnocline displacement.

One should not take too seriously the details of this simple linear theory flow pattern. The main point is that the adjustment of the isopycnals to a steady flow pattern may allow a transition from a quiescent interior (except for surface Ekman drift) to an active coastal boundary layer, over a distance range of order R_i. In particular, a longshore pycnocline tilt may develop, generating a pressure gradient in opposition to the wind in the top layer alone. Dissipative processes would presumably spread out the trapped density field implied by this model to a scale larger than R_i, but the basic character of the pycnocline tilts might well remain as predicted by the model. Along continental slopes, in particular, a considerable range of depth is available for the development of longshore isopycnal tilt.

5.7. Mean Circulation of the Mid-Atlantic Bight

The remaining sections of this article will attempt to connect the simple steady-state models of circulation to observational evidence. Circulation features more persistent than weather-related events are sometimes brought to light by averaging fixed-point current meter data for periods of the order of a month or longer, or by Lagrangian tracer studies involving horizontal displacements of drifters or of water masses over similar periods. A mean circulation pattern determined by such methods is not necessarily relatable to simple dynamic models of the kind discussed above, because unquantifiable cumulative effects of transient flow events may swamp those of steady, low-level forcing. Furthermore, Eulerian and Lagrangian mean flow patterns may differ significantly. If, however, the velocities associated with the long-term circulation component are not too much smaller than storm and tidal currents, these complications are less likely to be serious.

The circulation of the east coast shelf of North America north of Cape Hatteras, specifically that of the Mid-Atlantic Bight, is characterized by a persistent southwestward drift of an amplitude only somewhat less than tidal currents (Bumpus, 1973; Beardsley *et al.*, 1976; Mayer *et al.*, 1979). Upon time-averaging data over periods of one month or longer a clear and consistent pattern of longshore and cross-shore velocities emerges that is reasonably regarded as a steady-state flow field and compared with frictional equilibrium flow models.

A successful steady flow model of the winter circulation of the Mid-Atlantic Bight may be constructed by taking into account forcing by wind stress (both longshore and cross-shore), freshwater influx, and a larger-scale oceanic pressure field (Csanady, 1976b). It is reasonable to suppose that each of the component forces produces a pattern of its own and that the resultant pattern is a simple linear superposition of the component patterns. A parallel flow model as discussed above is adequate to describe wind-driven flow locally, even close to shore where the longshore pressure gradient may be affected by coastline orientation etc. Over the middle and outer shelf the pressure gradient may be supposed to be a deep ocean effect, i.e., constant with distance from shore. The freshwater influx is supposed to generate a two-dimensional density field, $\rho = \rho(x, z)$, represented over midshelf by $\partial \epsilon / \partial x = L^{-1}$ = constant, where L is an offshore scale of density variations. The parallel flow model and thermohaline circulation due to a two-dimensional density field are mutually consistent.

Such model calculations yielded "typical" magnitudes of the different velocity contributions, appropriate for conditions in the Mid-Atlantic Bight, at the surface, midcolumn, and bottom, at two different mid- and outer shelf locations of different depths. The magnitude of the longshore pressure gradient used was inferred from a comparison with observed currents (Scott and Csanady, 1976; Csanady, 1976b).

The observed mean circulation of the Mid-Atlantic Bight, over the middle and outer portions of the continental shelf, conforms remarkably closely to this parallel-flow thermohaline model (Flagg, 1977). Figure 29 shows some mean velocities observed over the New England shelf (averaging period a little over a month, March 1974, taken from Flagg, 1977). A larger body of similar evidence was summarized by Beardsley *et al.* (1976). Other current meter studies have shown comparable results, demonstrating the persistence of this pattern in time, and its spatial extension over the entire Mid-Atlantic Bight. The principal mean flow features derived from current meter studies characterize motion in the middle of the water column, below the surface, and above the bottom frictional layer. They are the following:

(1) Longshore (long-isobath, more accurately) flow toward the southwest at an intensity of 3–10 cm sec^{-1}, increasing noticeably with increasing distance from shore.

(2) Onshore (cross-isobath) flow over most of the water column, at an amplitude of 1–3 cm sec^{-1}.

Where reliable current meter measurements close to the surface or close to the bottom are available, they confirm the conclusions of Bumpus (1973) derived on the basis of surface and bottom drifter studies. These are the following:

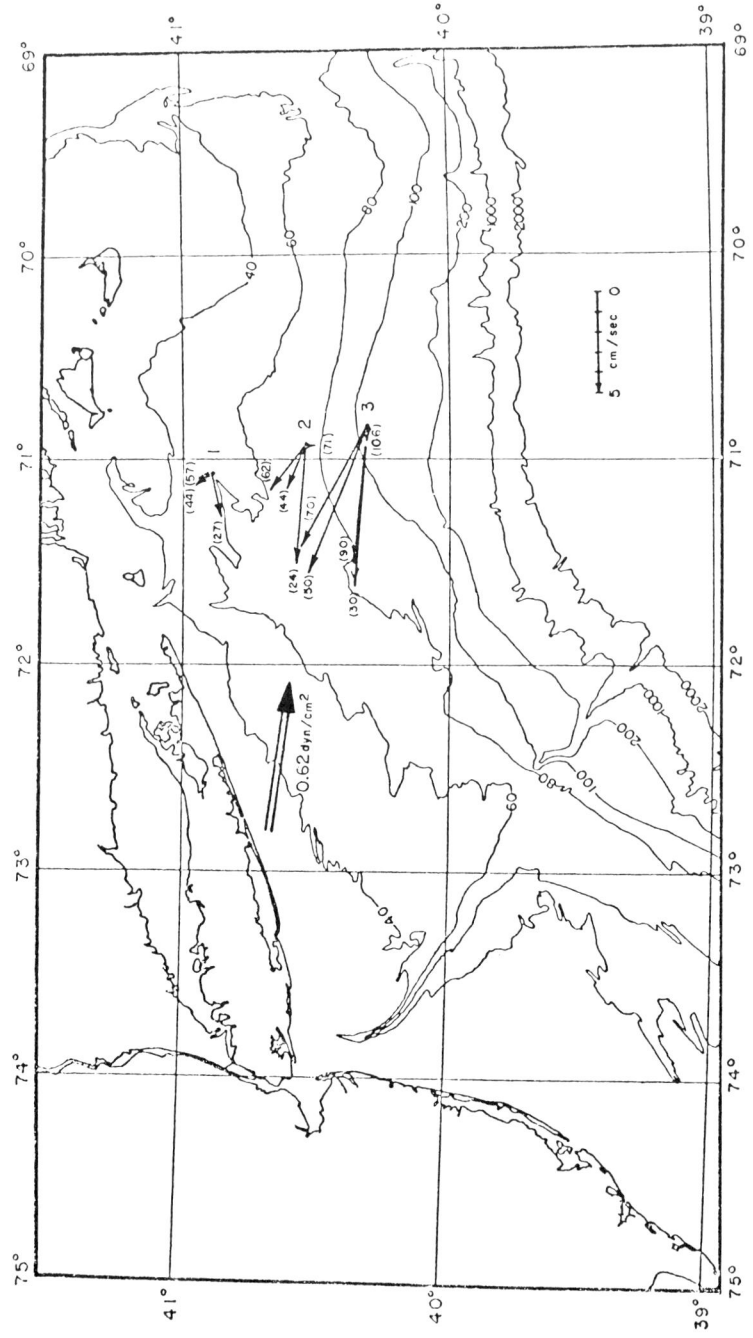

FIG. 29. Mean currents and wind stress during the 1974 MIT New England Shelf Dynamics Experiment: Feb. 27–April 3, 1974. Numbers in parentheses are the instrument depths in meters. Isobaths are in meters. Note that mean flow opposes wind stress and that velocities increase significantly in deeper water. (From Flagg, 1977.)

(3) Surface waters move to the southwest at mean speeds of 10–30 cm sec^{-1}, and in an offshore direction, at 3–10 cm sec^{-1}.

(4) Bottom waters diverge at about the 60 m isobath, moving in an onshore direction at 0–3 cm sec^{-1} in shallower water, offshore at similar speeds in deeper water.

These observed facts may be understood in the framework of the model described above as being a consequence of four mean circulation components. Specifically, the interaction of longshore wind stress and opposing sea level gradient is responsible for the increase of longshore velocity with distance from shore, according to the model discussed above. This also explains the divergence of the bottom boundary layer at a specific depth. The high offshore velocities at the surface result from the two windstress-related circulation components (due to longshore and cross-shore wind, respectively) and from the thermohaline circulation.

The observational evidence suggests that no significant trapped cells affect the mean flow beyond the 30 m isobath or so. The key driving force, the longshore pressure gradient, is then very likely a deep-water effect, impressed upon the shelf by offshore oceanic gyres (Csanady, 1978c; Beardsley and Winant, 1979). At the edge of the shelf this longshore gradient is certainly as large as it is closer to shore.

The magnitude of the longshore pressure gradient is not constant in time, however, but is subject to clear seasonal—and perhaps longer-term—variation (Chase, 1979). An interesting aside is that at the time of the Argo Merchant oil spill off Nantucket Island (December 1976), the usual southwestward driving longshore pressure gradient was fortuitously absent, and the water column moved eastward under strong northwest winds, taking the oil spill eastward and out to sea (Grose and Mattson, 1977).

One factor contributing to longshore sea level gradients, at least in the northern portion of the Mid-Atlantic Bight, appears to be freshwater influx further north, notably in the Gulf of Maine and the Gulf of St. Lawrence. According to the boundary layer model discussed above, longshore variations of freshwater influx over such a long range (order 1000 km) affect more or less the entire width of the shelf, which would make their effects difficult to distinguish from deepwater gyre effects. The magnitude of the longshore pressure gradient due to observed density variations may be estimated to be 10 cm in 1000 km (10^{-7}) during the spring runoff period only, and a much lower slope at other times of the year (Csanady, 1979). During the spring–early summer period the longshore sea level gradient due to freshwater sources is therefore of the same order of magnitude as required to explain the observed southwestward drift of shelf waters. This effect may explain the season variation of the longshore gradient. It should be added, however, that some uncertainty attaches to the estimation of a long-term mean density field, so that these conclusions must be regarded as tentative.

Over the inner shelf, evidence for trapped cells affecting the long-term mean circulation pattern comes from nearshore studies in different locations, arriving at different magnitudes of the longshore pressure gradient (e.g., Bennett and Magnell, 1979). Near the "apex" of New York Bight such local variations are particularly clear (Hansen, 1977; Mayer *et al.*, 1979). The details of these trapped cells have not been elucidated so far, not at least in connection with a long-term mean circulation pattern. On the other hand, there is clear evidence showing trapped pressure fields accompanying storms, which will be discussed next.

5.8. Storm Currents over Atlantic Type Shelves

From an economical point of view, the most important problem in applied oceanography is the prediction of storm surges, which from time to time cause tremendous damage along coastlines adjacent to broad continental shelves, such as the North Sea, the U.S. Gulf Coast, and the East Coast. Consequently, numerical models are well developed for the prediction of coastal sea levels associated with hurricanes and extratropical storms (Jelesnianski, 1965, 1966; Heaps, 1969; Welander, 1961). These models have been calibrated empirically and today constitute a useful practical tool. They do not, however, give a particularly realistic description of storm-driven currents (Forristal *et al.*, 1977), not at least without considerable further development and calibration. In any event, the predictions of the models are almost as complex as the observational evidence, and it is desirable to understand the contribution of storms to the circulation problem in terms of simpler concepts.

Strong winds acting over shallow water rapidly establish frictional equilibrium flow so that this aspect of the circulation problem is best approached by way of steady-state models. The classical models of hurricane surge are of this kind (Freeman *et al.*, 1957; Bretschneider, 1966). Although wavelike "resurgences" are sometimes important (Redfield and Miller, 1957), the bulk of the coastal sea level rise attributable to storms can be explained as a steady-state, coastally trapped pressure field. Associated with this pressure field, intense longshore currents are generated by storms, presumably giving rise to large particle displacements. Boicourt and Hacker (1976) point out, for example, that most of the mean southwestward drift off Chesapeake Bay in the Mid-Atlantic Bight is generated by a few nor'easterlies, as illustrated in their article vividly by progressive vector diagrams of observed currents.

In a steady-state model, coastal sea level rise is due to two effects: setup in response to onshore wind and geostrophic adjustment to balance the Coriolis force of longshore currents. Freeman *et al.* (1957) refer to the resultant effect at the coast as the "bathystrophic tide," because both effects are related to the bathymetry of the continental shelf, principally the width of the region of shallow water. "Storm surge" remains a much more descriptive term, however. From the point of view of the circulation problem, coastal sea levels are incidental, but of course longshore gradients of sea level forming part of a trapped pressure field under a storm affect the intensity of longshore currents. Of great practical importance for mass exchange are the cross-shore motions in the open circulation cells associated with trapped pressure fields. The simple boundary layer models discussed above have shown that the longshore scale ("wavelength") of the forcing by wind is of key importance in determining the circulation pattern. Storms of small spatial scale may create a particularly intense circulation cell.

Extratropical storms have typical scales of 1000 km and more. Mooers *et al.* (1976b) show a "typical" extratropical storm of this size over the Mid-Atlantic Bight. Hurricanes are 3–10 times smaller in diameter, but their maximum winds are much higher (Cardone *et al.*, 1976). Maximum wind stress in a hurricane reaches values of 33 Pa and more.

The simplest estimate of the intensity of storm-driven currents is obtained from the "Bretschneider formula" (Bretschneider, 1966), which is the same as Eq. (24) for $t \to \infty$, i.e., $v_b = u_*/\sqrt{c_d}$. Physically, this estimate arises from supposing wind stress to be balanced by bottom stress. Using $\tau_y = 10$ Pa and $c_d = 3 \times 10^{-3}$, one finds $v_b = 1.8$ m sec^{-1}, which is of the correct order of magnitude for hurricane-driven currents (see, e.g., Forristal *et al.*, 1977). However, Forristal *et al.* also demonstrate that bottom stress sometimes exceeds wind stress by a considerable margin, i.e., that currents are partly driven by a longshore pressure gradient for some time during a storm, especially during the passage of the peak wind stress and afterwards. Smith (1978) also shows winds, currents, and sea level along the Texas coast at the fringe of hurricane Anita, passing within about 300 km of the measurement site. The longshore pressure gradient reverses rapidly at about the time of passage of the peak longshore wind stress. As in the case documented by Forristal *et al.* (1977), the intense current is much longer-lived than the wind-stress episode and is driven by the longshore pressure gradient. Associated with changes in the latter are marked reversals in the cross-isobath component of the velocity. Smith draws attention to the fact that the cross-isobath transport does not vanish at a site about 21.5 km from the coast.

Clear evidence for a trapped pressure field associated with hurricanes has already been presented by Redfield and Miller (1957) (see Fig. 30). The maximum longshore sea level gradient shown in Fig. 30 is about 8×10^{-6}, which, in water 50 m deep, would alone drive a longshore current with a near-bottom velocity of about 1.5 m sec^{-1}. Moreover, this gradient drives the water in the same direction as the longshore stress ahead of the eye of the hurricane, just prior to landfall, and so presumably increases the maximum longshore current above what the Bretschneider formula predicts.

Intuitively, the sea level distribution shown in Fig. 30 is not difficult to understand as an effect of the cross-shore component of the wind, causing a setup to one side of the eye, a set-down to the other side. A similar effect is revealed by the boundary-layer shelf model, acted upon by variable cross-shore wind, the transport streamline pattern for which was shown in Fig. 24. Figure 31 shows the corresponding sea level distribution, characterized by a drop in levels near the maximum offshore wind ($y = 0$) and a rise near the onshore wind maximum. The coastal elevation field has an offshore extent of order $L = (2\kappa k^{-1})^{1/2}$, which should typically range from 10–30 km. Outside this range the cross-shore wind stress is balanced by Ekman drift in the surface layer. The equivalent of the eye of a storm is located near $ky = \pi/2$, where the cyclonic curl of the wind stress causes substantial onshore transport (see Fig. 24). The coastally trapped pressure field is instrumental in deflecting the onshore flow into a longshore

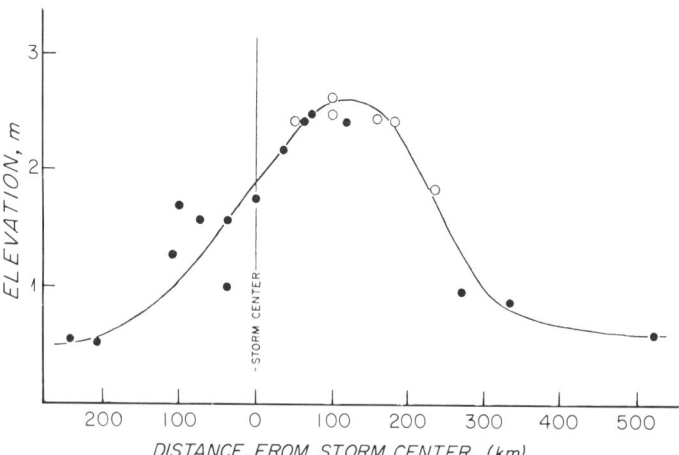

FIG. 30. Sea level rise associated with Atlantic coast hurricanes. (●) Tide gauge records; (○) field observations. (From Redfield and Miller, 1957.)

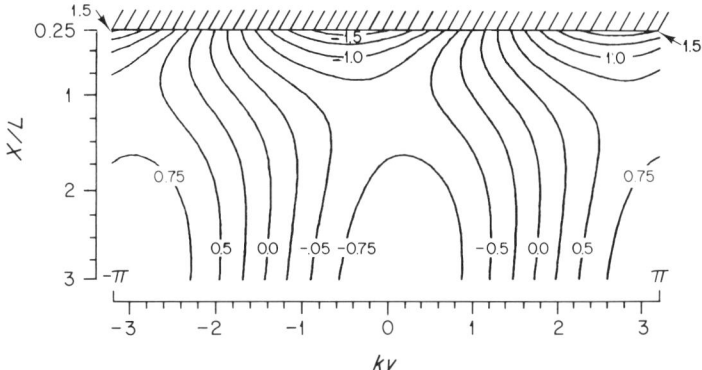

FIG. 31. Pressure field corresponding to pattern of transport streamlines in Fig. 24, generated by sinusoidally varying cross-shore wind. At $ky = 0$ wind is maximum offshore; at $ky = \pm\pi$ it is maximum onshore. Offshore (onshore) wind is seen to cause a set down (setup) over a coastal band of order L, while far from shore the surface stress is balanced by Ekman drift to the right. Cyclonic curl (which peaks at $ky = \pi/2$) is seen to induce strong onshore flow, geostrophic at large x/L (see also Fig. 24).

direction and causes strong longshore currents without any longshore wind. In an actual storm, of course, the important effects of longshore winds are superimposed on this pattern.

The longshore component of the wind in a storm generates a pressure field that generally opposes the wind stress and hence the longshore current close to the coast (cf. Fig. 22). Summing the effects of cross-shore and longshore winds causes partial cancellation of the longshore gradient in the case when the storm center is located offshore (Fig. 32). With the storm center over land, the pressure gradients add up, both opposing the direct driving force of the wind. The relative strength of the longshore gradients produced depends on the scale ratio kL. For a storm of hurricane size (large k), the cross-shore wind effect appears to dominate, as suggested by empirical data shown in Fig. 33. The longshore elevation gradient due to longshore wind is proportional to $(kL)^{-1}$ and should be relatively more important for an extratropical storm of large diameter.

The effect of cross-shore winds alone in generating longshore pressure gradients could recently be examined in isolation in some fortuitous observations south of Long Island (Csanady, 1980). A sharp impulse of offshore wind stress was found to generate a longshore sea level gradient of order 10^{-6}, superimposed on the tides, which drove a mean longshore current with a near-bottom velocity of order 0.1 m sec^{-1} in water 30 m deep. The effect could be related quantitatively to the simple model of Figs. 24 and 31.

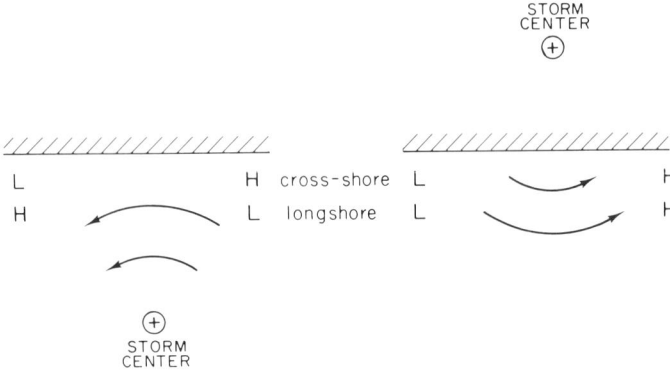

FIG. 32. Combination of cross-shore and longshore wind-stress effects on coastal elevation, for a hurricane or extratropical storm before and after landfall. Cross-shore winds produce the same effect: a high (H) to the right of the storm center, a low (L) to the left. Longshore winds, however, reverse on landfall, and with them the induced pressure gradient reverses.

Concerning the effects of larger scale extratropical storms on the Mid-Atlantic Bight, Beardsley and Butman (1974) have made some very illuminating comments. As did Boicourt and Hacker (1976), Beardsley and Butman point out that strong winter storms dominate the circulation and account for most of the observed net displacement. In addition, Beardsley and Butman report that storms with their center well to the south of the Bight produce vigorous southwestward flow, while storms with the center well to the north, although applying intense local wind stress directed to the northeast, produce little net flow because their effect is balanced mainly by an opposing sea level gradient. Scott and Csanady (1976) also comment on this phenomenon, as do Bennett and Magnell (1979), who estimate the transient longshore sea level gradients accompanying extratropical storms to be of order 10^{-7}. The effect may be explained by reference to Fig. 32 if one supposes that in such larger storms the cross-shore and the longshore wind effects are roughly equal as far as longshore pressure gradient generation is concerned. With the storm center offshore, nearly complete cancellation may then be expected, while a particularly strong gradient should oppose the longshore wind stress when the storm center is well to the north.

5.9. *Mean Circulation in Lake Ontario*

In enclosed basins, typical observationally determined mean velocities are about an order of magnitude weaker than wind-driven transient flow.

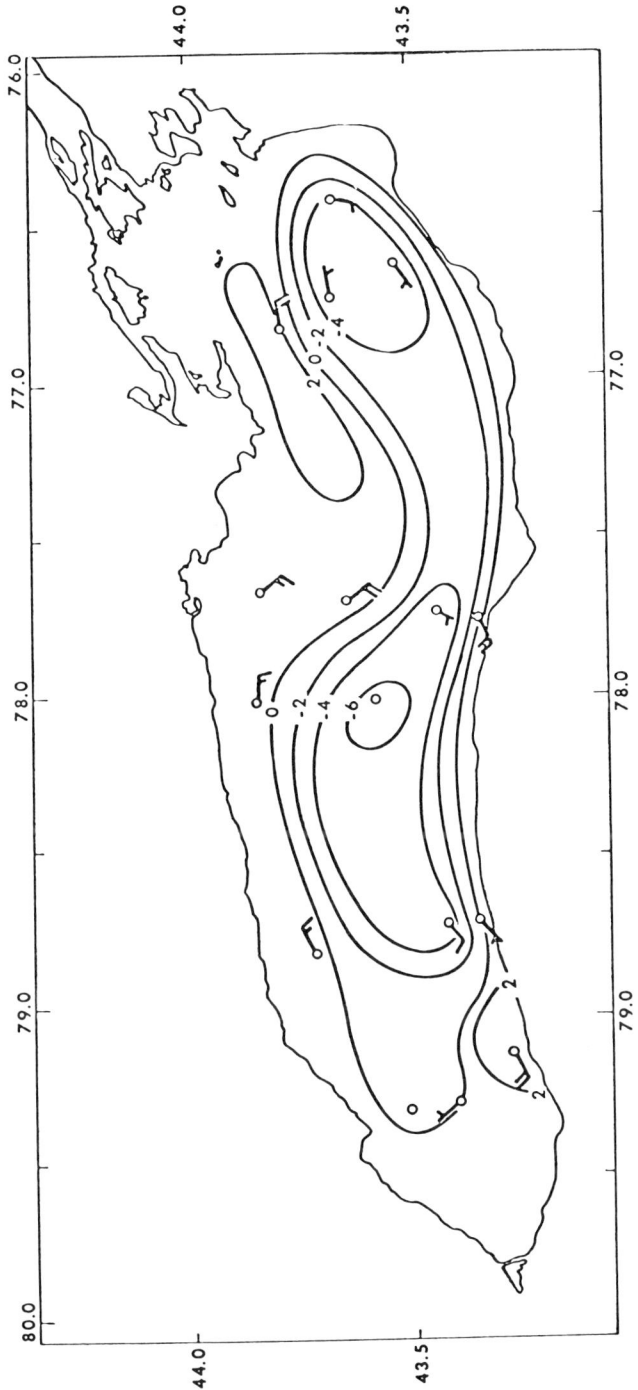

FIG. 33. Lake Ontario mean currents (barb 1 cm/sec) at −15 m in July 1972 and surface dynamic heights (mm). (From Pickett and Richards, 1975.)

One complication this gives rise to is a sampling problem: if a monthly average flow pattern is determined in a month that happens to contain one or two episodes of unusually strong wind in the same direction, the "mean" pattern will in fact be dominated by the response of the basin to those storms. Furthermore, the response of closed basins includes coastally trapped waves—in Lake Ontario of a period of 12–16 days (see earlier discussion). A monthly mean circulation pattern may therefore be expected to vary, depending on what precise phase of these waves is included in a month-long averaging period. Under summer conditions, storms bring about significant internal redistribution of mass in upwelling–downwelling episodes and subsequent wave or front propagation, the changes taking place again on a time scale much too close to a month not to distort monthly averages. When the location of a current meter mooring is alternately occupied by warm and cold water during the month (for periods of the order of a week at a time), the Eulerian average current is a composite of cold and warm water average velocities. Since temperature can be regarded as a particle tracer in a first approximation, this at once shows that Eulerian and Lagrangian averages are likely to be quite different.

Particle average (Lagrangian) velocities in enclosed basins generally tend to show a cyclonic circulation pattern (Emery and Csanady, 1973). For Lake Ontario, this was found to be the case by Harrington (1895) and was confirmed by recent evidence on mirex deposits (Pickett and Dossett, 1979). Eulerian observations, however, do not always show the same pattern. Pickett and Richards (1975) have shown a weak cyclonic mean flow pattern for July 1972 (Fig. 33); Pickett (1977) has shown a much more pronounced pattern of the same kind for November 1972. However, Pickett (1977) also illustrates the January 1973 mean flow pattern (Fig. 34), which is quite different. Sloss and Saylor (1976) and Saylor and Miller (1979) show cyclonic mean flow patterns in Lake Huron, both for winter and summer, based on current meter studies.

In our article on cyclonic mean flow patterns in enclosed shallow seas (Emery and Csanady, 1973), we proposed to explain cyclonic mean flow by an air–sea interaction mechanism, depending on the fact that the drag of wind is greater over warm water than over cold. Ekman drift at the surface coupled with upwelling and/or surface heating could give rise to significant temperature differences and increased drag on the right-hand side of a basin, looking downwind. The resulting cylonic curl of the wind stress would generate a cyclonic circulation component, which would add up from one wind-stress episode to the next, regardless of the wind direction. Observational data gained on Lake Ontario during IFYGL failed to provide specific support for this mechanism: the wind-stress curl was not found to be large enough. Alternative mechanisms that have been pro-

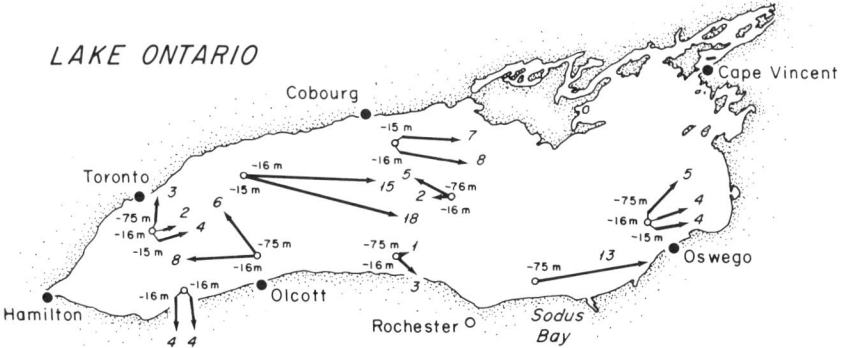

FIG. 34. Mean January current in Lake Ontario. (From Pickett, 1977.)

posed are (1) that the bottom-stress curl is cyclonic, due to the internal redistribution of mass under wind stress (Bennett, 1975) or (2) that the momentum input of the wind is redistributed in the water by advective motions, either in internal Kelvin waves (Wunsch, 1973), or (3) in frictional ageostrophic motions (Csanady, 1975), or finally (4) that the warm waters are internally redistributed to occupy preferentially the coastal zone and generate a cyclonic current by a thermohaline mechanism (Csanady, 1977c). Most of the proposed mechanisms fail to explain why a cyclonic pattern should arise in the Great Lakes in winter. As Saylor and Miller (1979) clearly point out, the cyclonic flow pattern seems to be a primary effect of the wind, and any inferred adjustment of density surfaces to geostrophic equilibrium must be regarded as secondary. Of the mechanisms proposed so far, only momentum advection in a homogeneous fluid (alternative 4 above) can account for winter cyclonic mean flow, but there is no specific evidence to support this mechanism and it must be regarded at this point as a hypothesis.

The Lake Ontario mean circulation pattern of January 1973 (Fig. 34) may, however, be satisfactorily related to the boundary layer model of coastal circulation, driven by eastward wind stress. During this month, alone of the winter months studied during IFYGL, the average wind was strong and westerly. Along both north and south shores this wind apparently generated an eastward longshore current. The current meter data are consistent with the interpretation that the longshore currents increased in width in the cyclonic direction and flowed around the ends of the basin. Specifically, the westward flow off Olcott may be interpreted as an extension of the north shore anticyclonic current around the western end of the lake, somewhat as a wrapped-around version of the theoretical pattern shown in Fig. 23.

Under summer stratified conditions the mean flow pattern is more complex and cannot be satisfactorily described without a detailed exploration of the coastal boundary layer (Csanady and Scott, 1980). The monthly average flow for July 15–Aug. 15, 1972, exhibited a coastal jet pattern similar to that suggested by the theory (Fig. 28), along the south shore of the lake only. Eastward and westward jets in warm water were associated with a thermocline tilt both cross-shore, for geostrophic equilibrium, and longshore. The longshore thermocline tilt corresponded to a pressure gradient not quite sufficient to balance the wind stress. Some momentum was apparently transferred downward by interface friction, which generated a cold water current especially around the eastern end of the lake, again somewhat as suggested by the boundary layer model (Fig. 23). The flow pattern along the north shore of the lake was quite different from the linear model of Fig. 28, however, because the return flow of warm water apparently took place some distance offshore. This is also suggested by the July 1972 circulation pattern reported by Pickett and Richards (Fig. 33), although that pattern applies to a period only partially overlapping. The July 15–Aug. 15 mean flow was not cyclonic, containing a clear anticyclonic loop in the western basin of the lake, although this was rather smaller than the cyclonic loop occupying the eastern $\frac{2}{3}$ of the basin.

The present status of understanding of the mean circulation in enclosed basins may perhaps be summed up as follows. There is a general tendency to cyclonic circulation, which shows up most clearly when the mean wind stress is weak. With a strong mean wind stress, a wind-driven two-gyre component overwhelms the cyclonic pattern, but the cyclonic half of the gyre remains stronger (Bennett, 1977). The boundary layer model and the stratified model of Fig. 28 partially account for the observed characteristics of the wind-driven two-gyre flow. A conspicuous departure from theory is, however, that the coastal jet sometimes separates from the coast. The tendency to a cyclonic circulation in the absence of wind, and the strengthening of the cyclonic cell with stronger wind stress acting, may tentatively be ascribed to momentum advection, although the theoretical basis of this is at present nonquantitative and unconfirmed. Attention should also be drawn to the fact that Eulerian mean currents may be physically meaningless in certain locations.

5.10. *Mean Summer Circulation over the Oregon Shelf*

Many observations of currents, densities, etc. are available during the summer (upwelling) season off the Oregon shelf and from these a seasonal mean circulation pattern may be pieced together. Because of the relatively large cross-shore excursions of the density front, which generally intersects the surface but sinks sometimes below the surface during this season, there are difficulties in the determination of mean particle velocities from fixed-point records. At the same time, surface heating and near-surface and nearshore mixing play an important role in determining the temperature and salinity structure of the coastal water mass, so that temperature and salinity anomalies can only be used with considerable discretion as short-term Lagrangian particle tracers. The main features of the mean summer circulation are nevertheless clear, and have been discussed by Smith et al. (1971), Mooers et al. (1976a), Smith (1974), Huyer et al. (1975), Bryden (1978), Halpern et al. (1978), and others.

The longshore components of mean velocities are southward at the surface, strong above the inclined density front, and northward below the surface, at depths of 100 m or more. Mooers et al. (1976a) give a schematic illustration of longshore velocity distribution in terms of a coastal jet above the upwelled pycnocline and a poleward undercurrent. The mean velocity of the coastal jet is given by Mooers et al. (1976a) as about 20 cm sec^{-1}, but this is probably an underestimate if jet velocity is defined as the average (Lagrangian) velocity of the warm layer above the (usually) upwelled density front. The northward flow below is referred to as the poleward undercurrent and this appears to be trapped over the upper slope, perhaps at depths less than 500 m. Mooers et al. also illustrate the typical appearance of the mean isopycnals, although these move about considerably, as already mentioned, and are not easily described in terms of a "mean" field.

Cross-shore velocities present a more complex picture and have been the subject of considerable controversy. Some of this was no doubt caused by a confusion of Eulerian and Lagrangian means, an acute problem in an upwelling zone where some fixed-point current meters sample widely different water masses in the course of upwelling–downwelling events. What is not in doubt is that over most of the water column the cross-shore velocity is directed shoreward most of the time and has an amplitude of about 2 cm sec^{-1}. Across the 100 m isobath this implies onshore transport of about 2 m^2 sec^{-1}, or about three times more than the offshore Ekman transport at the surface associated with the mean wind stress (Bryden, 1978). There is also some offshore Ekman transport in the bottom boundary layer associated with the poleward undercurrent (Kundu, 1977), but it is very unlikely that this is sufficient to maintain two-dimensional mass balance by transporting away most of the onshore flow arriving throughout the water column. Smith et al. (1971), Stevenson et al. (1974), and Mooers et al. (1976a) also convincingly demonstrate that some of the water drawn from deeper levels is heated at the surface and sinks along isopycnals of the pycnocline when the latter intersects the free surface. This implies offshore motion along some isopycnals, but it is not clear whether the quantity is sufficient to affect the cross-shore mass balance significantly.

A long-term mean onshore velocity of about 2 cm sec^{-1}, constant with depth, implies a longshore sea level gradient of 2×10^{-7}, driving northward. Reid and Mantyla (1976) have demonstrated that a longshore gradient of this sign and magnitude may also be inferred from the density field of the North Pacific. In the yearly average this gradient is confined to latitudes south of 38°N, but in the summer it extends to 44°N and thus encompasses the Oregon shelf. The cross-shore sea surface slope associated with the density field extends to about 100 km from the shore, i.e., it coincides with the poleward undercurrent. The longshore sea level slope is associated with a corresponding pycnocline slope. The longshore momentum balance of the poleward undercurrent is thus dominated by a northward-driving pressure gradient, balanced in the frictionless interior by onshore flow. Where the onshore flow runs into the continental slope, a northward current develops, with associated bottom friction. The vorticity-tendency balance of this current is as discussed in connection with the boundary layer model of shelf circulation, first pointed out by Pedlosky (1974b). The discussion of Reid and Mantyla (1976) also suggests that the longshore isopycnal slope along the west coast of North America is part of a larger-scale response of the North Pacific to wind stress. Halpern *et al.* (1978) confirm this by showing that the undercurrent transports relatively warm and saline water northwards over a considerable range of latitude.

In the surface layers, offshore wind-driven Ekman drift is compensated by onshore flow in a layer of about 30 m depth, i.e., essentially above the main pycnocline. Onshore flow in the deeper layers turns seaward partly in the bottom boundary layer below the undercurrent, but this circulation is confined only to the trapping width of the boundary current (~ 50 km) and is in any case insufficient for two-dimensional mass balance. The rest of the inflow is presumably accommodated in a broadening of the boundary current. The need to view the Oregon shelf circulation problem in three dimensions has been pointed out already by O'Brien and his collaborators (O'Brien and Hurlburt, 1972; Thompson and O'Brien, 1973).

In contrast with the case of an enclosed basin, or an Atlantic type shelf discussed above, the mean circulation of the Pacific type shelf is seen to be dominated by deep ocean influences.

6. Conclusion

The later sections of this article have increasingly focused on long-term, larger scale flow phenomena that are at best partially understood.

The mean circulation problem clearly requires further study in all three environments discussed here, as well as in even more complex cases, such as semienclosed basins. To quote another conspicuous gap in knowledge, the summer circulation over the east coast continental shelf shows complexities that at present appear completely puzzling, including a bottom-trapped cold band of water flowing along the outer shelf, the water supply for which seems to originate in the Gulf of Maine (Hopkins and Garfield, 1979). On the other hand, we seem to possess a fair understanding of short-term transient flow events in the coastal zone, even if the influence of dissipative processes is rather obscure, especially in the presence of stratification. Hopefully, the synthesis of knowledge attempted above will prove to be a spur to further progress, recognizing that parts of the conceptual framework erected here are likely to require major additions and significant modification.

Acknowledgments

This work has been supported by the Department of Energy under a contract entitled Coastal-Shelf Transport and Diffusion.

References

Allen, J. S. (1973). Upwelling and coastal jets in a continuously stratified ocean. *J. Phys. Oceanogr.* **3**(3), 245–257.

Allen, J. S. (1980). Models of wind-driven currents on the continental shelf. *Annu. Rev. Fluid Mech.* **12**, 389–433.

Allen, J. S., and Kundu, P. K. (1978). On the momentum, vorticity and mass balance of the Oregon shelf. *J. Phys. Oceanogr.* **8**, 13–27.

Ayers, J. C., Chandler, D. C., Lauff, G. H., Power, C. F., and Henson, E. B. (1958). Currents and water masses of Lake Michigan. *Publ.—Great Lakes Res. Div., Univ. Mich.* **3**, 1–169.

Beardsley, R. C., and Boicourt, W. C. (1980). On estuarine and continental shelf circulation in the Middle Atlantic Bight. *In* "Evolution of Physical Oceanography, Scientific Surveys in Honor of Henry Stommel" (B. A. Warren and C. Wunsch, eds.), pp. 198–233. MIT Press, Cambridge, Massachusetts.

Beardsley, R. C., and Butman, B. (1974). Circulation on the New England Continental Shelf: Response to strong winter storms. *Geophys. Res. Lett.* **1**, 181–184.

Beardsley, R. C., and Winant, C. D. (1979). On the mean circulation in the Mid-Atlantic Bight. *J. Phys. Oceanogr.* **9**, 612–619.

Beardsley, R. C., Boicourt, W. C., and Hansen, D. V. (1976). Physical oceanography of the Middle Atlantic Bight. *Spec. Symp.—Am. Soc. Limnol. Oceanogr.* **2**, 20–34.

Bennett, E. B., and Saylor, J. H. (1975). IFYGL water movement program: A post field work review. *Proc. IFYGL Symp., 55th Annu. Meet., Am. Geophys. Union* pp. 102–127.

Bennett, J. R. (1971). Thermally driven lake currents during the spring and fall transition periods. *Proc.—Conf. Great Lakes Res.* **14**, 535–544.

Bennett, J. R. (1973). A theory of large-amplitude Kelvin waves. *J. Phys. Oceanogr.* **3**, 57–60.
Bennett, J. R. (1974). On the dynamics of wind-driven lake currents. *J. Phys. Oceanogr.* **4**, 400–414.
Bennett, J. R. (1975). Nonlinearity of wind-driven currents. *Rep. Ser.—Inland Waters Dir. (Can.)* **43**.
Bennett, J. R. (1977). A three-dimensional model of Lake Ontario's summer circulation. *J. Phys. Oceanogr.* **7**, 591–601.
Bennett, J. R. (1978). A three-dimensional model of Lake Ontario's summer circulation. II. A diagnostic study. *J. Phys. Oceanogr.* **8**, 1095–1103.
Bennett, J. R., and Magnell, B. A. (1979). A dynamical analysis of currents near the New Jersey Coast. *J. Geophys. Res.* **84**, 1165–1175.
Bigelow, H. B., and Sears, M. (1935). Studies of the waters on the continental shelf, Cape Cod to Chesapeake Bay. II. Salinity. *Pap. Phys. Oceanogr. Meteorol.* **4**(1), 1–94.
Birchfield, G. E. (1969). The response of a circular model Great Lake to a suddenly imposed wind stress. *J. Geophys. Res.* **74**, 5547–5554.
Birchfield, G. E. (1972). Theoretical aspects of wind-driven currents in a sea or lake of variable depth with no horizontal mixing. *J. Phys. Oceanogr.* **2**, 355–362.
Birchfield, G. E., and Davidson, D. R. (1967). A case study of coastal currents in Lake Michigan. *Proc. Conf.—Great Lakes Res.* **10**, 264–273.
Birchfield, G. E., and Hickie, B. P. (1977). The time-dependent response of a circular basin of variable depth to a wind stress. *J. Phys. Oceanogr.* **7**, 691–701.
Blanton, J. O. (1974). Some characteristics of nearshore currents along the north shore of Lake Ontario. *J. Phys. Oceanogr.* **4**, 415–424.
Blanton, J. O. (1975). Nearshore lake currents measured during upwelling and downwelling of the thermocline in Lake Ontario. *J. Phys. Oceanogr.* **5**, 111–124.
Boicourt, W. C., and Hacker, P. W. (1976). Circulation on the Atlantic continental shelf of the United States, Cape May to Cape Hatteras. *Mem. Soc. R. Sci. Liege* **5**(X), 187–200.
Bowden, K. F. (1970). Turbulence. II. *Oceanogr. Mar. Biol.* **8**, 11–32.
Boyce, F. M. (1974). Some aspects of Great Lakes physics of importance to biological and chemical processes. *J. Fish. Res. Board Can.* **31**, 689–730.
Boyce, F. M. (1977). Response of the coastal boundary layer on the north shore of Lake Ontario to a fall storm. *J. Phys. Oceanogr.* **7**, 719–732.
Bretschneider, C. L. (1966). Engineering aspects of the hurricane surge. In "Estuarine and Coastline Hydrodynamics" (■■ Ippen, ed.), pp. 231–256. McGraw-Hill, New York.
Brooks, D. A. (1979). Coupling of the Middle and South Atlantic Bights by forced sea level oscillations. *J. Phys. Oceanogr.* **9**, 1304–1311.
Bryden, H. L. (1978). Mean upwelling velocities on the Oregon continental shelf during summer 1973. *Estuarine Coastal Mar. Sci.* **7**, 311–327.
Bumpus, D. F. (1973). A description of the circulation on the continental shelf of the east coast of the United States. *Prog. Oceanogr.* **6**, 111–159.
Cardone, V. J., Pierson, W. J., and Ward, E. G. (1976). Hindcasting the directional spectra of hurricane-generated waves. *J. Pet. Technol.* **28**, 385–396.
Charney, J. G. (1955). Generation of oceanic currents by wind. *J. Mar. Res.* **14**, 477–498.
Chase, R. R. P. (1979). The coastal longshore pressure gradient: Temporal variations and driving mechanisms. *J. Geophys. Res.* **84**, 4898–4904.
Church, P. E. (1945). The annual temperature cycle of Lake Michigan. II. Spring warming and summer stationary periods, 1942. *Misc. Rep.—Univ. Chicago, Inst. Meteorol.* **18**, 1–100.
Clarke, A. J. (1977). Observational and numerical evidence for wind-forced coastal trapped long waves. *J. Phys. Oceanogr.* **7**, 231–247.

Collins, C. A., and Patullo, J. G. (1970). Ocean currents above the continental shelf off Oregon as measured with a single array of current meters. *J. Mar. Res.* **28,** 51–68.

Collins, C. A., Mooers, C. N. K., Stevenson, M. R., Smith, R. L., and Patullo, J. G. (1968). Direct current measurements in the frontal zone of a coastal upwelling region. *J. Oceanogr. Soc. Jpn.* **24,** 295–306.

Crépon, M. (1967). Hydrodynamique marine en regime impulsionnel. *Cah. Oceanogr.* **19,** 847–880.

Crépon, M. (1969). Hydrodynamique marine en régime impulsionnel. *Cah. Oceanogr.* **21,** 333–353, 863–877.

Csanady, G. T. (1968a). Wind-driven summer circulation in the Great Lakes. *J. Geophys. Res.* **73,** 2579–2589.

Csanady, G. T. (1968b). Motions in a model Great Lake due to a suddenly imposed wind. *J. Geophys. Res.* **73,** 6435–6447.

Csanady, G. T. (1970). Dispersal of effluents in the Great Lakes. *Water Res.* **4,** 79–114.

Csanady, G. T. (1971). On the equilibrium shape of the thermocline in a shore zone. *J. Phys. Oceanogr.* **1,** 263–270.

Csanady, G. T. (1972). Response of large stratified lakes to wind. *J. Phys. Oceanogr.* **2,** 3–13.

Csanady, G. T. (1973). Wind-induced barotropic motions in long lakes. *J. Phys. Oceanogr.* **3,** 429–438.

Csanady, G. T. (1974a). Barotropic currents over the continental shelf. *J. Geophys. Res.* **4,** 357–371.

Csanady, G. T. (1974b). Spring thermocline behavior in Lake Ontario during IFYGL. *J. Phys. Oceanogr.* **4,** 425–445.

Csanady, G. T. (1974c). Mass exchange episodes in the coastal boundary layer, associated with current reversals. *Rapp. P.-V. Reun., Cons. Int. Explor. Mer.* **167,** 41–45.

Csanady, G. T. (1975). Lateral momentum flux in boundary currents. *J. Phys. Oceanogr.* **5,** 705–717.

Csanady, G. T. (1976a). Topographic waves in Lake Ontario. *J. Phys. Oceanogr.* **6,** 93–103.

Csanady, G. T. (1976b). Mean circulation in shallow seas. *J. Geophys. Res.* **81,** 5389–5399.

Csanady, G. T. (1977a). The coastal jet conceptual model in the dynamics of shallow seas. *In* "The Sea" (E. D. Goldberg, I. N. McCave, J. J. O'Brien, and J. H. Steele, eds.), Vol. 6, pp. 117–144. Wiley (Interscience), New York.

Csanady, G. T. (1977b). Intermittent 'full' upwelling in Lake Ontario. *J. Geophys. Res.* **82,** 397–419.

Csanady, G. T. (1977c). On the cyclonic mean circulation of large lakes. *Proc. Natl. Acad. Sci. U.S.A.* **74,** 2204–2208.

Csanady, G. T. (1978a). The arrested topographic wave. *J. Phys. Oceanogr.* **8,** 47–62.

Csanady, G. T. (1978b). Wind effects on surface to bottom fronts. *J. Geophys. Res.* **83,** 4633–4640.

Csanady, G. T. (1978c). Water circulation and dispersal mechanisms. *In* "Lakes, Their Physics and Chemistry" (A. Lerman, ed.), pp. 21–64. Springer-Verlag, Berlin and New York.

Csanady, G. T. (1979). The pressure field along the western margin of the North Atlantic. *J. Geophys. Res.* **84,** 4905–4914.

Csanady, G. T. (1980). Longshore pressure gradients caused by offshore wind. *J. Geophys. Res.* **85,** 1076–1084.

Csanady, G. T., and Scott, J. T. (1974). Baroclinic coastal jets in Lake Ontario during IFYGL. *J. Phys. Oceanogr.* **4,** 524–541.

Csanady, G. T., and Scott, J. T. (1980). Mean summer circulation in Lake Ontario within the coastal zone. *J. Geophys. Res.* **85,** 2797–2812.

Cutchin, D. L., and Smith, R. L. (1973). Continental shelf waves: Low frequency variations in sea level and currents over the Oregon continental shelf. *J. Phys. Oceanogr.* **3,** 73–82.

Defant, A. (1961). "Physical Oceanography," Vol. 2. Pergamon, Oxford.

Ekman, V. W. (1905). On the influence of the earth's rotation on ocean currents. *Ark. Mat, Astron., Fys.* **2**(11), 1–52.

Emery, K. O., and Csanady, G. T. (1973). Surface circulation of lakes and nearly landlocked seas. *Proc. Natl. Acad. Sci. U.S.A.* **70,** 93–97.

Fischer, H. B. (1980). Mixing processes on the Atlantic continental shelf, Cape Cod to Cape Hatteras. *Limnol. Oceanogr.* **25,** 114–125.

Flagg, C. N. (1977). The kinematics and dynamics of the New England continental shelf and shelf/slope. Ph.D. Thesis, MIT-WHOI Joint Program in Oceanography, Woods Hole, Massachusetts.

Forristal, G. Z., Hamilton, R. C., and Cardone, V. J. (1977). Continental shelf currents in tropical storm Delia: Observations and theory. *J. Phys. Oceanogr.* **7,** 532–546.

Freeman, J. C., Baer, L., and Jung, G. H. (1957). The bathystrophic storm tide. *J. Mar. Res.* **16,** 12–22.

Gill, A. E., and Clarke, A. J. (1974). Wind-induced upwelling, coastal currents and sea-level changes. *Deep-Sea Res.* **21,** 325–345.

Gill, A. E., and Schumann, E. H. (1974). The generation of long shelf waves by the wind. *J. Phys. Oceanogr.* **4,** 83–90.

Grant, W. D., and Madsen, O. S. (1979). Continued wave and current interaction with a rough bottom. *J. Geophys. Res.* **84,** 1797–1808.

Greenspan, H. P. (1968). "The Theory of Rotating Fluids." Cambridge Univ. Press, London and New York.

Grose, P. L., and Mattson, J. S., eds. (1977). "The Argo Merchant Oil Spill," a preliminary scientific report. National Oceanic and Atmospheric Administration, Department of Commerce, Rockville, Maryland.

Halpern, D. (1974). Variations in the density field during coastal upwelling. *Tethys* **6,** 363–374.

Halpern, D. (1976). Structure of a coastal upwelling event observed off Oregon during 1973. *Deep-Sea Res.* **23,** 495–508.

Halpern, D., Smith, R. L., and Reed, R. K. (1978). On the California undercurrent over the continental slope off Oregon. *J. Geophys. Res.* **83,** 1366–1372.

Hamon, B. V. (1962). The spectrums of mean sea level at Sydney, Coff's Harbour and Lord Howe Island. *J. Geophys. Res.* **67,** 5147–5155.

Hansen, D. V. (1977). Circulation. *MESA N.Y. Bight Atlas Monogr.* **3,** 1–23.

Harrington, M. W. (1895). "Surface Currents of the Great Lakes, as Deduced from the Movements of Bottle Papers During the Season of 1892, 1893 and 1894," Bull. B (rev.). *U.S. Weather Bur.,* Washington, D.C.

Heaps, N. S. (1969). A two-dimensional numerical sea model. *Philos. Trans. R. Soc. London* **265,** 93–137.

Hendershott, M., and Rizzoli, P. (1976). The winter circulation of the Adriatic Sea. *Deep-Sea Res.* **23,** 353–370.

Hickey, B. M., and Hamilton, P. (1980). A spin-up model as a diagnostic tool for interpretation of current and density measurements on the continental shelf of the Pacific Northwest. *J. Phys. Oceanogr.* **10,** 12–24.

Hopkins, T. S., and Garfield, N. (1979). Gulf of Maine Intermediate Water. *J. Mar. Res.* **37,** 103–139.

Hsueh, Y., and Peng, C. Y. (1978). A diagnostic model of continental shelf circulation. *J. Geophys. Res.* **83,** 3033–3041.

Hsueh, Y., Peng, C. Y., and Blumsack, S. L. (1976). A geostrophic computation of currents over a continental shelf. *Mem. Soc. R. Sci. Liege* **10**, 315–330.

Huang, J. C. K. (1971). The thermal current in Lake Michigan. *J. Phys. Oceanogr.* **1**, 105–122.

Huthnance, J. M. (1978). On coastal trapped waves: Analysis and numerical calculation by inverse iteration. *J. Phys. Oceanogr.* **8**, 74–92.

Huyer, A., and Pattullo, J. G. (1972). A comparison between wind and current observations over the continental shelf off Oregon, summer 1969. *J. Geophys. Res.* **77**, 3215–3220.

Huyer, A., and Smith, R. L. (1974). A subsurface ribbon of cool water over the continental shelf off Oregon. *J. Phys. Oceanogr.* **4**, 381–391.

Huyer, A., Smith, R. L., and Pillsbury, R. D. (1974). Observations in a coastal upwelling region during a period of variable winds (Oregon coast, July 1972). *Tethys* **6**, 391–404.

Huyer, A., Hickey, B. M., Smith, J. D., Smith, R. L., and Pillsbury, R. D. (1975). Alongshore coherence at low frequencies in currents observed over the continental shelf off Oregon and Washington. *J. Geophys. Res.* **80**, 3495–3505.

Huyer, A., Smith, R. L., and Sobey, E. J. C. (1978). Seasonal differences in low-frequency current fluctuations over the Oregon continental shelf. *J. Geophys. Res.* **83**, 5077–5089.

Huyer, A., Sobey, E. J., and Smith, R. L. (1979). The spring transition in currents over the Oregon continental shelf. *J. Geophys. Res.* **84**, 6995–7011.

Irbe, J. G., and Mills, R. J. (1976). Aerial surveys of Lake Ontario water temperature and description of regional weather conditions during IFYGL—January, 1972 to March, 1973. CLI 1-76, Atmos. Env. Service, Canada, 151 pp.

Jeffreys, H. (1923). The effect of a steady wind on the sea level near a straight shore. *Philos. Mag.* [6] **45**, 114–125.

Jelesnianski, C. P. (1965). Numerical computations of storm surges without bottom stress. *Mon. Weather Rev.* **93**, 343–358.

Jelesnianski, C. P. (1966). Numerical computations of storm surges with bottom stress. *Mon. Weather Rev.* **94**, 379–394.

Kundu, P. K. (1977). On the importance of friction in two typical continental waters: Off Oregon and Spanish Sahara. *In* "Bottom Turbulence" (J. C. G. Nihoul, ed.), pp. 187–208. Elsevier, Amsterdam.

Kundu, P. K., and Allen, J. S. (1976). Some three-dimensional characteristics of low-frequency current fluctuations near the Oregon coast. *J. Phys. Oceanogr.* **6**, 181–199.

Kundu, P. K., Allen, J. S., and Smith, R. L. (1975). Modal decomposition of the velocity field near the Oregon coast. *J. Phys. Oceanogr.* **5**, 683–704.

Leblond, P. H., and Mysak, L. A. (1977). Trapped coastal waves and their role in shelf dynamics. *In* "The Sea" (E. D. Goldberg, I. N. McCave, J. J. O'Brien, and J. H. Steele, eds.), Vol. 6, pp. 459–495. Wiley (Interscience), New York.

Longuet-Higgins, M. S. (1965). Some dynamical aspects of ocean currents. *Q. J. Roy. Meteorol. Soc.* **91**, 425–451.

Malone, F. D. (1968). An analysis of current measurements in Lake Michigan. *J. Geophys. Res.* **73**, 7065–7081.

Marmorino, G. O. (1978). Inertial currents in Lake Ontario, Winter 1972–73 (IFYGL). *J. Phys. Oceanogr.* **8**, 1104–1120.

Marmorino, G. O. (1979). Low frequency current fluctuations in Lake Ontario, Winter 1972–73 (IFYGL). *J. Geophys. Res.* **84**, 1206–1214.

Mayer, D. A., Hansen, D. V., and Ortman, D. A. (1979). Long term current and temperature observations on the Middle Atlantic Shelf. *J. Geophys. Res.* **84**, 1776–1792.

Monin, A. S., and Yaglom, A. M. (1971). "Statistical Fluid Mechanics." MIT Press, Cambridge, Massachusetts.

Mooers, C. N. K., and Smith, R. L. (1968). Continental shelf waves off Oregon. *J. Geophys. Res.* **73,** 549–557.

Mooers, C. N. K., Collins, C. A., and Smith, R. L. (1976a). The dynamic structure of the frontal zone in the coastal upwelling region off Oregon. *J. Phys. Oceanogr.* **6,** 3–21.

Mooers, C. N. K., Fernandez-Partages, J., and Price, J. F. (1976b). "Meteorological Forcing Fields of the New York Bight," Tech. Rep. 76-8. Rosenstiel School of Marine and Atmospheric Science, University of Miami, Coral Gables, Florida.

Mortimer, C. H. (1963). Frontiers in physical limnology with particular reference to long waves in rotating basins. *Publ.—Great Lakes Res. Div., Univ. Mich.* **10,** 9–42.

Murthy, C. R. (1970). An experimental study of horizontal diffusion in Lake Ontario. *Proc.—Conf. Great Lakes Res.* **13,** 477–489.

Mysak, L. A. (1980). Topographically trapped waves. *Annu. Rev. Fluid Mech.* **12,** 45–76.

O'Brien, J. J., and Hurlburt, H. E. (1972). A numerical model of coastal upwelling. *J. Phys. Oceanogr.* **2,** 14–26.

Pedlosky, J. (1974a). On coastal jets and upwelling in bounded basins. *J. Phys. Oceanogr.* **4,** 3–18.

Pedlosky, J. (1974b). Longshore currents, upwelling, and bottom topography. *J. Phys. Oceanogr.* **4,** 214–226.

Pedlosky, J. (1974c). Longshore currents and the onset of upwelling over bottom slope. *J. Phys. Oceanogr.* **4,** 310–320.

Pedlosky, J., and Greenspan, H. P. (1976). A simple laboratory model for oceanic circulation. *J. Fluid Mech.* **27,** 291–304.

Pickett, R. L. (1977). The observed winter circulation of Lake Ontario. *J. Phys. Oceanogr.* **7,** 152–156.

Pickett, R. L., and Dossett, D. A. (1979). Mirex and the circulation of Lake Ontario. *J. Phys. Oceanogr.* **9,** 441–445.

Pickett, R. L., and Richards, F. P. (1975). Lake Ontario mean temperatures and mean currents in July 1972. *J. Phys. Oceanogr.* **5,** 775–781.

Platzman, G. W. (1963). The dynamic prediction of wind tides on Lake Erie. *Meteorol. Monogr.* **4,** No. 26, 1–44.

Pritchard-Carpenter (1965). "Drift and Dispersion Characteristics of Lake Ontario's Nearshore Waters, Rochester, N.Y. to Sodus Bay, N.Y." Pritchard-Carpenter, 208 MacAlpine Rd., Elliott City, Maryland (unpublished report).

Proudman, J. (1953). "Dynamical Oceanography." Wiley, New York.

Redfield, A. C., and Miller, A. R. (1957). Water levels accompanying Atlantic coast hurricanes. *Meteorol. Monogr.* 2(10), 1–23.

Reid, J. L., Jr., and Mantyla, A. W. (1976). The effects of the geostrophic flow upon coastal sea elevations in the northern Pacific Ocean. *J. Geophys. Res.* **81,** 3100–3110.

Reid, R. O. (1958). Effect of Coriolis force on edge waves. (1). Investigation of the normal modes. *J. Mar. Res.* **16,** 109–144.

Robinson, A. R. (1964). Continental shelf waves and the response of sea level to weather systems. *J. Geophys. Res.* **69,** 367–368.

Rodgers, G. K. (1965). The thermal bar in the Laurentian Great Lakes. *Proc.—Conf. Great Lakes Res.* **8,** 358–363.

Rossby, C. G. (1938). On the mutual adjustment of pressure and velocity distribution in certain simple current systems. II. *J. Mar. Res.* **1,** 239–263.

Sato, G. K., and Mortimer, C. H. (1975). Lake currents and temperatures near the western shore of Lake Michigan. *Spec. Rep.—Univ. Wis.-Milwaukee, Cent. Great Lakes Stud.* **22.**

Saylor, J. H., and Miller, G. S. (1979). Lake Huron winter circulation. *J. Geophys. Res.* **84,** 3237-3252.

Scott, J. T., and Csanady, G. T. (1976). Nearshore currents off Long Island. *J. Geophys. Res.* **81,** 5401-5409.

Simons, T. J. (1973). Comparison of observed and computed currents in Lake Ontario during Hurricane Agnes, June 1972. *Proc.—Conf. Great Lakes Res.* **16,** 831-844.

Simons, T. J. (1974). Verification of numerical models of Lake Ontario. Part I. Circulation in spring and early summer. *J. Phys. Oceanogr.* **4,** 507-523.

Simons, T. J. (1975). Verification of numerical models of Lake Ontario. II. Stratified circulations and temperature changes. *J. Phys. Oceanogr.* **5,** 98-110.

Simons, T. J. (1976). Verification of numerical models of Lake Ontario. Part III. Long term heat transports. *J. Phys. Oceanogr.* **6,** 372-378.

Sloss, P. W., and Saylor, J. H. (1976). Large-scale current measurements in Lake Huron. *J. Geophys. Res.* **81,** 3069-3078.

Smith, J. D. (1977). Modelling of sediment transport on continental shelves. *In* "The Sea" Vol. 6 (E. D. Goldberg, I. N. McCave, J. J. O'Brien, and J. H. Steele, eds.), pp. 539-577. Wiley (Interscience), New York.

Smith, N. P. (1978). Longshore currents on the fringe of Hurricane Anita. *J. Geophys. Res.* **83,** 6047-6051.

Smith, R. L. (1974). A description of current, wind and sea-level variations during coastal upwelling off the Oregon coast, July-August 1972. *J. Geophys. Res.* **79,** 435-443.

Smith, R. L., Pattullo, J. G., and Lane, R. K. (1966). Investigation of the early stage of upwelling along the Oregon coast. *J. Geophys. Res.* **71,** 1135-1140.

Smith, R. L., Mooers, C. N. K., and Enfield, D. B. (1971). Mesoscale studies of the physical oceanography in two coastal upwelling regions: Oregon and Peru. *In* "Fertility of the Sea" (G. D. Costlow, ed.), Vol. 2, pp. 513-535. Gordon & Breach, New York.

Stevenson, M. R., Garvine, R. W., and Wyatt, B. (1974). Lagrangian measurements in a coastal upwelling zone off Oregon. *J. Phys. Oceanogr.* **4,** 321-336.

Stommel, H., and Leetmaa, A. (1972). Circulation on the continental shelf. *Proc. Natl. Acad. Sci. U.S.A.* **69,** 3380-3384.

Thompson, J. D., and O'Brien, J. J. (1973). Time-dependent coastal upwelling. *J. Phys. Oceanogr.* **3,** 33-46.

Verber, J. L. (1966). Inertial Currents in the Great Lakes. *Publ.—Great Lakes Res. Div., Univ. Mich.* **15,** 375-379.

Walsh, J. J., Whitledge, T. E., Barvenik, F. W., Wirick, C. D., Howe, S. O., Esaias, W. E., and Scott, J. T. (1978). Wind events and flood chain dynamics within the New York Bight. *Limnol. Oceanogr.* **23,** 659-683.

Wang, D. P. (1979). Low frequency sea level variability in the Middle Atlantic Bight. *J. Mar. Res.* **37,** 683-697.

Wang, D. P., and Mooers, C. N. (1976). Coastally-trapped waves in a continuously stratified ocean. *J. Phys. Oceanogr.* **6,** 853-863.

Weatherly, G. L., and Van Leer, J. C. (1977). On the importance of stable stratification to the structure of the bottom boundary layer on the western Florida shelf. *In* "Bottom Turbulence" (J. C. G. Nihoul, ed.), pp. 103-122. Elsevier, Amsterdam.

Welander, P. (1961). Numerical prediction of storm surges. *Adv. Geophys.* **8,** 316-379.

Wunsch, C. (1973). On the mean drift in large lakes. *Limnol. Oceanogr.* **18,** 793-795.

Yamagata, T. (1980). On cyclonic propagation of a warm front in a bay of a northern hemisphere. *Tellus* **32,** 73-76.

MESOSCALE NUMERICAL MODELING

Roger A. Pielke

Department of Environmental Sciences
University of Virginia
Charlottesville, Virginia

1.	Introduction.	186
2.	Basic Set of Equations	187
3.	Simplification of the Basic Equations.	189
	3.1 Conservation of Mass Approximation.	190
	3.2 Conservation of Heat, Water, and Other Gaseous and Aerosol Material Approximations	191
	3.3 Conservation of Motion Approximations	191
4.	Averaging the Conservation Relations	198
5.	Types of Models.	204
	5.1 Physical Models.	204
	5.2 Linear Models.	209
6.	Coordinate Representation	221
7.	Planetary Boundary-Layer Parameterization	226
	7.1 Laminar Sublayer	226
	7.2 The Surface Layer	227
	7.3 The Transition Layer	231
8.	Radiation Parametrization.	239
	8.1 Short-Wave Radiation	241
	8.2 Long-Wave Radiation.	245
9.	Moist Thermodynamics.	248
	9.1 Convectively Stable Atmosphere.	248
	9.2 Convectively Unstable Atmosphere	249
	9.3 Types of Parameterizations	250
10.	Methods of Solution.	253
	10.1 Advection Terms	255
	10.2 Diffusion Terms.	261
	10.3 Other Terms	262
	10.4 Other Calculations	264
	10.5 Nonlinear Effects	265
11.	Boundary and Initial Conditions.	269
	11.1 Grid and Domain Structure.	269
	11.2 Grid Staggering.	272
	11.3 Initialization.	272
	11.4 Initialization with Sparse Data	274
	11.5 Spatial Boundaries	276
12.	Model Evaluation.	285
	12.1 Kinetic Energy Evaluation	286
	12.2 Observational Validation.	288
	12.3 Model Intercomparisons.	289
	12.4 Model Logic	290
13.	Examples of Mesoscale Models.	290
	13.1 Terrain-Induced Mesoscale Systems.	291
	13.2 Synoptically-Induced Mesoscale Systems	313
14.	Conclusions.	317
	References.	321

1. INTRODUCTION

In order to effectively develop and utilize mesoscale dynamical simulations of the atmosphere, it is necessary to understand the basic physical and mathematical foundations of the models, and to have an appreciation of how the particular atmospheric system of interest works. This article provides such an overview of the field and should be of use to the operational forecaster as well as the researcher of mesoscale phenomena. This work also provides an update and expansion of many of the ideas discussed by Gutman (1972). Since the material starts from fundamental concepts, it should be possible to evaluate the scientific basis of any simulation model that has been or will be developed.

For the purposes of this article, mesoscale is defined as having a temporal and a horizontal spatial scale smaller than the conventional rawinsonde network, but significantly larger than individual cumulus clouds. In more quantitative terms, this implies that the horizontal scale is on the order of a few to 100 km or so, with a time scale of about 1–12 hr. The vertical scale extends from tens of meters or so to the depth of the troposphere. Clearly, this is a somewhat arbitrary limit; however, the shorter scale corresponds to atmospheric features that, for weather forecasting purposes, can only be described statistically, while the larger limit corresponds to the smallest features we can generally distinguish on a synoptic weather map. As will be shown later in the article, mesoscale can also be defined by those systems that are sufficiently large so that the hydrostatic approximation to the vertical pressure distribution is valid, yet small enough that the geostrophic and gradient winds are inappropriate approximations to the actual wind circulation above the planetary boundary layer. This scale of interest, then, along with computer resource limitations, defines the domain and grid sizes of mesoscale models.

In this article, the outline of material is as follows. In Sections 2 and 3 the fundamental conservation relations are introduced and appropriate simplifications given. In Section 4 the equations are averaged in order to conform to a mesoscale model grid mesh. In Section 5 types of models are discussed and their advantages and disadvantages for proper simulation of mesoscale phenomena are presented. The transformation of the equations to a generalized coordinate representation is given in Section 6, while the parameterizations in a mesoscale model of the planetary boundary layer, electromagnetic radiation, and moist thermodynamics are introduced in Sections 7, 8, and 9, respectively. In Section 10 methods of solution are illustrated with boundary and initial conditions, and grid structure is discussed in Section 11. Then the procedure for evaluating models is given in Section 12. Finally, examples of mesoscale simulations of particular mesoscale phenomena are provided in Section 13 and conclusions are presented in Section 14.

In this article, each section provides an overview of the subject matter. A text entitled "Mesoscale Dynamic Modelling—An Introductory Survey" is in preparation that provides more detailed information, including relevant derivations in each of the topical areas given in this article. Also, an extensive but not exhaustive reference list is given at the end of this article. The latest references available are given with the expectation that interested readers will seek additional source materials listed in those publications.

2. Basic Set of Equations

The foundation of any model based on fundamental physical principles is a set of conservation relations. For mesoscale atmospheric models these relations are the following:

- conservation of mass
- conservation of heat
- conservation of motion
- conservation of water
- conservation of other gaseous and aerosol materials.

These principles form a coupled set of relations that must be satisfied simultaneously and that include sources and sinks in the individual expressions.

In tensor notation, one form of these conservation relations can be written as

$$\frac{\partial \rho}{\partial t} = \frac{\partial}{\partial x_j} \rho u_j \tag{1}$$

$$\frac{\partial \theta}{\partial t} = -u_j \frac{\partial \theta}{\partial x_j} + S_\theta \tag{2}$$

$$\frac{\partial u_i}{\partial t} = -u_j \frac{\partial u_i}{\partial x_j} - \frac{1}{\rho} \frac{\partial p}{\partial x_i} - g\delta_{i3} - 2\epsilon_{ijk}\Omega_j u_k \tag{3}$$

$$\frac{\partial q_n}{\partial t} = -u_j \frac{\partial q_n}{\partial x_j} + S_q \qquad (n = 1, 2, 3) \tag{4}$$

(5) $$\frac{\partial \chi_m}{\partial t} = -u_j \frac{\partial \chi_m}{\partial x_j} + S_{\chi_m} \quad (m = 1, 2, \ldots, M)$$

with

(6) $$\theta = T_v \left[\frac{1000}{p(\text{in mbar})}\right]^{R_d/c_p}$$

(7) $$p = \rho R_d T_v$$

and

(8) $$T_v = T(1 + 0.61 q_3)$$

Mathematically, Eqs. (1)–(8) represent $11 + M$ nonlinear simultaneous partial differential equations in the $11 + M$ dependent variables: ρ, density; θ, potential temperature; T_v virtual temperature; T, temperature; p, pressure; u_i, velocity ($u_1 = u$, $u_2 = v$, $u_3 = w$); q_n, amounts of water ($q_1 = q_{\text{solid water}}$, $q_2 = q_{\text{liquid water}}$, $q_3 = q_{\text{water vapor}}$); χ_m, amounts of other gaseous and aerosol atmospheric materials (e.g., $\chi_1 = \chi_{SO_2}$, $\chi_2 = \chi_{SO_4}$). The variables q_n and χ_m have units of mass of substance to mass of air in the same volume. The independent variables are time t and the three space coordinates $x_1 = x$, $x_2 = y$ and $x_3 = z$. R_d is the gas constant for dry air.

In deriving Eqs. (1)–(8) the assumptions utilized included the following:

- The atmosphere is considered an ideal gas [Eq. (7)] in thermodynamic equilibrium.
- Mass is assumed to have no sources or sinks.
- External work performed by chemical reactions, phase changes, and electromagnetism is neglected in the first law of thermodynamics [Eq. (2)].
- Gravity and the rotation rate of the Earth are treated as constants, and electromagnetism is not considered as an external force in the equation of motion [Eq. (3)]. In this equation the dissipation of velocity by molecular interactions (called an internal force) is also ignored.
- The transfer of heat, water, and other gaseous and aerosol atmospheric materials by molecular diffusion is also neglected in Eqs. (2), (4), and (5).
- Changes in water mass caused by chemical reactions are ignored in Eq. (4).

The conservation relations are discussed in detail in Dutton (1976) and Iribarne and Godsen (1973).

The source/sink terms in Eqs. (2), (4), and (5) represent the following processes:

$$S_\theta = \begin{bmatrix} +\text{freezing} \\ -\text{melting} \end{bmatrix} + \begin{bmatrix} +\text{condensation} \\ -\text{evaporation} \end{bmatrix} + \begin{bmatrix} +\text{deposition (vapor to solid)} \\ -\text{sublimation (solid to vapor)} \end{bmatrix}$$

$$+ \begin{bmatrix} +\text{exothermic chemical reactions} \\ -\text{endothermic chemical reactions} \end{bmatrix}$$

$$+ \begin{bmatrix} +\text{net radiative flux convergence} \\ -\text{net radiative flux divergence} \end{bmatrix}$$

$$+ [\text{dissipation of kinetic energy by molecular motions}];$$

$$S_{q_1} = \begin{bmatrix} +\text{freezing} \\ -\text{melting} \end{bmatrix} + \begin{bmatrix} +\text{deposition (vapor to solid)} \\ -\text{sublimation (solid to vapor)} \end{bmatrix}$$

$$+ \begin{bmatrix} +\text{fallout from above} \\ -\text{fallout to below} \end{bmatrix};$$

$$S_{q_2} = \begin{bmatrix} +\text{melting} \\ -\text{freezing} \end{bmatrix} + \begin{bmatrix} +\text{condensation} \\ -\text{evaporation} \end{bmatrix} + \begin{bmatrix} +\text{fallout from above} \\ -\text{fallout to below} \end{bmatrix};$$

$$S_{q_3} = \begin{bmatrix} +\text{evaporation} \\ -\text{condensation} \end{bmatrix} + \begin{bmatrix} +\text{sublimation (solid to vapor)} \\ -\text{deposition (vapor to solid)} \end{bmatrix};$$

while S_{χ_m} can be written to include changes of state (analogous to that performed for water), plus chemical transformations from one compound to another, as well as precipitation scavenging and dry fallout. The mathematical representation of these source/sink terms can be quite complex and some of them are overviewed in Sections 8 and 9.

Equations (1), (2), (3), (4), and (5) represent the conservation of mass, heat, motion, water, and other gaseous and aerosol atmospheric materials, respectively. Equation (7) is the ideal gas law, while Eqs. (6) and (8) are definitions of dependent variables. In the remainder of this article methods of simplification and solution of these relations are discussed, as well as examples presented of mesoscale model results.

3. Simplification of the Basic Equations

Equations (1)–(8) can be simplified for specific mesoscale meteorological simulations, and by mathematical operations some of these relations can also be changed in form. In this section, commonly made assumptions will be overviewed and the resultant equations presented. In all cases the general equations given by (1)–(8) are altered in form and/or simplified in order to permit their solution in an easier and/or more economical fashion.

The method of *scale analysis* is used to determine the relative importance of the individual terms in the conservation relations. This technique involves the estimation of their order of magnitude through the use of representative values of the dependent variables and constants that make up these terms.

3.1. Conservation of Mass Approximation

Applying Dutton and Fichtl's (1969) results, it can be shown that the conservation of mass relation [Eq. (1)] can be written as

$$\frac{\partial}{\partial x_j} \bar{\rho} u_j = 0 \tag{9}$$

if density fluctuations are much smaller than the average value of density over a region of interest [$\bar{\rho}$ is defined formally in Section 4 by Eq. (19), or by (86) in Section 6]. In addition, as shown by Dutton and Fichtl (1969), if the depth of the circulation of interest is much less than the density scale height of the atmosphere, defined as

$$H_\rho^{-1} = -\frac{1}{\rho}\frac{\partial \rho}{\partial z} \tag{10}$$

then the conservation of mass relation can be simplified even further to

$$\frac{\partial u_j}{\partial x_j} = 0 \tag{11}$$

Dutton and Fichtl refer to Eqs. (9) and (11) as the *deep* and *shallow continuity equations*, respectively. Equation (11) is also referred to as the *incompressible* form of the conservation of mass relation since the same form of the equation results from Eq. (1) if the atmosphere is assumed homogeneous ($\rho = \mathring{\rho} = $ constant).

Because time changes in density are not included in (9) and (11), these equations are also referred to by Ogura and Phillips (1962) and Ogura (1963) as the *anelastic* or *soundproof* approximations to the conservation of mass relation. As shown by Ogura and Charney (1961), and as will be briefly mentioned in Section 5, sound waves are excluded as a possible solution to Eqs. (1)–(8) when the diagnostic equation (i.e., no time tendency terms) (9) or (11) is used in lieu of the more complete prognostic conservation of mass equation given by (1).

Many mesoscale models use Eq. (11) to represent the conservation of mass. There is a certain irony in its use, of course, because although air closely follows the ideal gas law, its behavior is also accurately approximated by the incompressible assumption given by Eq. (11) when the atmospheric circulations have a limited vertical extent. This apparent discrepancy is explained by realizing that, in the atmosphere, air is not physically constrained in its movement. When air moves into one side of a volume of atmosphere, the density can either increase by compression, or an equivalent mass of air can move out of the other side of this parcel. As long as this atmospheric parcel is not restricted in size by rigid boundaries, however, the creation of a pressure gradient between the two sides of the parcel as a result of the different velocities will force air out of one side so that mass conservation is closely approximated by Eq. (11).

3.2. Conservation of Heat, Water, and Other Gaseous and Aerosol Material Approximations

The conservation relations for heat, water, and other gaseous and aerosol atmospheric materials represented by Eqs. (2), (4), and (5), respectively, are simplified through the chosen form for the source/sink terms S_θ, S_{q_n}, and S_{χ_m}. The development of simplified mathematical representations for any of these terms is called *parameterization*. The most stringent assumption for the conservation of potential temperature relation is to require that all motions be *adiabatic* so that $S_\theta = 0$. This condition is most closely fulfilled when the following conditions are met:

- No phase changes of water occur.
- Motion occurs over comparatively short time periods so that radiational heating or cooling of the air is relatively small.
- The heating or cooling of the lowest levels of the atmosphere by the Earth's surface is comparatively small.

S_{q_n} and S_{χ_m} can also be neglected relative to the local time tendency and advection if contributions due to phase changes and chemical transformations are small. More specific discussions of S_θ, S_{q_n}, and S_{χ_m} are given in Sections 7–9.

3.3. Conservation of Motion Approximations

In developing simplified forms of the conservation of motion relation, it is useful to consider the vertical and horizontal components separately, since the gravitational acceleration is only included in the vertical equation.

3.3.1. The Horizontal Equation of Motion. The horizontal equations of motion in scalar form, obtained from Eq. (3), are given as

$$\frac{\partial u}{\partial t} = -u\frac{\partial u}{\partial x} - v\frac{\partial u}{\partial y} - w\frac{\partial u}{\partial z} - \frac{1}{\rho}\frac{\partial p}{\partial x} + fv - \hat{f}w \tag{12}$$

and

$$\frac{\partial v}{\partial t} = -u\frac{\partial v}{\partial x} - v\frac{\partial v}{\partial y} - w\frac{\partial v}{\partial z} - \frac{1}{\rho}\frac{\partial p}{\partial y} - fu \tag{13}$$

where the *Coriolis parameter* $f = 2\Omega \sin \phi$, $\hat{f} = 2\Omega \cos \phi$, ϕ is latitude, and Ω is the rotation rate of the Earth (2π/day). The procedure for determining the relative importance of individual terms in these two expressions is to use scale analysis to estimate the magnitude of the variables and constants that comprise each term. Using this method [details of the analysis are presented by Pielke (1982)], the ratios are given in Table 1, where

- L_x and L_z are used to represent the horizontal and vertical wavelengths of the mesoscale system;
- $C \sim L/t_u$ represents the speed of movement of a mesoscale system. If the mesoscale atmospheric feature of interest propagates predominantly in the horizontal direction, $L \cong L_x$ is assumed. If horizontal advective changes are most important, $C \cong U$, while C is equal to the group velocity of the system if wave propagation dominates (C could be in the vertical as well as horizontal direction). The time t_u represents the period over which significant changes in u and v occur at a fixed point;
- U represents the horizontal advecting velocity. The relation $W/L_z \sim U/L_x$, where W is a representative vertical advecting velocity, has been used to obtain the ratios and is derived from a scale analysis representation of Eq. (11);
- δT is obtained from the scale analysis representation to the pressure gradient force given as

$$\frac{\partial p}{\partial x} = \rho R \frac{\partial T}{\partial x} + TR \frac{\partial \rho}{\partial x} \cong \rho R \frac{\partial T}{\partial x},$$

where the ideal gas law has been used. For most mesoscale phenomena the quantity $T(\partial \rho/\partial x)$ is less than $\rho(\partial T/\partial x)$ (e.g., $|T \delta\rho/\rho \delta T| \sim 0.3$ with $T = 300°K$, $\delta T = 1°K$, $\delta p = 1$ mbar, and $\rho = 1.0$ kg m^{-3});
- δD represents the relation between horizontal pressure gradient and horizontal gradients in the depth of a homogeneous atmosphere (e.g., $P_{z=0} = \dot{\rho}gD$ so that $\delta P/L_x \sim \dot{\rho}g \, \delta D/L_x$);
- $R_0 = U/|f|L_x$ is called the *Rossby number*.

TABLE I. Ratio of Individual Terms in the Horizontal Equation of Motion, Obtained Using Scale Analysis

$$\frac{\text{advective term (e.g., } |u\,\partial u/\partial x|)}{\text{local tendency term (e.g., } |\partial u/\partial t|)} \cong \frac{U}{C}$$

$$\frac{\text{pressure gradient force (e.g., } |(1/\rho)\,\partial p/\partial x|)}{\text{local tendency term}} \cong \frac{R\delta T}{UC} \sim \frac{g\delta D}{UC}$$

$$\frac{\text{pressure gradient force}}{\text{advective term}} \cong \frac{R\delta T}{U^2} \sim \frac{g\delta D}{U^2}$$

$$\frac{f\,\text{Coriolis term (e.g., } fv)}{\text{local tendency term}} \cong \frac{U}{CR_0}$$

$$\frac{f\,\text{Coriolis term}}{\text{advective term}} \cong \frac{1}{R_0}$$

$$\frac{f\,\text{Coriolis term}}{\text{pressure gradient force}} \cong \frac{|f|UL_x}{R\delta T} \sim \frac{|f|UL_x}{g\delta D} \sim \frac{U^2}{g\delta D R_0}$$

$$\frac{\hat{f}\,\text{Coriolis term } (\hat{f}w)}{\text{local tendency term}} \cong \frac{U}{C}\frac{L_z}{L_x R_0}$$

$$\frac{\hat{f}\,\text{Coriolis term}}{\text{advective term}} \cong \frac{L_z}{L_x R_0}$$

$$\frac{\hat{f}\,\text{Coriolis term}}{\text{pressure gradient force}} \cong \frac{|f|UL_z}{R\delta T} \sim \frac{|f|UL_z}{g\delta D}$$

$$\frac{\hat{f}\,\text{Coriolis term}}{f\,\text{Coriolis term}} \cong \frac{L_z}{L_x}$$

Some of the interpretations that can be made from such a scale analysis are the following:

- The local tendency terms can be neglected if the movement of the mesoscale system is much less than the advecting wind speed (e.g., $C \ll U$). Such a system is said to be *steady state* if $C = 0$ and *quasi-steady*, otherwise.
- The ratio of the pressure gradient force and advective terms is inversely proportional to the square of the wind speed and proportional to the horizontal temperature gradient. Since the horizontal pressure gradient is approximately a linear term (e.g., $(1/\rho)(\partial p/\partial x) \cong (1/\hat{\rho})(\partial p/\partial x)$ if density fluctuations are small), while the advective terms are nonlinear, the solution to the conservation relations is greatly simplified if the ratio is large and this nonlinear contribution to the equations can be ignored. Equations (12) and (13) are then linearized (as long as $\rho \cong \hat{\rho}$).
- If the horizontal scale of the mesoscale circulation is much larger than the vertical scale, then the $\hat{f}$ Coriolis term can be ignored relative to the f Coriolis contribution.
- If the Rossby number is much less than unity, the advective terms can be neglected relative to the Coriolis terms. This is the assumption used to derive the gradient and geostrophic wind approximations used in synoptic meteorology.
- If the Rossby number is much greater than unity, the Coriolis terms can be neglected relative to the advective terms. This is the assumption used in cumulus and other smaller-scale models.

3.3.2. The Vertical Equation of Motion. The vertical equation of motion can be similarly treated using scale analysis. From Eq. (3) with $i = 3$,

$$\frac{\partial w}{\partial t} = -u\frac{\partial w}{\partial x} - v\frac{\partial w}{\partial y} - w\frac{\partial w}{\partial z} - \frac{1}{\rho}\frac{\partial p}{\partial z} - g + 2\Omega u \cos \phi$$

or, equivalently,

(14) $$\frac{dw}{dt} = -\frac{1}{\rho}\frac{\partial p}{\partial z} - g + 2\Omega u \cos \phi$$

To estimate the magnitude of the vertical acceleration let

$$|dw/dt| \sim W/t_w$$

where t_w is a time scale corresponding to the period required for significant changes in vertical velocity to occur. From the scale analysis of Eq. (11), W is related to U through the ratio of L_z to L_x. Thus, since

$$W \sim (L_z/L_x)U$$

a representative estimate for t_w is $t_w \sim L_z/W \sim L_x/U$ if advective changes dominate, or $t_w \sim L_z/C_z \sim L_x/C_x$ [using a scale analysis from Dutton (1976; Fig. 12.5)] if the group velocity of internal gravity waves is the most important cause of vertical acceleration. Thus,

$$\left|\frac{dw}{dt}\right| \sim \frac{WC_z}{L_z} \sim \frac{L_z}{L_x^2} UC_x$$

The magnitude of the vertical pressure gradient is estimated from

$$\left|\frac{1}{\rho}\frac{\partial p}{\partial z}\right| = \left|R\frac{\partial T}{\partial z} + \frac{RT}{\rho}\frac{\partial \rho}{\partial z}\right| \sim \left|\frac{RT}{\rho}\frac{\partial \rho}{\partial z}\right| = \frac{RT}{H_\rho}$$

where to simplify the analysis, the temperature lapse rate was assumed isothermal. For convenience the subscripts "d" and "v" have been deleted from R and T. H_ρ is defined by Eq. (10).

The ratios of the individual terms in Eq. (14) are then given by

$$g \cdot \left|\frac{1}{\rho}\frac{\partial p}{\partial z}\right|^{-1} \sim \frac{gH_\rho}{RT} = R_g$$

$$|2\Omega u \cos \phi| \cdot \left|\frac{1}{\rho}\frac{\partial p}{\partial z}\right|^{-1} \sim \frac{2\Omega UH_\rho}{RT} = R_f$$

and

(15) $$\left|\frac{dw}{dt}\right| \cdot \left|\frac{1}{\rho}\frac{\partial p}{\partial z}\right|^{-1} \sim \frac{H_\rho L_z}{L_x^2} \frac{UC_x}{RT} = R_w$$

If Eq. (9) is used as the conservation of mass relation, then H_ρ and L_z are the same order of magnitude so that Eq. (15) can also be written as

$$R_w = \frac{H_\rho^2}{L_x^2} \frac{UC_x}{RT}$$

This scale analysis shows that R_w, the ratio of the vertical accelerations to the vertical pressure gradient, becomes less as the horizontal wavelength increases and/or the vertical wavelength decreases, while it increases for stronger winds. According to this scale analysis, therefore, to neglect the vertical acceleration term in Eq. (14), R_w must be much less than unity. Unfortunately, however, this analysis is not complete, since we have only shown that the magnitude of the gravitational and pressure gradient accelerations are separately much larger than the vertical acceleration. Of more significance is the magnitude of the difference between these two terms. To better examine this relationship, it is convenient to define a large-scale averaged atmosphere that has an exact balance between the gravitational and pressure gradient terms. This can be defined as

$$\partial p_0/\partial z = -\rho_0 g$$

where the subscript "0" is used to indicate a large-scale average [such an average could be defined as given by Eq. (30)]. If any dependent variable is defined to be equal to such an average value plus a deviation from that average (i.e., $\phi = \phi_0 + \hat{\phi}$, where ϕ is any one of the dependent variables), then Eq. (14) can also be written as

$$\frac{dw}{dt} = -\frac{1}{\rho_0}\frac{\partial \hat{p}}{\partial z} + g\frac{\hat{\alpha}}{\alpha_0} + 2\Omega u \cos \phi$$

In this expression, the first two terms on the right-hand side are much smaller in magnitude than the first two terms on the right-hand side of Eq. (14).

The magnitude of the perturbation vertical pressure gradient can be estimated from

$$\left|\frac{1}{\rho_0}\frac{\partial \hat{p}}{\partial z}\right| \cong \frac{1}{\rho_0}\left|T_0 R\frac{\partial \hat{\rho}}{\partial z} + R\hat{T}\frac{\partial \rho_0}{\partial z} + R\rho_0\frac{\partial \hat{T}}{\partial z}\right|$$

$$\sim \left|\frac{R\,\delta T}{L_z} + \frac{R\,\delta T}{H_\rho} + \frac{R\,\delta T}{L_z}\right| \sim \frac{R|\delta T|}{L_z}$$

where for the purposes of this scale analysis the large-scale atmosphere is assumed isothermal and the linearized ideal gas law of the form $|\delta\rho/\rho| \cong |\delta T/T + \delta p/p| \cong |\delta T/T|$ has been used. From this analysis the term with the vertical gradient of the large-scale density ρ_0 can be neglected if $L_z \ll H_\rho$.

The ratio of the vertical acceleration to this perturbation pressure gradient term is thus given as

$$\hat{R}_w = \frac{L_z^2}{L_x^2}\frac{UC_x}{R\,\delta T} = \frac{H_\rho^2}{L_x^2}\frac{UC_x}{R\,\delta T}$$

where $L_z \sim H_\rho$ has been used in the last expression on the right. This relationship is more restrictive than R_w since δT is in the denominator, rather than T (note that $\hat{R}_w$ does not increase without bound because as δT goes to zero, UC_x should go to zero faster). If $\hat{R}_w \ll 1$, $g\hat{\alpha}/\alpha_0$ must be of the same order of magnitude as the perturbation vertical pressure gradient (since $R_f \ll 1$ under all expected atmospheric conditions as will be discussed shortly).

To illustrate the magnitude of $\hat{R}_w$ for representative values on the mesoscale, δT and U are set equal to 10°C and 10 m sec^{-1} (based on observed values), and C_x is set equal to U. R is equal to 287 J kg^{-1} deg^{-1}. This yields

$$\hat{R}_w = 0.03 H_\rho^2/L_x^2$$

where $H_\rho \cong 8$ km (Wallace and Hobbs, 1977). If the depth of the circulation is less than the scale height ($L_z < H_\rho$), then $\hat{R}_w$ is proportionally smaller (e.g., if $L_z = 0.1 H_\rho$, $\hat{R}_w = 0.0003$). Thus, from this analysis, a conservative estimate for neglecting vertical accelerations relative to the vertical pressure gradient term is given by

(16) $$\boxed{H_\rho/L_x \gtrsim 1}$$

The ratios R_g and $R_{\hat{f}}$ are independent of wavelength. For any expected values of T and U in the Earth's atmosphere

$$R_g \sim 1, \qquad R_{\hat{f}} \sim 10^{-4} \ll 1$$

so that the Coriolis term can be neglected relative to the vertical pressure gradient, whereas gravitational acceleration cannot.

If (16) is valid for a particular mesoscale system, Eq. (14) can be rewritten as

(17) $$\boxed{\partial p/\partial z = -\rho g}$$

and is called the *hydrostatic equation*. This relation replaces the prognostic equation for vertical velocity with a diagnostic relation for pressure. In utilizing this relationship it must be emphasized that the results of the scale analysis imply only that the magnitude of the vertical acceleration is much less than the magnitude of the pressure gradient force, *not* that the magnitude of the vertical acceleration is identically zero (i.e., $|dw/dt| \ll |(1/\rho)(\partial p/\partial z)|$, *not* $|dw/dt| \equiv 0$).

Using the scale analysis in this section a more formal definition of mesoscale can be given than that presented in the Introduction. The criteria are as follows:

- *The horizontal scale is sufficiently large so that the hydrostatic equation can be applied.*
- *The horizontal scale is sufficiently small so that the Coriolis term is small relative to the advective and pressure gradient forces, resulting in a flow field that is substantially different from the gradient wind relation above the planetary boundary layer.*

Scales of motion in which vertical accelerations become important can be referred to as the microscale and correspond to the meso-γ and smaller scale as defined by Orlanski (1975). This scale of motion somewhat smaller than mesoscale has also been referred to as the cumulus scale, with the smallest sizes referenced as the turbulence scale. Scales larger than the mesoscale, where the Coriolis effect becomes of the same magnitude as the pressure gradient, are termed the synoptic scale and correspond to Orlanski's meso-α and larger.

In a mesoscale model as defined here, Eq. (17) will replace the vertical equation of motion ($i = 3$) from Eq. (3), while Eq. (9) or (11) will be used in lieu of Eq. (1) to represent the conservation of mass. In performing these simplifications, mesoscale circulations, therefore, are defined to be *anelastic, hydrostatic,* and *significantly nongradient wind* meteorological systems.

4. Averaging the Conservation Relations

The conservation relations given by Eqs. (1)–(5), or simplified forms such as Eq. (9) or (11), are defined in terms of the differential operators ($\partial/\partial t$, $\partial/\partial x_i$) and thus, in terms of mathematical formalism are valid only in the limit when δt, δx, δy, and δz approach zero. In terms of practical application, however, they are usable only when the spatial increments δx, δy, and δz are much larger than the spacing between molecules (so that only the statistical characteristics of molecular motion, rather than the movement of individual molecules themselves, are important), but are small enough so that the differential terms over these distances and over the time interval δt can be represented accurately by a constant. If the variation of the dependent variables within these intervals is significant, however, the differential prognostic relations must be integrated over the distance and time interval in which they are being applied.

Stated more formally, if

$$l_m \ll \delta x, \delta y, \text{ and } \delta z$$

where l_m is the representative spacing between molecules, and if

$$\frac{\partial u}{\partial x} \gg \frac{\delta x}{2} \frac{\partial^2 u}{\partial x^2}; \quad \frac{\partial u}{\partial t} \gg \frac{\delta t}{2} \frac{\partial^2 u}{\partial t^2}; \quad \text{etc.}$$

then the use of differential forms of the conservation relations is appropriate.

In the atmosphere, however, these criteria limit the application of the differential forms to space scales on the order of a centimeter, and to time scales of a second or so. Therefore, to accurately represent the atmosphere, these relations must be evaluated over those space and time intervals. Since mesoscale circulations have horizontal scales on the order of 10–100 km with a vertical size of approximately 10 km, these equations would have to be solved at 10^{18}–10^{20} locations. This amount of information, unfortunately, far exceeds the capability of any existing or foreseeable computer system.

To circumvent this problem, the conservation relations are integrated over specific spatial and temporal scales, whose sizes are determined by the available computer capacity, including its speed of operation. For a specific mesoscale system, the smaller these scales, the better the *resolution* of the circulation.

In performing this integration, it is convenient to decompose the dependent variables by

$$(18) \quad \phi = \bar{\phi} + \phi''$$

where ϕ represents any one of the dependent variables and

$$(19) \quad \bar{\phi} = \int_{t}^{t+\Delta t} \int_{x}^{x+\Delta x} \int_{y}^{y+\Delta y} \int_{z}^{z+\Delta z} \phi \, dz \, dy \, dx \, dt / (\Delta t \, \Delta x \, \Delta y \, \Delta z).$$

Thus, $\bar{\phi}$ represents the average of ϕ over the finite time increment Δt and space intervals Δx, Δy, and Δz. The variable ϕ'' is the deviation of ϕ from this average and is often called the *subgrid scale perturbation*. In a numerical model Δt is called the *time step* and Δx, Δy, and Δz represent the *grid intervals*.

The decomposition given by Eq. (18) can be performed for each dependent variable and substituted into Eqs. (2)–(5), and (9) or (11).*
These equations can then be integrated by

$$(20) \quad \overline{(\quad)} = \int_{t}^{t+\Delta t} \int_{x}^{x+\Delta x} \int_{y}^{y+\Delta y} \int_{z}^{z+\Delta z} \times (\quad) \, dz \, dy \, dx \, dt / (\Delta t \, \Delta x \, \Delta y \, \Delta z)$$

which is the same as that used to obtain $\bar{\phi}$. This operation is often called *grid-volume averaging* since it is performed over the spatial increments Δx, Δy, and Δz.

In order to simplify the equations that result from this integration, it is customary to assume that the averaged dependent variables change much more slowly in time and space than do the deviations from the average. This *scale separation* between the average and the perturbation implies that $\bar{\phi}$ is approximately constant across the distances Δx, Δy, and Δz, and time interval Δt. In addition, the grid intervals and time increment are also

---

* The expressions for θ and T_v (Eqs. (6) and (8)) and the ideal gas law (Eq. (7)) are not represented in this fashion since they are definitions and can be defined separately for the subgrid scale perturbations and grid-volume averaged variables.

presumed not to be a function of location or time so that derivatives (e.g., $\partial/\partial t$, $\partial/\partial x_j$) can be removed from the integrals.

Using these assumptions, it follows that

$$\bar{\bar{\phi}} = \bar{\phi}; \quad \overline{\phi''} = 0; \quad \overline{\frac{\partial \phi}{\partial t}} = \frac{\partial \bar{\phi}}{\partial t}; \quad \overline{\frac{\partial \phi}{\partial x_j}} = \frac{\partial \bar{\phi}}{\partial x_j}$$

The stipulation that the average of the deviations is zero ($\overline{\phi''} = 0$) is commonly called the *Reynold's assumption*.

Using the assumption that $\rho''/\bar{\rho} \ll 1$, performing the integral operation given by (20), simplifying, and rearranging yields the anelastic equations given as

(21) $$\frac{\partial}{\partial x_j} \bar{\rho} \bar{u}_j = 0 \quad \text{or} \quad \frac{\partial \bar{u}_j}{\partial x_j} = 0$$

(22) $$\bar{\rho} \frac{\partial \bar{\theta}}{\partial t} = -\frac{\partial}{\partial x_j} \bar{\rho} \bar{u}_j \bar{\theta} - \frac{\partial}{\partial x_j} \overline{\rho u_j'' \theta''} + \bar{\rho} \bar{S}_\theta$$

(23) $$\bar{\rho} \frac{\partial \bar{u}_i}{\partial t} = -\frac{\partial}{\partial x_j} \bar{\rho} \bar{u}_j \bar{u}_i - \frac{\partial}{\partial x_j} \overline{\rho u_j'' u_i''} - \frac{\partial \bar{p}}{\partial x_i} - \bar{\rho} g \delta_{i3} - 2\epsilon_{ijk} \bar{\rho} \Omega_j \bar{u}_k$$

(24) $$\bar{\rho} \frac{\partial \bar{q}_n}{\partial t} = -\frac{\partial}{\partial x_j} \bar{\rho} \bar{u}_j \bar{q}_n - \frac{\partial}{\partial x_j} \overline{\rho u_j'' q_n''} + \bar{\rho} \bar{S}_{q_n}$$

(25) $$\bar{\rho} \frac{\partial \bar{\chi}_m}{\partial t} = -\frac{\partial}{\partial x_j} \bar{\rho} \bar{u}_j \bar{\chi}_m - \frac{\partial}{\partial x_j} \overline{\rho u_j'' \chi_m''} + \bar{\rho} \bar{S}_{\chi_m}$$

where $\bar{S}_\theta$, $\bar{S}_q$, and $\bar{S}_{\chi_m}$ represent the integrated contributions of the source/sink terms. The terms $\overline{\rho u_j'' \theta''}$, $\overline{\rho u_j'' u_i''}$, $\overline{\rho u_j'' q_n''}$, and $\overline{\rho u_j'' \chi_m''}$, which, of course, do not appear in the unaveraged equations, are called the *subgrid scale correlation terms* and represent the contribution of scales smaller than the resolvable grid on the mesoscale. Such terms, often of order the same as or even larger than the terms that involve only resolved dependent variables, must be *parameterized* in terms of the resolved variables in order to assure that the number of unknowns is equal to the number of equations. The proper specification of these subgrid scale correlations as a function of resolvable, averaged quantities is referred to as the *closure* problem and is discussed further in Section 7.

If the motions are assumed hydrostatic, the vertical equation of motion

($i = 3$) from Eq. (23) is replaced with

(26) $$\partial \bar{p}/\partial z = -\bar{\rho} g$$

Equations (21), (22), (24), and (25) together with (23), or (25) together with (23), where $i = 1, 2$, are often called the *primitive equations* because they are derived straightforwardly from the original conservation laws presented in Section 2. As evident from the assumptions required to obtain them, however, they are not the most fundamental form of the conservation laws as implied by the word "primitive."

In using these equations in a mesoscale model, the horizontal and vertical pressure gradients are often written in terms of a scaled pressure defined as

$$\pi = C_p \left[\frac{P \text{ (in mbar)}}{1000 \text{ mbar}} \right]^{R/C_p} = C_p T_v/\theta$$

where π is called the *Exner function*. Since the ideal gas law [Eq. (7)] and the definition of potential temperature [Eq. (6)] must apply separately to the subgrid-scale and resolvable quantities, the pressure gradient term in Eq. (23) can be rewritten as

$$\frac{1}{\bar{\rho}} \frac{\partial \bar{p}}{\partial x_i} = \bar{\theta} \frac{\partial \bar{\pi}}{\partial x_i}$$

Equation (23) then becomes

(27) $$\frac{\partial \bar{u}_i}{\partial t} = -\bar{u}_j \frac{\partial \bar{u}_i}{\partial x_j} - \frac{1}{\bar{\rho}} \frac{\partial}{\partial x_j} \overline{\rho u_j'' u_i''} - \bar{\theta} \frac{\partial \bar{\pi}}{\partial x_i} - g\delta_{i3} - 2\epsilon_{ijk}\Omega_j \bar{u}_k$$

If the hydrostatic assumption is used, the vertical equation of motion ($i = 3$) from Eq. (27) is replaced with

(28) $$\partial \bar{\pi}/\partial z = -g/\bar{\theta}$$

Equations (23) and (27) can also be written in a different form if

(29) $$\bar{\phi} = \phi_0 + \phi'$$

where

(30) $$\phi_0 = \int_x^{x+D_x} \int_y^{y+D_y} \bar{\phi} \, dx \, dy / D_x D_y$$

ϕ_0 is referred to as the *layer-domain-averaged variable*. The symbols D_x

and D_y represent distances that are large compared with the mesoscale system of interest [perhaps the horizontal size (*domain*) of the mesoscale model representation], so that ϕ_0 can be considered to represent the synoptic scale. The variable ϕ', then, represents the mesoscale deviations from this larger scale.

Using the decomposition given by Eq. (29) in Eqs. (23) and (27) with $\rho''/\rho_0 \ll 1$, then the horizontal and vertical components of (23) and (27) can be written as

$$(31) \quad \boxed{\begin{aligned} \frac{\partial \bar{u}_i}{\partial t} &= -\bar{u}_j \frac{\partial \bar{u}_i}{\partial x_j} - \frac{1}{\bar{\rho}} \frac{\partial}{\partial x_j} \overline{\rho u_j'' u_i''} - \frac{1}{\rho_0} \frac{\partial \bar{p}}{\partial x_i} \\ &\quad - 2\epsilon_{ijk} \Omega_j \bar{u}_k \quad (i = 1, 2) \end{aligned}}$$

or in an equivalent form

$$(32) \quad \boxed{\begin{aligned} \frac{\partial \bar{u}_i}{\partial t} &= -\bar{u}_j \frac{\partial \bar{u}_i}{\partial x_j} - \frac{1}{\bar{\rho}} \frac{\partial}{\partial x_j} \overline{\rho u_j'' u_i''} - \bar{\theta} \frac{\partial \bar{\pi}}{\partial x_i} \\ &\quad - 2\epsilon_{ijk} \Omega_j \bar{u}_k \quad (i = 1, 2) \end{aligned}}$$

while

$$(33) \quad \boxed{\frac{\partial \bar{w}}{\partial t} = -\bar{u}_j \frac{\partial \bar{w}}{\partial x_j} - \frac{1}{\bar{\rho}} \frac{\partial}{\partial x_j} \overline{\rho u_j'' w''} - \frac{1}{\rho_0} \frac{\partial p'}{\partial z} - \frac{\rho'}{\rho_0} g + 2\bar{u}\Omega \cos \phi}$$

or

$$(34) \quad \boxed{\frac{\partial \bar{w}}{\partial t} = -\bar{u}_j \frac{\partial \bar{w}}{\partial x_j} - \frac{1}{\bar{\rho}} \frac{\partial}{\partial x_j} \overline{\rho u_j'' w''} - \bar{\theta} \frac{\partial \pi'}{\partial z} + \frac{\theta'}{\theta_0} g + 2\bar{u}\Omega \cos \phi}$$

In deriving these expressions, it has been assumed that the synoptic scale pressure field is hydrostatic ($\partial p_0/\partial z = -\rho_0 g$; $\partial \pi_0/\partial z = -g/\theta_0$) so that, using the requirement that $\rho'/\rho_0 \ll 1$,

$$\frac{1}{\rho_0 + \rho'} \frac{\partial}{\partial z}(p_0 + p') + g \cong \frac{1}{\rho_0}\left(1 - \frac{\rho'}{\rho_0}\right) \frac{\partial(p_0 + p')}{\partial z} + g$$

$$\cong \frac{1}{\rho_0} \frac{\partial p'}{\partial z} + \frac{\rho'}{\rho_0} g$$

and

$$(\theta_0 + \theta') \frac{\partial(\pi_0 + \pi')}{\partial z} + g = \bar{\theta} \frac{\partial \pi'}{\partial z} - \frac{\theta'}{\theta_0} g$$

(Note that the condition $\rho'/\rho_0 \ll 1$ is not needed in deriving the $\theta \, \partial\pi/\partial z$ expression.)

The hydrostatic relations obtained from Eqs. (33) and (34), corresponding to Eq. (17), are

(35) $$\frac{\partial p'}{\partial z} = -\rho' g$$

and

(36) $$\frac{\partial \pi'}{\partial z} = \frac{\theta'}{\theta_0 \bar{\theta}} g$$

The inequality given by (16) defines when either one of these hydrostatic equations or Eq. (26) or (28) can be used in lieu of the third equation of motion. Since, as shown in Section 10, the smallest-sized horizontal feature that can be accurately resolved in a mesoscale model has a size corresponding to $4\,\Delta x$, then

(37) $$\frac{H_\rho}{L_x} = \frac{H_\rho}{4\,\Delta x} \gtrsim 1$$

defines the necessary restriction for the use of the hydrostatic equation. If $\Delta x = 2$ km, for example, the hydrostatic approximation appears to be appropriate using the results of scale analysis.

If (37) is applicable, and $\bar{\rho} \cong \rho_0$, then Eqs. (21), (22), (24), (25), (28), and (32) represent $8 + M$ equations as a function of the $8 + M$ dependent variables $\bar{u}_i$, $\bar{\theta}$, $\bar{\pi}$, $\bar{q}_n$, and $\bar{\chi}_m$ (providing the subgrid correlation terms and source/sink terms are parameterized in terms of known quantities). Similarly, Eqs. (21), (22), (24)–(26), (31), along with (6)–(8), defined using $\bar{T}_v$, $\bar{\theta}, \bar{\rho}, \bar{p},$ and $\bar{q}_3$, represent $11 + M$ equations in the $11 + M$ dependent variables $\bar{u}_i$, $\bar{\theta}$, $\bar{p}$, $\bar{q}_n$, $\bar{\chi}_m$, $\bar{\rho}$, $\bar{T}_v$, and $\bar{T}$. Equations identical to these, or closely equivalent, form the foundation of all mesoscale models based on fundamental principles.

Finally, it is possible to use a vorticity equation instead of the primitive equations to represent the conservation of motion. Such an expression can be obtained by taking the curl (e.g., $\epsilon_{pqi}\,\partial/\partial x_q$) of Eq. (3), or some equivalent averaged form. Unfortunately, the differential operation (i.e., $\partial/\partial x_q$), required to obtain vorticity, magnifies errors that may occur in an initial imposed velocity field. Second derivatives of the subgrid scale correlation terms in the equation of motion appear when the equations are averaged over a grid volume and are more difficult to represent accurately than in the primitive equation format. Realistic boundary conditions are also more difficult to apply. In addition, the computation procedure is more complicated than with the primitive equations since velocity must be mathematically recovered from the vorticity field in order to integrate the conservation equations.

Previous use of the vorticity equation to represent meteorological flows has been limited to one of its components. Early synoptic models such as the Equivalent Barotropic Model (Haltiner, 1971) used a simplified form of the vertical component of vorticity in order to simulate the movement of large-scale troughs and ridges in the middle troposphere. In mesoscale and cumulus cloud models, one of the horizontal components of vorticity is used to simulate a vertical cross section of the circulation. Examples of such models include those of Pearson (1973), Orville (1965), Orville and Sloan (1970), and Murray (1970).

In recent years, meteorological investigators have increasingly deferred using a vorticity equation to represent the conservation of motion, and instead have relied on the primitive equation form because of its computational and conceptual simplicity. No three-dimensional mesoscale model utilizes the vorticity form, nor is such an approach likely because of its mathematical complexity in three dimensions. For this reason, the remainder of this article will focus on the primitive equation format.

5. Types of Models

There are two fundamental methods of simulating mesoscale atmospheric flows—*physical models* and *mathematical models*. With the first technique, scale model replicas of observed ground surface characteristics (e.g., topographic relief, buildings) are constructed and inserted into a chamber such as a wind tunnel. The flow of air, or other gases or liquids, in this chamber is adjusted so as to best represent the larger scale, observed atmospheric conditions. Mathematical modeling, on the other hand, utilizes such basic analysis techniques as algebra and calculus in order to solve all or a subset of the conservation relations. As will be discussed in this section, certain idealized subsets of the conservation equations can be solved exactly, while the more complete mesoscale simulations require the approximate solution technique called *numerical modeling*.

5.1. Physical Models

As discussed in Section 3, scale analysis can be used to estimate the magnitude of individual terms in the conservation relations. In utilizing a physical model, it is therefore desirable that each term be proportionately changed in magnitude so that the relative relation between each term is retained.

To illustrate this scaling [based on the discussions by Cermak (1971,

1975)], Eqs. (32), (34), and (22), for example, can be rewritten as

$$\frac{\partial \hat{\bar{u}}_i}{\partial \hat{t}} = -\hat{\bar{u}}_j \frac{\partial \hat{\bar{u}}_i}{\partial \hat{x}_j} - \left[\frac{e_{u_i}^2}{S^2}\right] \frac{\partial}{\partial \hat{x}_j} \overline{u''_j u''_i} - \left[\frac{R\delta\theta}{S^2}\right] \hat{\theta}_0 \frac{\partial \hat{\pi}'}{\partial \hat{x}_i} \tag{38}$$

$$- \left[\frac{R\delta\theta}{S^2}\right] \hat{\theta}_0 \left\{\frac{\partial \hat{\pi}_0}{\partial \hat{x}} \delta_{i1} + \frac{\partial \hat{\pi}_0}{\partial \hat{y}} \delta_{i2}\right\} - [\text{Ri}_{\text{bulk}}]\hat{\theta}'\delta_{i3}$$

$$- \left[\frac{1}{R_0}\right] 2\epsilon_{ijk}\hat{\Omega}_j\hat{\bar{u}}_k - \left[\frac{1}{\text{Re}}\right] \frac{\partial^2 \hat{\bar{u}}_i}{\partial \hat{x}_j^2}$$

and

$$\frac{\partial \hat{\bar{\theta}}}{\partial \hat{t}} = -\hat{\bar{u}}_j \frac{\partial \hat{\bar{\theta}}}{\partial x_j} - \left[\frac{e_\theta e_u}{\delta\theta S}\right] \frac{\partial}{\partial \hat{x}_j} \overline{u''_j \theta''} \tag{39}$$

$$+ \left[\frac{k_\theta}{\rho_0 C_p \nu}\right]\left[\frac{\nu}{LS}\right] \frac{\partial^2 \hat{\bar{\theta}}}{\partial \hat{x}_j^2} + \hat{S}_\theta$$

where the last term on the right-hand side of (38) and the next-to-last term on the right-hand side of (39) have been added to represent the molecular dissipation of motion and molecular conduction of heat. The significance of these two terms, which are not necessary to describe actual mesoscale motions, will become evident in the discussion that follows.

For this analysis L and S are the representative length and velocity scale of the circulation of interest (no distinction is made here between U, V, and W, nor L_x, L_y, and L_z). The coefficients k_θ and ν are the molecular conduction coefficient and kinematic viscosity of air. A circumflex (ˆ) over a dependent or independent variable indicates that it is nondimensional. The remaining symbols in Eqs. (38) and (39) are defined as follows:

$e_{u_i} = [\overline{u''^2_i/2}]^{1/2}$ (subgrid scale kinetic energy);
$e_\theta = [\overline{\theta''^2/2}]^{1/2}$;
$R_0 = S/\Omega L$ (*a Rossby number* analogous to that defined in Section 3);
$\text{Re} = LS/\nu$ (*Reynolds number*);
$\text{Ri}_{\text{bulk}} = gL\delta\theta/S^2\theta_0$ (bulk Richardson number; $\delta\theta$ represents the potential temperature perturbation and is the same order as δT used in Section 3).

In order to use a physical model to accurately represent the conservation of motion and of heat relations given by Eqs. (38) and (39), it is essential that

- the individual bracketed terms be equal in the model and in the atmosphere, or
- the bracketed terms that are not equal must be much less in magnitude than the individual remaining bracketed terms.

When these conditions are met in Eqs. (38) and (39), the actual and modeled atmospheres are said to have *dynamic* and *thermal similarity*.

Thus, from Eqs. (38) and (39), to properly represent all of the terms in these equations, the two types of similarity include the following requirements:

- The ratio of turbulent kinetic energy to mean kinetic energy must be kept constant.
- Reducing the length scale L in the physical model necessitates the following: (*i*) an increase in the magnitude of the horizontal temperature perturbation $\delta\theta$ and/or a reduction in the simulated flow speed S; (*ii*) an increase in the rotation rate Ω and/or a reduction in S; (*iii*) a decrease in the viscosity ν and/or an increase in S.
- An increase in $\delta\theta$ and a decrease in S in the pressure gradient terms necessitates that θ_0 decrease.
- The partitioning of heat transport between the subgrid scale and resolvable fluxes must be the same.

Unfortunately, it is impossible to achieve all of these requirements simultaneously in existing physical models of mesoscale atmospheric circulations. Such physical models are constructed inside of buildings, which limits the dimensions of the simulated circulations to the size of meters, whereas actual mesoscale circulations extend over kilometers. Moreover, although the nondimensional source/sink term $\hat{S}_\theta$ (equivalent expressions could be developed for $\hat{S}_{\chi_m}$ and $\hat{S}_{q_n}$ as well) is included in the analysis, the mathematical symbolism used to represent this term masks its physical complexity. As is discussed in Sections 8 and 9, this term includes such effects as radiative flux divergence and phase changes of water, and is an involved function of the dependent variables. Thus it is extremely difficult to evaluate this term using scale analysis, and, in practice, physical modelers exclude it in their representation of mesoscale atmospheric flows.

To illustrate the difficulty of obtaining dynamic similarity in a mesoscale physical model for all the terms in Eq. (38), let the horizontal scale of a mountain ridge be 10 km, while the model utilizes a 1 m representation. Therefore, the scale reduction is 10^4. Thus, if $S = 10$ m sec^{-1} in the real atmosphere and air is used in the scaled model atmosphere, then the simulated wind speed would have to be 10^5 m sec^{-1} to maintain identical Reynold's number similarity. Similarly, in order to have the same Rossby number for this example, the physical model must rotate 10,000 times more rapidly than the Earth or the wind speed must be reduced by 10,000. Reducing the speed, of course, is contrary to what is required to obtain

Reynolds number similarity. Only if the results are relatively insensitive to changes in these nondimensional quantities, as suggested, for example, by Cermak (1975) for large values of the Reynolds number in simulations of the atmospheric boundary layer, can one ignore large differences in the nondimensional parameters.

Other types of similarity conditions must also be considered. These include the following:

- *kinematic similarity,* which requires that the ratio of the vertical to horizontal scales of the mesoscale and simulated circulations be the same;
- *geometric similarity,* which specifies that the vertical to horizontal representation of the terrain and other physical features of the ground surface not be exaggerated;
- *boundary similarity,* which requires that flow entering the physical model have profiles that have dynamic and thermal similarity and are in near-equilibrium [i.e., $\partial \hat{\bar{u}}_i/\partial \hat{t}$ and $\partial \hat{\bar{\theta}}/\partial \hat{t}$ are small relative to the remaining terms in Eqs. (38) and (39)]. In addition, such bottom boundary conditions as surface temperature and aerodynamic roughness conditions must be scaled so as to produce kinematic, thermal, and dynamic similarity in the lowest levels of the physical model.

From this discussion, it should be clear that it is impossible to obtain exact similarity between mesoscale atmospheric features and the physical model when all the terms in the conservation relations are included. Nonetheless, investigators who use physical models have proposed a type of similarity between actual atmospheric circulations in which $e_{u_i}^2/S^2 \gg \nu/LS$ and physical model representations in which $e_{u_i}^2/S^2 \ll \nu/LS$. With this type of physical model simulation it is assumed that the mixing by molecular motion, proportional to $\partial^2 \hat{\bar{u}}_i/\partial \hat{x}_j^2$, acts in the same manner as the mixing by air motions as represented by the turbulence flux divergence term, $(\partial/\partial \hat{x}_j)\overline{u_j'' u_i''}$. In its dimensional form this latter term can be approximated by

$$(40) \qquad \frac{\partial}{\partial x_j} \overline{u_j'' u_i''} \cong \frac{\partial}{\partial x_j}\left(-K \frac{\partial \bar{u}_i}{\partial x_j}\right)$$

(the accuracy of this approximation is discussed in Section 7), where K is called the *turbulent exchange coefficient* and is analogous to the kinematic viscosity ν. If a *turbulent Reynolds number* $\mathrm{Re}_{\mathrm{turb}}$ is defined as the ratio of the advective terms to the subgrid scale correlation terms, then by scale analysis,

$$\mathrm{Re}_{\mathrm{turb}} = L_{\mathrm{meso}} S_{\mathrm{meso}}/K$$

where the subscript "meso" refers to the mesoscale. Similarity of flow between the atmosphere and the physical model are then assumed to occur when

$$\mathrm{Re}_{\mathrm{turb}} = L_{\mathrm{meso}} S_{\mathrm{meso}}/K = \mathrm{Re} = L_{\mathrm{model}} S_{\mathrm{model}}/\nu$$

where the subscript "model" refers to the scaled physical representation.

If both the actual and simulated wind speeds are equal, $S_{\mathrm{meso}} = S_{\mathrm{model}}$; then if $L_{\mathrm{meso}} = 10^4 L_{\mathrm{model}}$, for example, then the condition

$$K = 10^4 \nu$$

is used to justify similarity between the mesoscale and the physical model, when subgrid scale mixing is the dominant forcing term in Eq. (38). If air is used in the physical model, then $\nu \cong 1.5 \times 10^{-5}$ m² sec⁻¹ so that K must be equal to about 1.5×10^{-1} m² sec⁻¹ — a condition that may be fulfilled near the ground when the air is very stably stratified.

Using this analysis, physical modelers assume that turbulent mesoscale atmospheric circulations are accurately simulated by laminar laboratory models providing that the appropriate ratio between the eddy exchange coefficient and kinematic viscosity is obtained. With all the other similarity requirements, however, physical modeling of the mesoscale has been limited to stably stratified flows over irregular terrain. Even for this class of problem, however, such observed features of the real atmosphere as the veering of the wind with height, radiational cooling, and condensation cannot be reproduced.

The main advantage of physical models of the mesoscale, therefore, has been to provide qualitative estimates of the airflow over terrain obstacles, during dense overcast or nighttime conditions. Cermak (1971), for example, illustrates such a simulation for a scale model of Port Arguello, California, where helium is released as a tracer in order to represent pollution dispersal. The influence of the model topography on the flow was quite marked and corresponded well with the observed trajectories and concentrations. A number of other physical model simulations have been performed for the mesoscale, including Meroney *et al.* (1978), Cermak (1970), SethuRaman and Cermak (1973), Chaudhry and Cermak (1971), Yamada and Meroney (1971), and Hunt *et al.* (1978).

Because physical models are severely limited in their applicability to the mesoscale, however, it is necessary to utilize the techniques of mathematical modeling. The remainder of this article will be devoted to this methodology.

5.2. Linear Models

As mentioned in Section 2, the conservation relations represent a simultaneous set of nonlinear partial differential equations. They are termed *simultaneous* because each conservation relation must be satisfied at any given time, and they involve *partial derivatives* because four independent variables, x, y, z, and t, are involved. The *nonlinear* character of the equations occurs because products of the dependent variables (e.g., $\bar{u}_j \, \partial \bar{u}_i / \partial x_j$, $\bar{u}_j \, \partial \bar{\theta} / \partial x_j$) are included in the relationship.

Over the last several hundred years, mathematical techniques have evolved that permit exact solutions of a range of algebraic and differential equations; however, except for a few highly simplified and idealized situations, the exact solution of general sets of nonlinear equations such as those expressed by the conservation relations is impossible. To solve these nonlinear equations, the differential operators must be *approximated*, as discussed in Section 10, so that the results obtained are *not exact*.

In order to obtain exact solutions to the conservation relationships, it is necessary to remove the nonlinearities in the equations. Although an idealization of real mesoscale systems, results from such simplified linear models are useful for the following reasons:

- The exact solution of simplified linear differential equations gives some indication as to the physical mechanisms involved in specific atmospheric circulations. Because precise solutions are obtained, an investigator can be certain the results are not due to computational errors as can be true with numerical models.
- The use of results obtained with the linear equations can be contrasted with those obtained from a numerical model in which the magnitude of the nonlinear terms are small relative to the linear terms. An accurate nonlinear numerical model must be able to closely reproduce the linear results when the products of the dependent variables are small.

Linear representations of the conservation relations have been used to investigate expected wave motions in the atmosphere, as well as to represent actual mesoscale circulations. Kurihara (1976), for example, used a linear analysis to investigate spiral bands in a tropical cyclone, while Klemp and Lilly (1975) used such an approach to study wave dynamics in downslope wind storms to the lee of large mountain barriers.

To illustrate the first use of these criteria, Eqs. (1)–(3) and (7) can be rewritten using Eqs. (18) and (29) to decompose the dependent variables, and Eqs. (20) and (30) to average the resultant expressions. Assuming that

- the layer-domain-averaged fields (u_0, v_0, w_0, θ_0, p_0, ρ_0) are hydrostatic, horizontally homogeneous, and unchanging in time;
- the layer-domain-averaged velocity fields are identically zero ($u_0 \equiv v_0 \equiv w_0 \equiv 0$);
- all subgrid scale correlation terms are ignored;
- gradients in the y direction ($i = 2$) are neglected;
- the vertical gradients of θ_0 and ρ_0 are constant throughout the atmosphere;
- the Coriolis term is neglected in the vertical equation of motion;
- all motion is adiabatic ($\bar{S}_\theta = 0$);
- variations of density ρ', pressure p', and potential temperature θ' are assumed to be much less than the magnitude of the domain-averaged variables ρ_0, p_0, and θ_0;
- products of the dependent variables are removed (e.g., $u_j' \partial u_i'/\partial x_j$; $u_j' \partial \theta'/\partial x_j$);
- moisture and effects of other gaseous and aerosol atmospheric materials are ignored,

then these equations can be rewritten as

$$\rho_0 \frac{\partial u'}{\partial t} = -\frac{\partial p'}{\partial x} + \rho_0 f v' \tag{41}$$

$$\frac{\partial v'}{\partial t} = -f u' \tag{42}$$

$$\lambda_1 \rho_0 \frac{\partial w'}{\partial t} = -\frac{\partial p'}{\partial z} - \rho' g \tag{43}$$

$$\frac{\partial \theta'}{\partial t} = -w' \frac{\partial \theta_0}{\partial z} \tag{44}$$

$$\lambda_2 \frac{\partial \rho'}{\partial t} = -\rho_0 \left[\frac{\partial u'}{\partial x} + \frac{\partial w'}{\partial z} \right] - w' \frac{\partial \rho_0}{\partial z} \tag{45}$$

$$\rho' = \rho_0 \frac{C_v}{C_p} \frac{p'}{p_0} - \rho_0 \frac{\theta'}{\theta_0} \tag{46}$$

Note that (41) and (42), (43), (44), and (45) have the same form as (31), (33), (22), and (21) (left-hand expression); if $\lambda_2 = 0$ the assumptions listed above are used; and (21) is not used to write $-\overline{\rho u_j} \frac{\partial \bar{\theta}}{\partial x_j} = -\frac{\partial}{\partial x_j} \overline{\rho u_j \theta}$ in (22). Equation (46) is the linearized form of the ideal gas law obtained by substituting for T_v in Eq. (7) using the definition of potential temperature (6), logarithmically differentiating, and then assuming that the perturbations in density, pressure, and potential temperature are small (e.g., $dp \sim$

$p' \ll p_0$; $d\theta \sim \theta' \ll \theta_0$; $d\rho \sim \rho' \ll \rho_0$). The parameters λ_1 and λ_2 are introduced to identify the contributions of compressibility and vertical accelerations in the linearized solutions ($\lambda_1 = 0$ and $\lambda_2 = 0$ correspond to the hydrostatic and anelastic assumptions, respectively; λ_1 and λ_2 equal unity, otherwise).

Equations (41)–(46) are six simultaneous linear algebraic and differential equations in the six unknowns u', v', w', p', θ', and ρ'. Using a Fourier transformation (e.g., see Richtmyer and Morton, 1967; Churchill, 1963) given by

(47) $\quad \phi'(x, z, t) = \tilde{\phi}(k_x, k_z, \omega) \exp(i(k_x x + k_z z + \omega t))$

where ϕ' is any one of the dependent variables, Eqs. (41)–(46), after rearranging, can be written as the six linear homogeneous algebraic relations

(48) $\quad \rho_0 i\omega \tilde{u} + ik_x \tilde{p} - \rho_0 f \tilde{v} = 0$

(49) $\quad i\omega \tilde{v} + f\tilde{u} = 0$

(50) $\quad \lambda_1 \rho_0 i\omega \tilde{w} + ik_z \tilde{p} + g\tilde{\rho} = 0$

(51) $\quad i\omega\tilde{\theta} + \tilde{w}\dfrac{\partial \theta_0}{\partial z} = 0$

(52) $\quad \lambda_2 i\omega\tilde{\rho} + \rho_0 ik_x \tilde{u} + \rho_0 ik_z \tilde{w} + \tilde{w}\dfrac{\partial \rho_0}{\partial z} = 0$

(53) $\quad \tilde{\rho} + \dfrac{\rho_0}{\theta_0}\tilde{\theta} - \dfrac{\rho_0}{p_0}\dfrac{C_v}{C_p}\tilde{p} = 0$

In the transformation $\tilde{\phi}(k_x, k_z, \omega)$ is the amplitude as a function of horizontal and vertical wavenumbers k_x and k_z and frequency ω. $\tilde{\phi}$ is a complex number in general. In the nonlinear equivalent to Eqs. (41)–(46) the Fourier transformation of $\phi'(x, z, t)$ would involve the integral of $\tilde{\phi}(k_x, k_z, \omega) \exp(i(k_x + k_z + \omega t))$ over all possible wavenumbers. Because Eqs. (41)–(46) are linear, however, any single term (also called a *harmonic*) in the Fourier transformation is a solution. Any linear combination is also a solution with the complete representation given by adding together all of the harmonics. In a nonlinear system, on the other hand, products of integrals arise (e.g., from $u'_j \partial u_i / \partial x_j$, $u'_j \partial \theta' / \partial x_j$) reflecting the interactions between different scales of motion. No such interaction is possible, however, with a linear system.

The use of complex variables in the Fourier representation (47) is a convenient mathematical procedure to represent periodic waves. The exponential term in (47), for example, can be written as

$$\exp(i(k_x x + k_z z + \omega t)) = \cos(k_x x + k_z z + \omega t) + i \sin(k_x x + k_z z + \omega t)$$

which corresponds to a unit vector with components $\cos(k_x x + k_z z + \omega t)$ on the real axis and $\sin(k_x x + k_z z + \omega t)$ on the imaginary axis. The wavenumbers k_x and k_z and frequency ω can also be expressed as complex numbers. As an example, if $\omega = \omega_R + i\omega_i$, where ω_R and ω_i are real, then

$$e^{i\omega t} = e^{-\omega_i t} e^{i\omega_R t} = e^{-\omega_i t}(\cos \omega_R t + i \sin \omega_R t)$$

where $e^{-\omega_i t}$ indicates whether $\phi'(x, z, t)$ in Eq. (47) damps ($\omega_i > 0$) or amplifies ($\omega_i < 0$) with time, while the term $\cos \omega_R t + i \sin \omega_R t$ represents changes in $\phi'(x, z, t)$ due to propagation. A similar decomposition can be applied to wavenumbers k_x and k_z. In the solution of equations using complex representations of the dependent variables, only the real part of the right-hand side of Eq. (47) gives information concerning the magnitude of $\phi'(x, z, t)$.

Complex variables are used extensively in linear models, and in the analysis of approximate solution techniques to the nonlinear form of the conservation relations, as is shown in Section 10. Texts concerned with functions of complex variables (e.g., Hildebrand, 1962, Chapter 10) provide detailed discussions concerning this subject.

Equations (48)–(53) can be solved by writing them in matrix form:

$$\begin{bmatrix} \rho_0 i\omega & -f\rho_0 & 0 & 0 & 0 & ik_x \\ f & i\omega & 0 & 0 & 0 & 0 \\ 0 & 0 & \lambda_1 \rho_0 i\omega & 0 & g & ik_z \\ 0 & 0 & \partial\theta_0/\partial z & i\omega & 0 & 0 \\ \rho_0 ik_x & 0 & \rho_0 ik_z + \partial\rho_0/\partial z & 0 & \lambda_2 i\omega & 0 \\ 0 & 0 & 0 & \rho_0/\theta_0 & 1 & -\rho_0 C_v/p_0 C_p \end{bmatrix} \begin{bmatrix} \tilde{u} \\ \tilde{v} \\ \tilde{w} \\ \tilde{\theta} \\ \tilde{\rho} \\ \tilde{p} \end{bmatrix} = \begin{bmatrix} 0 \\ 0 \\ 0 \\ 0 \\ 0 \\ 0 \end{bmatrix}$$

This system of equations is homogeneous because no terms arise that are not coefficients of one of the dependent variables. Since it is homogeneous, as shown, for example, by Murdoch (1957), a nontrivial solution exists only if the determinant of the coefficients is zero. Expanding this large determinant in order to obtain frequency ω in terms of wavenumbers k_x and k_z is quite tedious (see Murdoch, 1957, or other texts in linear and matrix algebra for the procedure), but it eventually results in the equation

$$\left(\frac{\rho_0^2}{p_0} i \frac{C_v}{C_p} \lambda_2 \lambda_1\right) \omega^4 + \left\{-k_x^2 \lambda_1 \rho_0 i + \frac{\rho_0}{p_0} i \frac{C_v}{C_p} g \left(\rho_0 i k_z + \frac{\partial \rho_0}{\partial z}\right)\right. $$
$$\left. - k_z \left(\rho_0 i k_z + \frac{\partial \rho_0}{\partial z}\right) - f^2 \frac{\rho_0^2}{p_0} i \frac{C_v}{C_p} \lambda_2 \lambda_1 - \frac{\rho_0}{\theta_0} \frac{\partial \theta_0}{\partial z} \lambda_2 k_z\right\} \omega^2$$

$$+ f^2 \frac{\rho_0}{\theta_0} \frac{\partial \theta_0}{\partial z} \lambda_2 k_z + f^2 k_z \left(\rho_0 i k_z + \frac{\partial \rho_0}{\partial z} \right)$$

$$- f^2 \frac{\rho_0}{\rho_0} i \frac{C_v}{C_p} g \left(\rho_0 i k_z + \frac{\partial \rho_0}{\partial z} \right)$$

(54) $$+ g k_x^2 \frac{\rho_0}{\theta_0} i \frac{\partial \theta_0}{\partial z} = 0$$

Ogura and Charney (1961) obtained a similar fourth-order algebraic equation in frequency in the analysis of a similar set of the conservation relations. Despite the simplifying assumptions and linearization used to obtain Eqs. (41)–(46), Eq. (54) is not simple to evaluate. Nonetheless, it is useful to make further simplifying assumptions in order to reduce (54) to forms that can be interpreted in terms of atmospheric wave motions.

If no variations of the dependent variables in the x direction are permitted ($k_x = 0$), the Earth's rotation is neglected, no amplification or decay of the waves in the vertical or in time are permitted (k_z and ω are real variables), $\lambda_1 = \lambda_2 = 1$, and $(g/\rho_0)\,\partial\rho_0/\partial z$ is neglected relative to $(p_0/\rho_0)(C_p/C_v)k_z^2$ (as can be shown to be reasonable using scale analysis), then Eq. (54) reduces to

(55) $$\frac{\lambda_2 \lambda_1 \omega^2}{k_z^2} = \frac{\omega^2}{k_z^2} = C_{a_z}^2 = \frac{p_0}{\rho_0} \frac{C_p}{C_v} = RT_0 \frac{C_p}{C_v} \cong 402 T_0$$

(SI units—see Wallace and Hobbs, 1977, pp. xv–xvii) where $C_{a_z} = -\omega/k_z$ represents the speed of a vertically propagating sound (*acoustic*) wave. When either $\lambda_1 = 0$ (the hydrostatic assumption is applied) or $\lambda_2 = 0$ (the local tendency of density is neglected), vertically propagating sound waves do not appear as a solution to Eq. (54). For this reason Ogura and Phillips (1962) referred to the conservation of mass relation given by Eq. (9) as the anelastic assumption. For reasonable values of temperature ($T_0 \cong 300°K$) found in the Earth's lower troposphere, $|C_{a_z}| \cong 350$ m sec^{-1}.

Another type of wave motion can be isolated from Eq. (54) by requiring the flow to be anelastic ($\lambda_2 = 0$), neglecting terms associated with the rotation of the Earth, permitting amplification or decay in the vertical (k_z is complex) but not in the horizontal or in time (k_x and ω are real), and ignoring the terms $\frac{1}{4}(\rho_0^3/p_0^2)(C_v/C_p)^2 g^2$; $(1/4\rho_0)(\partial\rho_0/\partial z)^2$; and $\frac{1}{4}(\rho_0/p_0)(C_v/C_p)g\,\partial\rho_0/\partial z$ compared with $\rho_0(\lambda_1 k_x^2 + k_z^2)$, which arise in the solution for the frequency ω. This last criterion, using scale analysis as an estimate, results in an error of about 10% if the wavelength is around 5 km, while the terms are about equal for wavelengths of 50 km. Neglecting these terms simplifies the qualitative interpretation of the resultant equation, however, they should be retained in a quantitative analysis of this type of wave motion.

With these assumptions the solution to Eq. (54) is

$$\omega^2 = \frac{k_x^2}{\lambda_1 k_x^2 + k_{z_R}^2} \frac{g}{\theta_0} \frac{\partial \theta_0}{\partial z} \tag{56}$$

where the form of the dependent variables is given by

$$\phi(x, z, t) = \tilde{\phi}(k_x, k_z, \omega) \exp(-\beta z) \exp(i(k_x x + k_{z_R} z + \omega t))$$

with

$$\beta = \frac{\rho_0}{2 p_0} \frac{C_v}{C_p} g + \frac{1}{2\rho_0} \frac{\partial \rho_0}{\partial z}$$

The exponential term in βz occurs in this expression because k_z is permitted to be complex. The variable k_{z_R} is the real component of k_z. The phase speed, obtained from Eq. (56), is

$$C_{gc} = \pm \frac{k_x}{(k_x^2 + k_{z_R}^2)} \left(\frac{g}{\theta_0} \frac{\partial \theta_0}{\partial z} \right)^{1/2} \tag{57}$$

where C_{gc} is the speed of an *internal gravity wave* in a *continuously and uniformly stratified* fluid, while $[(g/\theta_0) \partial \theta_0/\partial z]^{1/2}$ is called the Brunt–Väisälä frequency. Using representative values of the parameters in Eq. (57) ($\theta_0 = 300°K$, $\partial \theta_0/\partial z = 1°C/100$ m, $g = 9.8$ m sec^{-2}), we get $|C_{gc}| \cong 4$ m sec^{-1} for $k_{z_R} = 2\pi/10^3$ m and $k_x = 2\pi/(7 \times 10^3$ m); and $|C_{gc}| \cong 29$ m sec^{-1} for $k_{z_R} = 2\pi/(7 \times 10^3$ m) and $k_x = 2\pi/10^3$ m. Since

$$C_{gc_x} = \frac{-\omega k_x}{k_x^2 + k_{z_R}^2}; \quad C_{gc_z} = \frac{-\omega k_{z_R}}{k_x^2 + k_{z_R}^2}$$

the direction of propagation is mostly vertical when $k_x^2 \ll k_{z_R}^2$ and mostly horizontal when $k_x^2 \gg k_{z_R}^2$.

If the same assumptions used to derive Eq. (57) are used in (54) except that motions are assumed hydrostatic ($\lambda_1 = 0$), the expression equivalent to Eq. (56) is written as

$$\omega^2 = \frac{k_x^2}{k_{z_R}^2} \frac{g}{\theta_0} \frac{\partial \theta_0}{\partial z} \tag{58}$$

Comparing (56) and (58), an alternative justification for the use of the hydrostatic assumption [Eq. (17)] when internal gravity waves in a continuous fluid are the dominant atmospheric circulation is that $k_x^2 \ll k_{z_R}^2$, or, since $k_x = 2\pi/L_x$ and $k_{z_R} = 2\pi/L_z$,

$$\boxed{L_x^2 \gg L_z^2} \tag{59}$$

is required. When this condition applies, (56) and (58) are the same without requiring that $\lambda_1 = 0$.

From Eq. (56) it is evident that a nonzero value of $\partial\theta_0/\partial z$ is needed to have an internal gravity wave. Thus, to isolate a third type of wave motion from Eq. (54), the vertical gradient of potential temperature is set to zero, all terms which include gravity are neglected, terms with $\partial\rho_0/\partial z$ are ignored (if $\partial\theta_0/\partial z = 0$, the atmosphere is well-mixed, implying that $\partial\rho_0/\partial z \cong 0$ also), and the influence of the rotation of the Earth is neglected. With these assumptions, the solution to Eq. (54), with $C_a^2 = \omega^2/(k_x^2 + k_z^2)$, is

(60) $$\frac{\lambda_2 \lambda_1 \omega^2}{k_x^2 + k_z^2} = \frac{\omega^2}{k_x^2 + k_z^2} = C_a^2 = \frac{RT_0 C_p}{C_v} \cong 402 T_0 \quad \text{(SI units)}$$

The phase speed given by this expression is identical to Eq. (55) and represents a *sound wave* that can propagate in the horizontal as well as vertical directions. As with Eq. (55), either the hydrostatic ($\lambda_1 = 0$) or anelastic ($\lambda_2 = 0$) assumption eliminates this wave form from the solutions.

Finally, a fourth wave form can be derived from Eq. (54) assuming that the lapse rate is adiabatic ($\partial\theta_0/\partial z = 0$), and either the hydrostatic assumption is used ($\lambda_1 = 0$) or no variations are permitted in the x direction. In this case (54) reduces to the simple frequency equation

$$\omega^2 = f^2$$

This oscillatory notation is referred to as an *inertial wave*.

Using the results of this analysis thus far, it is evident that the anelastic form of the conservation of mass relation given by (21) is sufficient to remove sound waves as a possible solution. The addition of the hydrostatic assumption, therefore, does not influence sound waves; however, internal gravity wave propagation, as defined by Eq. (57), will be affected if the horizontal wavelength of the wave is not much larger than its vertical wavelength.

If the smallest horizontal feature that can be resolved in a numerical model is $4\Delta x$, as used to obtain Eq. (37), and $H_\rho \cong 8$ km is used to estimate the largest expected vertical wavelength, then $\Delta x \gtrsim 6$ km is required to assure that the hydrostatic assumption is valid within about a 10% error for all internal waves formed in a continuously stratified medium. If the predominant horizontal wavelength of such waves is assumed to be determined by bottom surface variations (e.g., a mountain), then such forcings must have a horizontal scale of 25 km or more. This criterion for use of the hydrostatic assumption is more restrictive than that given by Eq. (37), but it must be emphasized that it only applies in mesoscale models in which internal gravity waves, propagating in a continuously stratified medium, are an important part of the physical solution.

A second type of internal gravity wave can occur in a mesoscale model—but which is not included as a possible solution to Eqs. (48)–(53)—when $\partial\theta_0/\partial z$ and $\partial\rho_0/\partial z$ are discontinuously stratified. As shown by Holton (1972) and others, if a less dense fluid overlies a more dense fluid, an internal gravity wave forms on the discontinuous interface between the two. For an idealized two-layer fluid such an internal gravity wave can be shown to have a phase speed of

$$C_{gd} = \pm ((\Delta\rho/\rho_0)gh_0)^{1/2} \tag{61}$$

where $\Delta\rho$ is the difference in density between the two fluids, ρ_0 is the density of the lower fluid, and h_0 is the height of the interface. With $\Delta\rho/\rho_0 = 0.1$ and $h_0 = 0.98$ km, for example, $C_{gd} \cong 31$ m sec^{-1}, which is close to the speed of wave motion observed on frontal interfaces in the atmosphere when cold, dense surface air is being overrun by warmer, less dense air aloft (see, e.g., Gedzelman and Donn, 1979).

Finally, when two or more waves are present and traveling with different propagation speeds, it is necessary to consider their *group velocity*. This is defined as the speed of movement of the region where the waves *constructively reinforce* one another and can be shown (see, e.g., Dutton, 1976) to be defined as

$$C_{\text{group}} = -\partial\omega/\partial k_i$$

where $\omega = \omega(k_x, k_y, k_z) = \omega(k_i)$, in general. As shown, for example, by Dutton (1976), the group velocity of internal gravity waves in a continuously stratified medium propagates at a 90° angle to the phase velocity of individual wave components. Thus, in a hydrostatic model the group velocity is almost horizontal, while the phase velocity is predominantly vertical.

In this section it was shown that the anelastic assumption eliminates sound waves as a possible solution, while the hydrostatic relation erroneously influences internal gravity wave propagation in a continuously stratified fluid if the ratio of the vertical to horizontal wavelength becomes too large. In addition, the rotation of the Earth introduces oscillatory motion that is a function of latitude.

All of the wave forms, of course, were derived after linearizing the conservation relations, as well as making additional simplifying assumptions in order to isolate idealized wave forms. In the atmosphere such limitations on the modes of interaction are not present, hence it is seldom possible to observe the idealized wave forms as have been described here. Moreover, even if the response of the atmosphere was linear, velocity shear and multiple temperature inversions, for example, will produce different phase and group velocities of internal gravity waves than are obtained from Eq. (54).

Such idealized linear models as given by Eqs. (41)–(46) are used only to ascertain types of possible wave motions in a mesoscale model, as well as the effect on these motions if simplified forms of the conservation relations are used. In the study of specific mesoscale systems, therefore, it is necessary to utilize somewhat different linear models. Linear models, of course, may be an accurate tool to represent real mesoscale circulations when the linear terms dominate over the nonlinear expressions in the conservation equations.

Defant (1950) presents a linearized model of the sea and land breezes. His model is used here to illustrate that linear models of this mesoscale phenomena reproduce some aspects of actual circulations, but fail to reproduce other important facets. To outline Defant's model (an analysis of which has been performed by Martin, 1981), Eqs. (31), (33), (22), and (21) (right-hand expression) can be written in the form

$$\frac{\partial u'}{\partial t} = -\frac{1}{\rho_0}\frac{\partial p'}{\partial x} + fv' - \sigma u' \tag{62}$$

$$\frac{\partial v'}{\partial t} = -fu' - \sigma v' \tag{63}$$

$$\frac{\partial w'}{\partial t} = -\sigma w' - \frac{1}{\rho_0}\frac{\partial p'}{\partial z} + \gamma \theta' \tag{64}$$

$$\frac{\partial u'}{\partial x} + \frac{\partial w'}{\partial z} = 0 \tag{65}$$

$$\frac{\partial \theta'}{\partial t} + w'\beta = K\left[\frac{\partial^2 \theta'}{\partial z^2} + \frac{\partial^2 \theta'}{\partial x^2}\right] \tag{66}$$

where $\beta = \partial \theta_0/\partial z$ and $\gamma = g/\theta_0$. In these expressions $u_0 \equiv v_0 \equiv w_0 \equiv 0$. θ_0 is a linear function of z; $\sigma u'$, $\sigma v'$, $\sigma w'$, and $K[\partial^2\theta'/\partial z^2 + \partial^2\theta'/\partial x^2]$ (where σ and K are constants) are introduced to represent the subgrid scale terms $(1/\bar{\rho})(\partial/\partial z)\overline{\rho u''w''}$, $(1/\bar{\rho})(\partial/\partial z)\overline{\rho v''w''}$, $(1/\bar{\rho})(\partial/\partial z)\overline{\rho w''^2}$, and $(1/\bar{\rho})\{(\partial/\partial z)\overline{\rho w''\theta''} + (\partial/\partial x)\overline{\rho u''\theta''}\}$; and the $\hat{f}$ Coriolis term is neglected. Except for the subgrid scale correlation terms, Eqs. (62)–(66) are the same as (41)–(46) if $p' = 0$ in (46), $\lambda_1 = 1$, $\lambda_2 = 0$, and $\partial \rho_0/\partial z = 0$. To solve Eqs. (62)–(66), Defant recognized that u' and v' must be 90° out of phase with w', p', and θ' since the first two dependent variables are expressed in terms of derivatives of the others. Moreover, the solutions should be a function of height above the ground surface, rather than simply a periodic function, since the sea and land breezes do not extend upward indefinitely.

For this reason, Defant assumed solutions, different than used to obtain Eqs. (48)–(53), as given by

$$w' = w(z)e^{i\omega t} \sin k_x x$$
$$p' = p(z)e^{i\omega t} \sin k_x x$$
$$\theta' = \theta(z)e^{i\omega t} \sin k_x x$$
$$u' = u(z)e^{i\omega t} \cos k_x x$$
$$v' = v(z)e^{i\omega t} \cos k_x x$$

These representations are in a form such that the circulations are periodic in time and in the x direction with a horizontal wavelength of $l_x = 2\pi/k_x$. $w(z)$, $p(z)$, $\theta(z)$, $u(z)$ and $v(z)$ are complex variables. Substitution of these expressions into Eqs. (62)–(66) [with $w(z) = w$, $p(z) = p$, etc.] yields

(67) $$i\omega u = -\frac{k_x}{\rho_0} p + fv - \sigma u$$

(68) $$i\omega v = -fu - \sigma v$$

(69) $$i\omega w = -\sigma w - \frac{1}{\rho_0} \frac{\partial p}{\partial z} + \gamma \theta$$

(70) $$-k_x u + \frac{\partial w}{\partial z} = 0$$

(71) $$i\omega \theta = -w\beta - K\theta k_x^2 + K\frac{\partial^2 \theta}{\partial z^2}$$

These equations are solved simultaneously with the boundary conditions

$$w(z = 0) = 0, \quad \theta(z = 0) = Me^{i\omega t} \sin k_x x$$
$$w(z \to \infty) = 0, \quad \theta(z \to \infty) = 0$$

where M is the amplitude of the maximum perturbation surface potential temperature.

The solution of these equations is tedious and is reported by Martin (1981). The final analytic result in terms of the original dependent variables u', v', w', p', and θ' is given by

(72) $$u' = -\frac{r}{(b^2 - a^2)k_x} M[ae^{az} + be^{-bz}]e^{i\omega t} \cos k_x x$$

(73) $$v' = -\frac{f}{i\omega + \sigma} u'$$

(74) $$w' = -\frac{r}{(b^2 - a^2)} M[e^{az} - e^{-bz}]e^{i\omega t} \sin k_x x$$

(75) $$p' = -\frac{\rho_0}{k_x} \frac{(i\omega + \sigma)^2 + f^2}{in + \sigma} u'$$

(76) $$\theta' = M[e^{-bz} + \frac{b^2 - s}{b^2 - a^2}(e^{az} - e^{-bz})]e^{i\omega t} \sin k_x x$$

where

$$a^2 = \frac{s + \eta^2}{2} - \frac{1}{2}((s + \eta^2)^2 - 4(s\eta^2 - r\epsilon))^{1/2}$$

$$b^2 = \frac{s + \eta^2}{2} + \frac{1}{2}((s + \eta^2)^2 - 4(s\eta^2 - r\epsilon))^{1/2}$$

$$a = -\sqrt{a^2}; \quad b = \sqrt{b^2}$$

$$\eta^2 = \frac{k_x^2(i\omega + \sigma)^2}{(i\omega + \sigma)^2 + f^2}; \quad r = \frac{\gamma k_x^2(i\omega + \sigma)}{(i\omega + \sigma)^2 + f^2}$$

$$\epsilon = \frac{\beta}{K}; \quad s = \frac{i\omega}{K} + k_x^2$$

It would be desirable to express Eqs. (72)–(76) in terms of real and imaginary components since only the real part has physical significance; however, Defant did not because the resultant expressions are quite complicated in form. It is much more convenient to solve these equations on a computer using complex arithmetic and then display the resultant real part of the solutions. Figure 1, reproduced from Defant, shows results for the vertical profile of vertical velocity with various values of σ with a value of $M = 1°C$ and $e^{i\omega t} \sin k_x x = 1$. As evident in Fig. 1 and as expected intuitively, the larger the friction (as represented by σ) the weaker the circulation for a particular value of temperature perturbation. In addition, the solution appears as a periodic function that damps with height, as required by the upper boundary condition. The direction of the wind reverses by 180° between the upper and lower levels of the circulation (about 400 m using the input to create Fig. 1) with the upper portion having a greater vertical depth.

Such solutions provide insight into the physical mechanisms that generate and influence the strength of the sea and land breezes. Unfortunately, in Defant's model physical interactions such as the following are inappropriately represented, or not even included:

- The subgrid-scale parameterizations for σ and K are assumed independent of time and space so that the intensity of the land breeze is equal to that of the sea breeze. As vertical mixing is known to be reduced at night over land, however, the land breeze is usually observed to be shallower and weaker than the sea breeze (e.g., Mahrer and Pielke, 1977b). Moreover, realistic parameterizations of the subgrid-scale mixing are nonlinear functions of the dependent variables as discussed in Section 7.

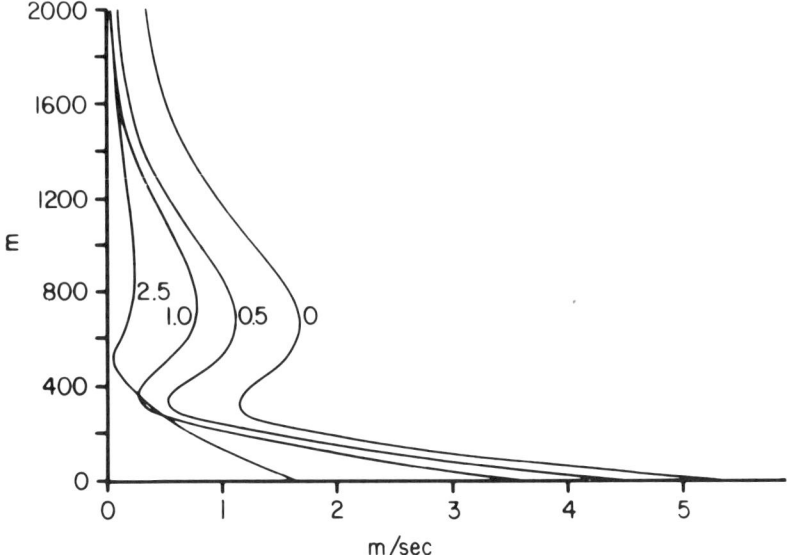

Fig. 1. The maximum amplitude of the sea-breeze wind speed normal to the coast as a function of height for values of $\sigma = 0$, 0.5×10^{-4}, 1.0×10^{-4}, and 2.5×10^{-4}. The other parameters used have values of $\eta = 7.273 \times 10^{-5}$ sec^{-1} = 2π/day; $\beta = 5°$C/km; $\alpha = 1/273.2$; $K = 22.5$ m^2 sec^{-1}; $f = 1.031 \times 10^{-4}$ sec^{-1}; $l = 2\pi/120$ km. (From Defant, 1950.)

- Advection of temperature and velocity are ignored. Even if the large-scale prevailing flow is zero, the marine air is known to move the region of maximum upward motion inland when a sea breeze occurs (e.g., Estoque, 1961).
- The vertical profile of the large-scale potential temperature is assumed linear. In general, such a condition does not exist and subgrid-scale mixing causes changes in potential temperature due to curvature in the large scale as well as in the perturbation field.
- The surface temperature perturbation is prescribed, whereas, in reality, it is a function of the mesoscale circulation (e.g., Physik, 1976).
- No interactions are permitted among the dependent variables. Although a necessary condition to obtain analytic results, if $u' \, \partial u'/\partial x$, for example, attains a magnitude on the order of $(1/\rho_0) \partial p'/\partial x$, nonlinear effects need to be considered.

While some of these shortcomings were eliminated in later linear models [e.g., Smith (1957) included a linear advection term], it is still impossible to analytically solve the conservation equations when one or more of the terms involve products of dependent variables. Such nonlinear terms arise in the representation of the subgrid-scale processes, the source/sink terms as well as in the expression for advection. Therefore, in order to obtain more realistic simulations of sea and land breezes, and other types of mesoscale flows, investigators have resorted to numerical models that provide approximate solutions to the conservation equations. Much of the remainder of this article will concentrate on the formulation of and results from this tool.

6. Coordinate Representation

Up to now the conservation equations have been expressed in the rectangular coordinate system (x, y, z) in which each axis is perpendicular to the other two and always oriented in the same direction. In the application of the conservation relations in a mesoscale model, however, it is not always desirable to utilize this coordinate representation. In synoptic meteorology, for example, pressure is often substituted for height z as the vertical coordinate since it is the quantity measured by radiosondes. When a different coordinate form is used, however, the conservation relations must be unchanged despite the different mathematical representation. Therefore, in transforming the conservation relations from one coordinate system to another, the equations must be written so that the physical representation is *invariant* in either system. The mathematical procedure developed to preserve this invariance requires some knowledge of tensor analysis and the reader is referred to Dutton (1976) for an explanation of this mathematical tool. A more detailed explanation and derivation of the material that follows is presented by Pielke and Martin (1981).

In mesoscale meteorology (and in meteorology, in general), the relationship between the rectangular coordinate system and the transformed representation can be given as

(77)
$$\begin{aligned} \tilde{x}^1 &= x & x &= \tilde{x}^1 \\ \tilde{x}^2 &= y & y &= \tilde{x}^2 \\ \tilde{x}^3 &= z^*(x, y, z) & z &= z(\tilde{x}^1, \tilde{x}^2, \tilde{x}^3) \end{aligned}$$

where z^* is called the *generalized vertical coordinate*. Examples of z^* used are $z^* = \theta$, $z^* = p$, and

(78) $$z^* = s(z - z_G)/(s - z_G)$$

where z_G is the ground elevation and s is the initial top of the model. z_G is a function of x and y, while s is assumed constant. The first two coordinates are referred to as *isentropic* and *isobaric* representations, while Eq. (78), used frequently in mesoscale modeling (e.g., Blondin, 1978; Colton, 1976; Yamada, 1978) and smaller-scale models (e.g., Gal-Chen and Sommerville, 1975a; Clark, 1977), is a *terrain-following* coordinate, usually called a *sigma* representation. Phillips (1957), using pressure as the original independent vertical coordinate, originated the concept whereby the lowest coordinate surface was coincident with the ground. Recently Uccellini *et al.* (1979), Friend *et al.* (1977), Deaven (1974, 1976) and Bleck (1978) have utilized an isentropic system well above the ground and a type of sigma representation below. Kasahara (1974) discusses various types of vertical coordinates, including the sigma system.

In a mesoscale model, equations developed from (1)–(5) and equivalent to (21) (left-hand part), (32), (28), (22), (24), and (25) can be written in contravariant form using tensor transformation procedures [shown in detail by Pielke (1981)] as

(79) $$\frac{\partial}{\partial \tilde{x}^j} \bar{\rho} \frac{s - z_G}{s} \tilde{u}^j = 0$$

(80) $$\frac{\partial \bar{\tilde{u}}^1}{\partial t} = -\bar{\tilde{u}}^j \frac{\partial \bar{\tilde{u}}^1}{\partial \tilde{x}^j} - \overline{\tilde{u}^{j''} \frac{\partial \tilde{u}^{1''}}{\partial \tilde{x}^j}} - \bar{\theta} \frac{\partial \bar{\pi}}{\partial \tilde{x}^1} + g \frac{z^* - s}{s} \frac{\partial z_G}{\partial \tilde{x}^1} - \hat{f}\bar{\tilde{u}}^3 + f\bar{\tilde{u}}^2$$

(81) $$\frac{\partial \bar{\tilde{u}}^2}{\partial t} = -\bar{\tilde{u}}^j \frac{\partial \bar{\tilde{u}}^2}{\partial \tilde{x}^j} - \overline{\tilde{u}^{j''} \frac{\partial \tilde{u}^{2''}}{\partial \tilde{x}^j}} - \bar{\theta} \frac{\partial \bar{\pi}}{\partial \tilde{x}^2} + g \frac{z^* - s}{s} \frac{\partial z_G}{\partial \tilde{x}^2} - f\bar{\tilde{u}}^1$$

(82) $$\frac{\partial \bar{\pi}}{\partial \tilde{x}^3} = -\frac{g}{\bar{\theta}} \frac{s - z_G}{s}$$

(83) $$\frac{\partial \bar{\theta}}{\partial t} = -\bar{\tilde{u}}^j \frac{\partial \bar{\theta}}{\partial \tilde{x}^j} - \overline{\tilde{u}^{j''} \frac{\partial \theta''}{\partial \tilde{x}^j}} + \bar{S}_\theta$$

(84) $$\frac{\partial \bar{q}_n}{\partial t} = -\bar{\tilde{u}}^j \frac{\partial \bar{q}_n}{\partial \tilde{x}^j} - \overline{\tilde{u}^{j''} \frac{\partial q_n''}{\partial \tilde{x}^j}} + \bar{S}_{q_n} \qquad (n = 1, 2, 3)$$

(85) $$\frac{\partial \bar{\chi}_m}{\partial t} = -\bar{\tilde{u}}^j \frac{\partial \bar{\chi}_m}{\partial \tilde{x}^j} - \overline{\tilde{u}_j'' \frac{\partial \chi_m''}{\partial \tilde{x}^j}} + \bar{S}_{\chi_m} \qquad (m = 1, 2, 3, \ldots, M)$$

where $\bar{\tilde{u}}^j$ are the contravariant components of velocity in the transformed system with z^* of Eq. (78) as the vertical coordinate. In obtaining these equations, a number of assumptions and definitions have been made.

(1) The averaging operator defined by (20) is not the correct form because the coordinate system is no longer rectangular. In the transformed system the appropriate grid volume averaging operator, denoted by an overbar in Eqs. (79)–(85), is

$$\text{(86)} \quad \overline{(\quad)} = \int_t^{t+\Delta t} \int_{\tilde{x}^1}^{\tilde{x}^1+\Delta \tilde{x}^1} \int_{\tilde{x}^2}^{\tilde{x}^2+\Delta \tilde{x}^2} \int_{\tilde{x}^3}^{\tilde{x}^3+\Delta \tilde{x}^3}$$
$$\times (\quad) \, d\tilde{x}^3 \, d\tilde{x}^2 \, d\tilde{x}^1 \, dt / (\Delta t)(\Delta \tilde{x}^1)(\Delta \tilde{x}^2)(\Delta \tilde{x}^3)$$

The averaging volume defined by (86) does not, in general, represent the same volume as given by (20). In addition, in performing the averaging operator defined by (86) on the conservation equations, it must be assumed that changes of the Jacobian of the transformation

$$\text{(87)} \quad \frac{\partial x^j}{\partial \tilde{x}^i} = \begin{bmatrix} 1 & 0 & 0 \\ 0 & 1 & 0 \\ \dfrac{s - z^*}{s} \dfrac{\partial z_G}{\partial x} & \dfrac{s - z^*}{s} \dfrac{\partial z_G}{\partial y} & \dfrac{s - z_G}{s} \end{bmatrix}$$

and its derivatives over the volume $\Delta \tilde{x}^1 \, \Delta \tilde{x}^2 \, \Delta \tilde{x}^3$ are small. This latter requirement of the Jacobian indicates that the variations of the terrain across the grid volumes in the transformed coordinate system must be small compared with the averaged values.

(2) The hydrostatic equation given by Eq. (82) does *not* result directly from the transformation of the hydrostatic equation, (28).† To obtain this expression, it is necessary to transform the complete conservation of motion relation given by Eq. (27) and to determine if the vertical acceleration, subgrid-scale correlation, and Coriolis terms that appear in the transformed representation of the equation for $\bar{u}^3$ are much less than the pressure gradient and gravitational accelerations. To estimate the contribution of the individual terms in this equation, the characteristic scale of the circulation in the terrain-following system can be projected onto a horizontal surface. If the resultant horizontal scale in the rectangular system, L_x, is such that (16) [and (59), if appropriate] is satisfied, then the equation for $\bar{u}^3$ (i.e., vertical velocity) reduces to

$$\text{(88)} \quad -\bar{\theta} \left\{ \frac{\partial z^*}{\partial x} \frac{\partial \bar{\pi}}{\partial \tilde{x}^1} + \frac{\partial z^*}{\partial y} \frac{\partial \bar{\pi}}{\partial \tilde{x}^2} + \left[\left(\frac{\partial z^*}{\partial x} \right)^2 + \left(\frac{\partial z^*}{\partial y} \right)^2 \right. \right.$$
$$\left. \left. + \left(\frac{\partial z^*}{\partial z} \right)^2 \right] \frac{\partial \bar{\pi}}{\partial \tilde{x}^3} \right\} = \frac{\partial z^*}{\partial z} g$$

Only if $\partial z^*/\partial x^1$ and $\partial z^*/\partial x^2$ are much less than $\partial z^*/\partial z$, and $\partial \bar{\pi}/\partial \tilde{x}^3$ has an equal or greater magnitude than $\partial \bar{\pi}/\partial \tilde{x}^1$ and $\partial \bar{\pi}/\partial \tilde{x}^2$ does Eq. (88) reduce to (82). Since, from (87),

---

† Unless the hydrostatic relation is assumed exactly valid in the x-y-z coordinate system.

$$\frac{\partial z^*}{\partial x}\left(\frac{\partial z^*}{\partial z}\right)^{-1} = \frac{z^* - s}{s}\frac{\partial z_G}{\partial x}; \quad \frac{\partial z^*}{\partial y}\left(\frac{\partial z^*}{\partial z}\right)^{-1} = \frac{z^* - s}{s}\frac{\partial z_G}{\partial y},$$

a sufficient condition to obtain Eq. (82) is that the terrain slope be much less than 45°. K. P. Hoinka (personal communication, 1979) obtained the same conclusion. This requirement for the hydrostatic equation is similar to the argument used to justify hydrostatic conditions as suggested by Long (1954). As reported by Hovermale (1965), Long maintained that hydrostatic conditions are well approximated if the radius of curvature of a barrier is large compared to the depth of the fluid.

The proper evaluation of the hydrostatic equation and horizontal pressure gradient force in the terrain-following coordinate is not trivial since slight inconsistencies in the approximate representation between the expressions can cause substantial errors. This problem arises because a portion of the pressure gradient in the vertical direction is evaluated on the $\tilde{x}^1-\tilde{x}^2$ surface in the transformed system, causing the resultant acceleration in Eqs. (80) and (81) to be the small difference between two large terms. Janjić (1977) and Mahrer and Pielke (1977b) discuss consistent representations of the terms in particular terrain-following coordinate representations.

The individual contravariant components of velocity in Eqs. (79)–(85) can be expressed in terms of the rectangular components as

(89)
$$\begin{aligned}\bar{\tilde{u}}^1 &= \bar{u} \\ \bar{\tilde{u}}^2 &= \bar{v} \\ \bar{\tilde{u}}^3 &= \bar{u}\frac{z^* - s}{s - z_G}\frac{\partial x_G}{\partial x} + \bar{v}\frac{z^* - s}{s - z_G}\frac{\partial z_G}{\partial y} + \bar{w}\frac{s}{s - z_G}\end{aligned}$$

It must be emphasized that the equations in the transformed coordinate system are obtained such that the physical invariance of the conservation laws are retained. If assumptions are made [such as those used to obtain Eq. (82)], the limitations of the simplified form can be demonstrated. Using the chain rule to obtain Eqs. (79)–(85), as performed by Kasahara (1974) and Haltiner and Williams (1980), however, is inappropriate unless the hydrostatic relation is assumed exact in the x-y-z system because, although derivatives of scalars can be transformed between coordinate systems properly using the chain rule, derivatives of vectors such as velocity and higher-order tensors cannot. It is only because of the two assumptions listed above that the resultant equations are of the same form.

Finally, using Eq. (79), the first two terms on the right-hand side of Eq. (80) can be multiplied by $\bar{\rho}(s - z_G)/s$ and written as

(90) $\quad \bar{\rho}\dfrac{s - z_G}{s}\left[\overline{\bar{\tilde{u}}^j\dfrac{\partial \bar{\tilde{u}}^1}{\partial x^j}} + \overline{\tilde{u}^{j''}\dfrac{\partial \tilde{u}^{1''}}{\partial x^j}}\right]$

$$= \bar\rho\,\frac{s - z_G}{s}\,\overline{\left[(\bar u^j + \bar u^{j\prime\prime})\,\frac{\partial}{\partial \tilde x^j}\,(\bar u^1 + \bar u^{1\prime\prime})\right]}$$

$$= \frac{\partial}{\partial \tilde x^j}\,\bar\rho\,\frac{s - z_G}{s}\,\overline{(\bar u^j + \bar u^{j\prime\prime})(\bar u^1 + \bar u^{1\prime\prime})}$$

$$= \frac{\partial}{\partial \tilde x^j}\left[\bar\rho\,\frac{s - z_G}{s}\,\bar u^j \bar u^1\right] + \frac{\partial}{\partial \tilde x^j}\left[\bar\rho\,\frac{s - z_G}{s}\,\overline{\bar u^{j\prime\prime}\bar u^{1\prime\prime}}\right]$$

where $\overline{u^{j\prime\prime}} = \overline{u^{1\prime\prime}} = 0$, using the Reynold's assumption discussed in Section 4. Similar terms can be obtained for the first two terms on the right of Eqs. (81) and (83)–(85). If $|\partial z_G/\partial z| \gg |\partial z_G/\partial x|$ and $|\partial z_G/\partial y|$ then it is reasonable to assume that the subgrid-scale fluxes in the $\tilde x^3$ and z directions are almost equal so that, for $j = 3$, the last term in the brackets on the right of Eq. (90) can be written as

$$\overline{\bar\rho w^{\prime\prime} u^{\prime\prime}}^R = \bar\rho\,\frac{s - z_G}{s}\,\overline{\bar u^{3\prime\prime}\bar u^{1\prime\prime}}$$

or

$$\overline{\bar u^{3\prime\prime}\bar u^{1\prime\prime}} = \frac{s}{s - z_G}\,\overline{w^{\prime\prime} u^{\prime\prime}}^R$$

where the superscript "R" after the overbar is added to emphasize that the averaging operators are different in the rectangular and transformed systems.

Moreover, as discussed in the next section, if $\overline{w^{\prime\prime} u^{\prime\prime}}^R$ is assumed equal to the product of the negative of an exchange coefficient which is a function of height above the ground h and the averaged velocity profile $\partial \bar u^R/\partial z$, then

$$\overline{\bar u^{3\prime\prime}\bar u^{1\prime\prime}} = \frac{s}{s - z_G}\,\overline{w^{\prime\prime} u^{\prime\prime}}^R = -\frac{s}{s - z_G}\,K\,\frac{\partial \bar u}{\partial z}$$

Since $\bar{\bar u}^1 = \bar u$, from (89), $h = z^*(s - z_G)/s$ from (78), and $\partial/\partial z = (s/(s - z_G))\,\partial/\partial \tilde x^3$ from (87), then

$$\overline{\bar u^{3\prime\prime}\bar u^{1\prime\prime}} = -\left(\frac{s}{s - z_G}\right)^2 K\,\frac{\partial \bar{\bar u}^1}{\partial \tilde x^3}$$

The subgrid-scale flux term in Eq. (80) can, therefore, be represented as

(91)
$$\overline{\bar u^{3\prime\prime}\,\frac{\partial \bar u^{1\prime\prime}}{\partial \tilde x^3}} = \left(\frac{s}{s - z_G}\right)^2\,\frac{\partial}{\partial \tilde x^3}\,K\,\frac{\partial \bar{\bar u}^1}{\partial \tilde x^3}$$

where K is a function of $z^*(s - z_G)/s$.

The vertical subgrid-scale fluxes in Eqs. (81), (83)–(85) can be similarly represented (e.g.,

$$\overline{\tilde{u}^{3''}\frac{\partial \theta''}{\partial \tilde{x}^3}} = \left(\frac{s}{s-z_G}\right)^2 \frac{\partial}{\partial \tilde{x}^3} K_\theta \frac{\partial \overline{\theta}}{\partial \tilde{x}^3}$$

where K_θ is an exchange coefficient for potential temperature).

The subgrid-scale fluxes in the $\tilde{x}^1$ and $\tilde{x}^2$ directions could be written in a similar form, but since essentially nothing is known about their functional form on the mesoscale in the rectangular coordinate representation, no purpose is served by writing them in a functional form here. In Section 10 it is shown that subgrid-scale fluxes in the horizontal direction are included in mesoscale models for computational reasons only.

Equations (79)–(85), along with the assumption that $\overline{\rho} \cong \rho_0$ in (79), represent $8 + M$ equations in the $8 + M$ variables $\tilde{\tilde{u}}^j$, $\overline{\pi}$, $\overline{\theta}$, $\overline{q}_n$, and $\overline{\chi}_m$ (providing the subgrid-scale and source/sink terms can be parameterized in terms of the dependent variables). In applying these equations to mesoscale modeling it is essential that the simplifications and assumptions used in deriving them be considered in interpreting the results.

7. Planetary Boundary-Layer Parameterization

The representation of the planetary boundary layer in mesoscale models is primarily handled through the subgrid-scale correlation terms since the model grid resolution is too large to explicitly resolve the small-scale fluxes found in this layer. The treatment of the influence of the planetary boundary layer in numerical models can be grouped into two classes:

- those that treat it as a single layer (e.g., Mahrt, 1974; Deardorff, 1972);
- those that resolve it into a number of discrete levels.

In mesoscale models, the second approach is the most common. As shown by Anthes *et al.* (1980), for example, detailed boundary-layer resolution is essential for accurate solutions when differential heating along complex terrain and across land–water boundaries is being represented.

With the discrete level approach, the planetary boundary layer can be divided into three sections: the laminar sublayer, the surface layer, and the transition layer.

7.1. Laminar Sublayers

The laminar sublayer is the level near the ground ($z < z_0$) where transfers of the dependent variables are performed by molecular motions;

z_0 usually varies from as low as 0.0015 cm over calm water to several meters over a forest or city. Zilitinkevich (1970) and Deardorff (1974) suggest relating temperature and specific humidity at the top of this layer, $\bar{\theta}_{z_0}$ and $\bar{q}_{z_0}$, to the surface values of these variables, $\bar{\theta}_G$, and $\bar{q}_G$, using expressions of the form

$$\bar{\theta}_{z_0} = \theta_G + 0.0962(\theta_*/k)(u_* z_0/\nu)^{0.45}$$
$$\bar{q}_{z_0} = q_G + 0.0962(q_*/k)(u_* z_0/\nu)^{0.45}$$
$$\bar{\chi}_{z_0} = \chi_G + 0.0962(\chi_*/k)(u_* z_0)/\nu)^{0.45}$$

where ν is the kinematic viscosity of air ($\sim 1.5 \times 10^{-5}$ m^2 sec^{-1}), k is von Karman's constant ($k \sim 0.35$) with θ_*, u_*, q_*, and χ_* defined according to Eqs. (92)–(95). Between $z = z_0$ and $z = z_G$, $\bar{u} = \bar{v} = \bar{w} = 0$, while variations of $\bar{p}$ and $\bar{\pi}$ across this depth are ignored. The height z_0 is called the *roughness height* and is a function of the aerodynamic roughness of the ground surface.

7.2. The Surface Layer

The surface layer is from z_0 to h_s, with h_s usually varying from about 10 to 100 m. In this layer the subgrid-scale fluxes are represented by values that are assumed to be independent of height, and the veering of the wind with height due to the Coriolis effect is neglected. With the assumption that the conditions in this layer are *steady* and *horizontally homogeneous*, investigators (e.g., Yamamoto, 1959; Yamamoto and Shimanuki, 1966; Shimanuki, 1969) have developed formulas that can be used to specify the relationship between the dependent variables and the subgrid-scale fluxes. Only a limited number of studies have been made over nonhomogeneous terrain (e.g., Taylor, 1977a,b; Taylor and Gent, 1981; Peterson, 1969) and this work has not yet been applied to mesoscale models.

One of the most common forms of horizontally homogeneous steady-state surface boundary-layer theory used in mesoscale models is that reported by Businger *et al.* (1971) and Businger (1973), in which

(92) $\quad u_* = k(\bar{u}^2 + \bar{v}^2)^{1/2}/[\ln z/z_0 - \psi_1(z/L)]$

(93) $\quad \theta_* = k(\bar{\theta}(z) - \theta_{z_0})/0.74[\ln z/z_0 - \psi_2(z/L)]$

(94) $\quad q_* = k(\bar{q}_3(z) - q_{z_0})/0.74[\ln z/z_0 - \psi_2(z/L)]$

(95) $\quad \chi_{*_m} = k(\bar{\chi}_m(z) - \chi_{z_0 m})/0.74[\ln z/z_0 - \psi_2(z/L)]$

where

(96) $\psi_1(z/L) = \int_{z_0/L}^{z/L} \frac{(1-\phi_1)}{\xi/L} d(\xi/L)$

$\cong \begin{cases} 2\ln[1 + \phi_1^{-1}/2] + \ln[(1 + \phi_1^{-2})/2] \\ \quad - 2\tan^{-1}\phi_1^{-1} + \pi/2 & \text{for } z/L \leq 0 \\ -4.7z/L & \text{for } z/L > 0 \end{cases}$

(97) $\psi_2(z/L) = \int_{z_0/L}^{z/L} \frac{(1-\phi_2)}{\xi/L} d(\xi/L)$

$\cong \begin{cases} 2\ln[(1 + 0.74\phi_2^{-1}/2] & \text{for } z/L \leq 0 \\ -6.35z/L & \text{for } z/L > 0 \end{cases}$

(98) $\phi_1 = \frac{kz}{u_*} \frac{\partial(\bar{u}^2 + \bar{v}^2)^{1/2}}{\partial z}$

$\cong \begin{cases} (1 - 15z/L)^{-1/4} & \text{for } z/L \leq 0 \\ 1 + 4.7z/L & \text{for } z/L > 0 \end{cases}$

(99) $\phi_2 = \frac{kz}{\theta_*} \frac{\partial \bar{\theta}}{\partial z} = \frac{\partial \bar{q}_3}{\partial z} = \frac{kz}{q_*} \frac{\partial \bar{q}_3}{\partial z} = \frac{kz}{\chi_{*_m}} \frac{\partial \bar{\chi}_m}{\partial z}$

$\cong \begin{cases} 0.74(1 - 9z/L)^{-1/2} & \text{for } z/L \leq 0 \\ 0.74 + 4.7z/L & \text{for } z/L > 0 \end{cases}$

and $L = \bar{\theta} u_*^2 / kg\theta_*$ (often called the Monin–Obukov stability length).

The empirical formulas on the right of Eqs. (96) and (97) were derived from observational data with the assumptions that $z_0/L \ll 1$, and that heat, water vapor, and other gaseous and aerosol atmospheric materials mix differently than velocity (it is expected that velocity changes can occur due to pressure effects in addition to those caused by the vertical transfer of air).

The parameters u_* and θ_* are called the *friction velocity* and *flux temperature*, respectively, while q_* and χ_{*_m} are similar variables except they have not been assigned labels. These parameters are related to the subgrid-scale fluxes [such as appear in Eqs. (22), (24), (25), and (32), for example] in the surface layer by

(100) $\overline{w''\theta''} = -u_*\theta_* = -K_\theta \frac{\partial \bar{\theta}}{\partial z}; \qquad \overline{w''u''} = -u_*^2 \cos\tau = -K_m \frac{\partial \bar{u}}{\partial z};$

$\overline{w''q_3''} = -u_*q_* = -K_\theta \frac{\partial \bar{q}_3}{\partial z}; \qquad \overline{w''v''} = -u_*^2 \sin\tau = -K_m \frac{\partial \bar{v}}{\partial z};$

$\overline{w''\chi_m''} = -u_*\chi_{*_m} = -K_\theta \frac{\partial \bar{\chi}_m}{\partial z}; \qquad \arctan\left(\frac{\bar{v}}{\bar{u}}\right) = \tau$

where K_m is called the *exchange coefficient for momentum* and K_θ is the *exchange coefficient for heat and other scalar quantities*.

The subgrid-scale flux of other gaseous and atmospheric materials in the surface layer can also be written as

$$\overline{w''\chi''_m} = -v_s\bar{\chi}_{z0m}$$

where v_s is called the *deposition velocity* and $\bar{\chi}_{z0m}$ is the mixing ratio of the gas or aerosol at level z_0. In the absence of scavenging by rain or snow (called *wet deposition*), this deposition velocity is used to estimate the dry deposition of materials with a negligible fall velocity onto the ground and vegetation surfaces. The value of v_s depends on the chemical species involved. For SO_2, for example, the deposition velocity is estimated to be on the order of 1 cm sec^{-1}, while sulfates with sizes of 0.1–1 μm are reported to have values ranging from around 0.01 to 1 cm sec^{-1} (Eliassen, 1980). Attempts to formulate dispersion representations of pollutants for use in three-dimensional models include those of Yamada (1977), while Sheih (1977) and others have considered the influence of thermal coagulation and gravitational sedimentation on pollution concentrations. Hane (1978) has simulated the wet deposition of pollutants over St. Louis using a two-dimensional squall line model.

Using Eqs. (98)–(100), the exchange coefficients in the surface layer can also be written as

(101) $$K_m = ku_*z/\phi_1; \quad K_\theta = ku_*z/\phi_2$$

Similar expressions could be derived for the solid and liquid water conservation relations (for $\bar{q}_1$ and $\bar{q}_2$); however, this has not been done for mesoscale models.

As evident from these relations, the subgrid-scale fluxes and profiles of the dependent variables are functions of the ground roughness and the stability of the surface layer as measured by z/L. When $\partial\bar{\theta}/\partial z < 0$ the parameter z/L is less than zero and equal to the gradient Richardson number,

$$\text{Ri} = \frac{g}{\bar{\theta}}\frac{\partial\bar{\theta}}{\partial z}\left[\frac{\partial}{\partial z}(\bar{u}^2 + \bar{v}^2)^{1/2}\right]^{-2}$$

(Pandolfo, 1966), while z/L is greater than zero when the surface air is stably stratified. When $\partial\bar{\theta}/\partial z = 0$, $z/L = \text{Ri} = 0$. Figure 2 illustrates schematically how the wind profile varies in the surface layer as a function of stability. With unstably stratified air, for example, this layer is well mixed and velocity changes with height are small, while reduced mixing in stable air tends to produce large velocity shears. Such diurnal variations in stability can contribute to large nocturnal wind maxima at the top of the planetary boundary layer as discussed by Hoxit (1975), Blackadar (1957), McNider and Pielke (1981), Zeman (1979), and others.

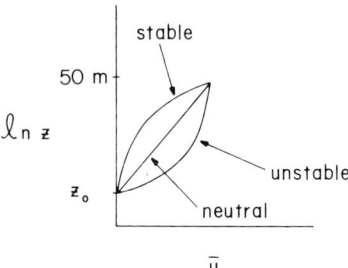

FIG. 2. Schematic illustration of wind profiles in the surface layer when the lower 50 m is unstably, neutrally, and stably stratified. The roughness z_0 and the wind at 50 m are assumed the same for the three situations.

Values of the roughness length z_0 for various types of ground surfaces have been reported by Oke (1978, pp. 48, 263), Rosenberg (1974, pp. 104, 105), and elsewhere. Pielke (1974a), for example, used a value of $z_0 = 4$ cm over south Florida to represent the saw grass, typical of much of that region. For specific locations z_0 is calculated by taking wind observations at several heights within the surface layer, when the mean wind speed is strong [so that $z/L = 0$ and $(\bar{u}^2 + \bar{v}^2)^{1/2} \cong (u_*/k) \ln z/z_0$ from Eq. (92)], plotting the winds as a function of the natural logarithm of height, and extrapolating the winds to the value $(\bar{u}^2 + \bar{v}^2)^{1/2} = 0$. The intersection of the ln z axis defines z_0. Over water Clarke (1970) suggests using the formula

$$z_0 = 0.032 u_*^2/g$$

with the condition that $z_0 \geq 0.0015$ cm to estimate roughness.

If the ground cover is high enough such that significant nonlaminar flow can occur below the top (e.g., within a pine forest, corn field), it is necessary to modify Eqs. (92)–(95) so that, for example, (92) would be rewritten as

$$(102) \quad u_* = k(\bar{u}^2 + \bar{v}^2)^{1/2} \bigg/ \left[\ln\left(\frac{z - D}{z_0}\right) - \psi_1\left(\frac{z - D}{L}\right) \right]$$

where D is called the *zero-plane displacement*. Values of z_0 that are then computed using this formula are displaced a distance D from the actual ground surface. The air flowing over this ground cover is thus more sensitive to the aerodynamic characteristics of its top, rather than what its morphology is below. (Within the ground cover, of course, the profiles of the dependent variables do depend on the physical structure of the vegetation or other obstacle.) Values of D are determined by plotting the wind speed as a function of $\ln(z - D)$ for strong winds, using different values of D. When Eq. (102) is well approximated, that particular value of D is used henceforth for that ground cover. Oke (1978, p. 48) and Rosenberg (1974, p. 104) give representative values of D for different types of surfaces.

7.3. The Transition Layer

The transition layer is from h_s to z_i, which ranges from 100 m or so to several kilometers or more. Above the surface layer, the mean wind tends to veer* with height and approach the *free-stream velocity* at z_j. This layer is also called the *Ekman layer*. The definition of z_i, the top of the planetary boundary layer, is *the lowest level in the atmosphere at which the ground surface no longer directly influences the dependent variables through turbulent mixing*. Tennekes (1974) gives a useful qualitative discussion of the atmospheric boundary layer. When thunderstorms are occurring, such a level can extend into the stratosphere; however, for most applications in mesoscale models, the planetary boundary layer is between a few hundred meters and several kilometers above the ground.

When the bottom surface is heated, the planetary boundary layer tends to be well mixed, particularly in potential temperature. Specific humidity is somewhat less well mixed, because the entrainment of dry air into a growing boundary layer permits a gradient in $\bar{q}_3$ to exist between the top of the planetary boundary layer and the, usualy, more moist surface (Mahrt, 1976). Because of horizontal pressure gradients, winds are the least well mixed. When the surface is cool relative to the overlying air, substantial vertical gradients in all of the dependent variables usually exist within the planetary boundary layer.

The parameterization of the subgrid-scale correlation terms in the planetary boundary layer can be grouped into four categories:

- drag coefficient representations
- local exchange coefficients
- exchange coefficients derived from profile functions
- explicit equations for the subgrid-scale fluxes.

The first three classes listed above are often called *first-order closure* representations because the subgrid-scale correlations are specified as functions of one or more of the averaged dependent variables (e.g., $\bar{u}_i$, $\bar{\theta}$, $\bar{q}_n$). The fourth category is referred to as *second-order closure* because prognostic equations are developed for the fluxes, which include triple correlation terms involving subgrid-scale variables $\left(\text{e.g., } \overline{u_k'' u_j'' \frac{\partial \theta''}{\partial x_j}}\right)$ that must be represented in terms of the double correlation terms and/or the averaged dependent variables. Mellor and Yamada (1974) discuss the dif-

---
* In the northern hemisphere.

ferent levels of complexity using various simplifications of the explicit representation of the subgrid-scale fluxes.

The drag coefficient form (also called the *bulk aerodynamic* formulation) is given, for example, by

$$\overline{u''w''} = -C_D(\bar{u}^2 + \bar{v}^2)\cos\tau \tag{103}$$

and is designed to represent the fluxes at the top of the surface layer. The parameter C_D is called the *drag coefficient*. Above this level a local exchange coefficient form is sometimes used if there is vertical resolution within the boundary layer. Lavoie (1972), Rosenthal (1970), and others have obtained realistic simulations using this form. Rosenthal used a value of $C_D = 3 \times 10^{-3}$ for velocity and $C_D = 0$ for heat and moisture. Lavoie used $C_D = 7 \times 10^{-3}$ over land and $C_D = 1.5 \times 10^{-3}$ over water for velocity, while for heat $C_D = 1.5 \times 10^{-3}$ was used if the surface layer was defined to be superadiabatic and $C_D = 0$, otherwise.

From Eqs. (92) and (100), however, it is straightforward to show that

$$C_D = k^2/[\ln z/z_0 - \psi_1(z/L)]^2 \tag{104}$$

so that, except for special cases, such as when the winds are strong [so that $\psi_1(z/L) \cong 0$] and the aerodynamic roughness of the surface is unchanging, it is inappropriate to treat the drag coefficient as a constant. For potential temperature, specific humidity, and other gaseous and aerosol materials, a form of the drag coefficient similar to that given by Eq. (104) can be derived from (92), (93), and (100). Using drag coefficients, fluxes in the boundary layer can be represented by requiring $C_D = 0$ at z_i with a specified functional form between the surface and z_i.

The use of exchange coefficients is of the form

$$\overline{u''w''} = -K_m \, \partial\bar{u}/\partial z \tag{105}$$

for example, where K_m is the exchange coefficient. If K_m is defined only in terms of local gradients, it is a local exchange coefficient, while it is a profile coefficient if it is derived from a vertical interpolation formula. The other subgrid-scale fluxes are of the same form except K_m is replaced by K_θ; and $\partial\bar{\theta}/\partial z$ or $\partial\bar{q}/\partial z$ etc., as appropriate, is substituted for $\partial\bar{u}/\partial z$. Blackadar (1979) suggested one form of local exchange coefficient, when the layer being simulated is stably stratified air ($\partial\bar{\theta}/\partial z > 0$), which is given (McNider and Pielhe, 1981) as

$$K_m = K_\theta = \begin{cases} \dfrac{1.1}{\text{Ri}_C}\left[(\text{Ri}_C - \text{Ri})l^2 \dfrac{\partial(\bar{u}^2 + \bar{v}^2)^{1/2}}{\partial z}\right] & \text{for } \text{Ri} \leq \text{Ri}_C \\ 0 & \text{for } \text{Ri} > \text{Ri}_C \end{cases} \tag{106}$$

where l is a mixing length given by

$$l = \begin{cases} 0.35z & \text{for } z < 200 \text{ m} \\ 70 \text{ m} & \text{for } z \geq 200 \text{ m} \end{cases}$$

The parameter Ri_C is the critical Richardson number, which should be equal to 0.25 (Dutton, 1976) in the limit as the vertical grid spacing approaches zero.

With such a representation, fluxes are always *down-gradient*. As shown by Deardorff (1966), however, *countergradient* fluxes are known to occur, and he suggested the vertical gradient of potential temperature used in the representation $\overline{w''\theta''} = -K_\theta \, \partial\bar{\theta}/\partial z$ be modified to

$$\frac{\partial \bar{\theta}_\text{C}}{\partial z} = \frac{\partial \bar{\theta}}{\partial z} - \gamma_\text{C}$$

where $\gamma_\text{C} = 0.65 \times 10^{-3}\,°\text{K m}^{-1}$ in order to permit fluxes of heat upgradient.

Another example of a local exchange coefficient representation is that of Orlanski *et al.* (1974), while Klemp and Lilly (1978) use a form of mixing equivalent to a local exchange coefficient that requires that the Richardson number always equal or exceed the critical Richardson number, 0.25. Such forms are appropriate when the vertical grid resolution is high, so that the gradients can be accurately approximated, and when the characteristic length scales of the subgrid-scale mixing are approximately the same size as the grid spacing.

If these requirements are not fulfilled, however, it is desirable to use the profile coefficient formulation. With this technique, the exchange coefficient is defined as a function of distance above the ground. The form suggested by O'Brien (1970a),

(107) $\quad K(z) = K_{z_\text{i}} + [(z_\text{i} - z)^2/(z_\text{i} - z_{h_\text{s}})^2]\{K_{h_\text{s}} - K_{z_\text{i}}$
$\quad + (z - h_\text{s})[\partial K_{h_\text{s}}/\partial z + 2(K_{h_\text{s}} - K_{z_\text{i}})/(z_\text{i} - h_\text{s})]\}, \qquad z_\text{i} \geq z \geq h_\text{s}$

where K_{h_s}, $\partial K/\partial z|_{h_\text{s}}$, and K_{z_i} are evaluated at the top of the surface layer h_s [defined by Eq. (109)] and at z_i, has been shown by Yu (1977), Pielke and Mahrer (1975), and others to produce realistic simulations for the growth of a boundary layer in which $\partial\bar{\theta}/\partial z \leq 0$ in the surface layer. $K(z)$ can refer to either $K_\text{m}(z)$ or $K_\theta(z)$. This cubic representation of the vertical exchange coefficient is illustrated schematically in Fig. 3. With this approach, mixing can occur throughout the boundary layer, even in regions where $\text{Ri} > \text{Ri}_\text{C}$, because K is not dependent on local gradients of the dependent variables above the surface layer. K_{h_s} and $\partial K/\partial z|_{h_\text{s}}$ are evaluated using Eq. (101) with $z = h_\text{s}$, while K_{z_i} is usually set equal to an arbitrarily small value [Pielke and Mahrer (1975) used $K_{z_\text{i}} = 0.0001 \text{ m}^2 \text{ sec}^{-1}$].

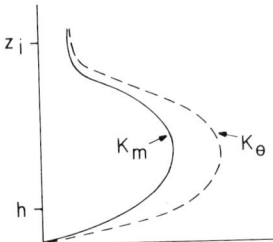

FIG. 3. Schematic illustration of the cubic representation of the exchange coefficients as a function of height. (From Pielke, 1974a.)

The depth of the planetary boundary layer z_i is usually associated with an inversion. As discussed by Oke (1978), there are three types of inversions:

- *inversions due to cooling:* (i) *radiational cooling* at night, or above stratiform clouds and smog layers; (ii) *evaporative cooling* over moist ground.
- *inversions due to warming:* (i) *synoptic subsidence;* (ii) *cumulus-induced subsidence.*
- *inversions due to advection:* (i) *frontal inversions;* (ii) *warm air over cold land, water,* or *snow;* (iii) *differential vertical temperature advection.*

In the absence of an inversion, when the air is neutrally stratified, Blackadar and Tennekes (1968) suggest that z_i is proportional to u_*/f, while Deardorff (1972) and Mahrt (1972), as reported by Moss (1978), suggest that the lifting condensation level is the appropriate height. Nocturnal surface inversion height formulations have also been suggested by Yamada (1979), Yu (1978), Nieuwstadt and Driedonks (1979), Zeman (1979), and others. In Yamada's formulation nocturnal long-wave radiational cooling is included.

Variations of the planetary boundary-layer depth due to subgrid-scale fluxes need not be parameterized when a local representation to the exchange coefficients is used, but will show up through changes in the vertical profile of the dependent variables. Using a profile form, however, such as Eq. (107), it is necessary to specify z_i. Deardorff (1974) has suggested such a parameterization following the work of Ball (1960) and others, which has subsequently been adopted by Pielke and Mahrer (1975) and others and shown to be quite realistic when the variation of z_i with time is strongly influenced by surface heating. This prognostic representation for the planetary boundary-layer height is given by

(108) $\dfrac{\partial z_i}{\partial t} = -\bar{u}_{z_i}\dfrac{\partial z_i}{\partial x} - \bar{v}_{z_i}\dfrac{\partial z_i}{\partial y} + \bar{w}_{z_i}$

$\qquad\qquad + 1.8(w_*^3 + 1.1u_*^3 - 3.3u_*^2 fz_i)\left(g\dfrac{z_i^2}{\theta_{h_s}}\dfrac{\partial\bar{\theta}^+}{\partial z} + 9w_*^2 + 7.2u_*^2\right)^{-1}$

where

$$w_* = \begin{cases}(-(g/\theta_h)\overline{u_*\theta_*}z_i)^{1/3} & \text{for } \theta_* \leq 0 \\ 0 & \text{for } \theta_* > 0\end{cases}$$

In Eq. (108) $\partial\bar{\theta}^+/\partial z$ is the potential temperature stratification immediately above z_i, $\bar{w}_z$ is the grid-scale vertical velocity at z_i (which can include synoptic mesoscale and averaged cumulus-induced subsidence), and θ_h is the potential temperature at the top of the surface layer. In this expression, the growth of z_i is directly proportional to the surface heat flux and mesoscale vertical velocity, and inversely proportional to the overlying stability.

Equation (108) can also be used to estimate $\bar{w}_{z_i}$ if it is assumed that the boundary-layer height is unchanging in time and horizontally homogeneous, $\theta_* = 0$, and that the net radiational flux divergence is zero. For this case (108) reduces to

$$\bar{w}_{z_i} = (1.98u_*^3 - 5.94u_*^2 fz_i)\left(g\dfrac{z_i^2}{\theta_{h_s}}\dfrac{\partial\bar{\theta}^+}{\partial z} + 7.2u_*^2\right)^{-1}$$

where z_i is obtained from a radiosonde or other observational platform.

The height of the surface layer h_s can be estimated from z_i as

(109) $\qquad\qquad\qquad h_s = 0.04 z_i$

This formulation was based on the results of A. K. Blackadar (personal communication, 1972) and Blackadar and Tennekes (1968), who found the best agreement between their predictions and observations, in a neutrally stratified boundary layer, when Eq. (109) was adopted.

An alternative representation of the planetary boundary layer in mesoscale models is that of Bush et al. (1976), in which a prognostic equation for mixing length is applied above the surface layer. This formulation, however, has not yet been tested against other techniques, such as has been performed by Yu (1977).

When air advects over heterogeneous ground surfaces (say, from forest to grassland) more than one boundary-layer height can result. At upper levels, subgrid-scale mixing is characteristic of the original surface, while mixing at lower levels is governed by the new surface. The interface between two such regions is called the *internal boundary layer*. Using a local exchange coefficient formulation, such internal boundary layers can be represented providing the horizontal grid resolution is adequate to resolve the heterogeneous ground surface. With the profile formulation, however, internal boundary layers of this type cannot be resolved.

Peterson (1969), Onishi (1968), and others have numerically simulated changes in surface-layer structure due to inhomogeneous terrain using two-dimensional, steady-state models. In Peterson's study, he showed that, in neutrally stratified air, the internal boundary layer grows about 1 unit upward for its first 10 units downstream of a change in surface characteristics. He further claims that the horizontal fetch must be 100 times its height in order for the new boundary layer to be in equilibrium. In an unstably stratified lower boundary layer the approach to its asymptotic equilibrium value would be much more rapid, however, since the surface fluxes are strongly coupled with the top of the growing boundary layer. Slower growth would occur if the layer was stably stratified. Since a local exchange coefficient representation is used when $\theta_* > 0$ (so that an internal boundary layer of this sort, if present, would be resolved), the major problem would arise when $\theta_* \leq 0$ and the profile representation is applied. When the surface layer is unstable, however, convective updrafts in the lower boundary layer are significant and a rapid adjustment to the new surface should occur. With light large-scale winds when $\theta_* < 0$, for example, the internal boundary layer would be much steeper than the 1:10 ratio reported by Peterson for $\theta_* = 0$. If Peterson's estimate is used as the slowest rate of growth for a heated internal boundary layer, a 10-km horizontal grid interval should be sufficient to generate planetary boundary-layer conditions close to equilibrium up to 1 km in height. Since the use of the hydrostatic assumption requires that the horizontal scale of the mesoscale circulation be larger than 8 km or so [from (16)], the use of Eq. (108) to represent the growth of the heated boundary layer in heterogeneous regions in a mesoscale model should give reasonable estimates.

Hsu's (1973) observations support the rapid growth of a heated boundary layer after a change in surface characteristics. Using two towers, one of 10 m elevation over a beach and one of 100 m, 10 km inland, he found the observations to closely agree with surface-layer theory developed for horizontally homogeneous steady conditions. A more detailed observation network, however, such as performed by Vugts (1980), is needed to validate this result because of Hsu's limited spatial resolution. In any case, lack of alternative formulations for the boundary layer in heterogeneous, nonsteady conditions requires that idealized theory be used in mesoscale models.

Various forms of the explicit representation of the subgrid-scale fluxes have been used by Donaldson (1973), Lee and Kao (1979), Brost and Wyngaard (1978), Deardorff (1974), Lumley and Khajeh-Nouri (1974), Burk (1977), Gambo (1978), Wyngaard and Coté (1974), André et al. (1978), and others. Mellor and Yamada (1974) categorize the level of complexity of those second-order representations. Although theoretically more satisfying, this more expensive approach with its greater degrees of freedom has not improved simulations of the evolution of the resolvable dependent variables in the planetary boundary layer over those obtained using the best first-order representations.

For example, days 33 and 34 of the Wangara Experiment (Clarke et al., 1971) have been used extensively to examine the accuracy of various parameterizations of the planetary boundary layer. Deardorff (1974), Wyngaard and Coté (1974), Pielke and Mahrer (1975), Yamada and Mellor (1975), and Dobosy (1979), for example, have attempted to simulate boundary-layer structure for all or a portion of these days. Figure 4 (solid line), reproduced from the sophisticated higher-order model of André et al. (1978), illustrates the evolution of the averaged virtual potential temperature in the boundary layer using a model that has an explicit representation of the subgrid-scale fluxes. Although useful understandings of turbulence can be derived from such models, their use is precluded in mesoscale models because of their expense. Figure 4 (dotted line) shows the results for the same period except Eqs. (107) and (108) are used to represent the vertical exchange coefficient when the surface layer is unstable or neutrally stratified ($\theta_* \leq 0$) and Eq. (106) when it is not ($\theta_* > 0$). Both results closely correspond to the observed profile [Fig. 4 (dashed line)]. The profiles of the other dependent variables produced by the two models also closely agree. Yu's (1977) results support this part of this conclusion. He found that use of Eqs. (107) and (108) produced accurate simulations of the growth of the mixed layer when compared against a range of other schemes, including a simplified second-order representation.

Zeman (1979) claims his economical one-layer model compares favorably against second-order closure models in simulating the evolution of a nocturnal boundary layer, while Klöppel et al. (1978) show that the development and decay of ground-based inversions can be satisfactorily simulated by simple models. In addition, Chang (1979) and Blackadar (1980) produced accurate results using simple representatives for heat and momentum exchanges and concluded that their methods provide an economic, realistic alternative to higher order closure schemes.

Based on these studies, the optimal representation of the planetary boundary layer in mesoscale models is a first-order closure representation

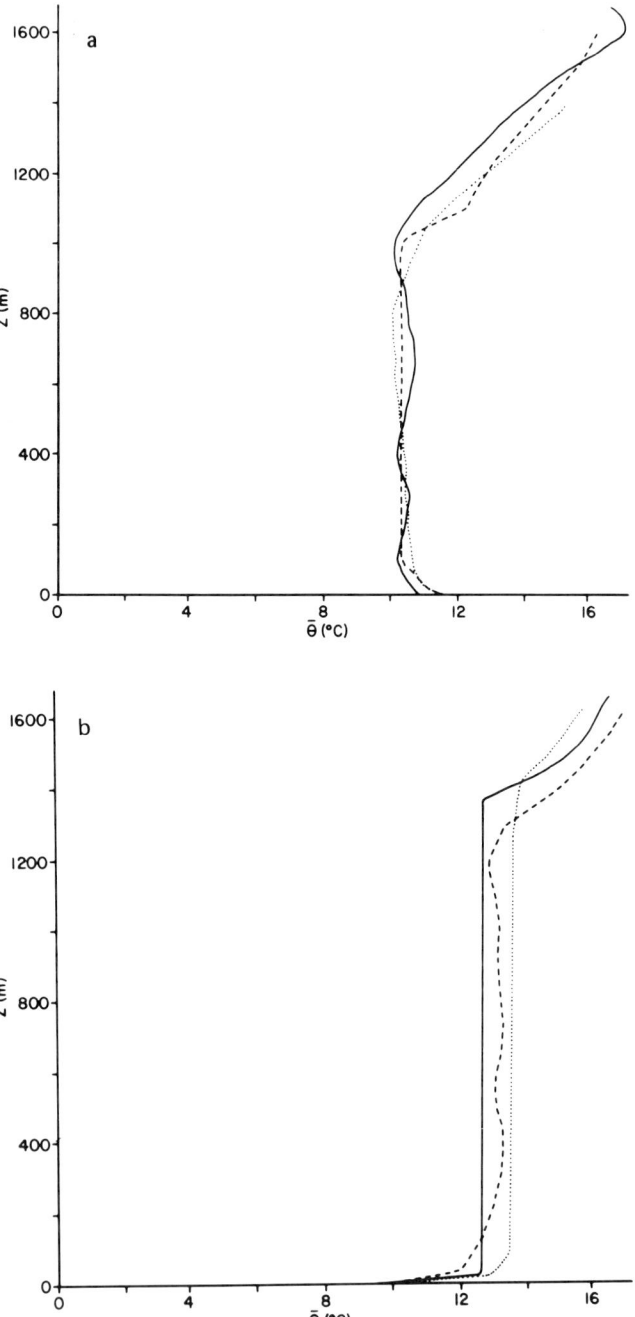

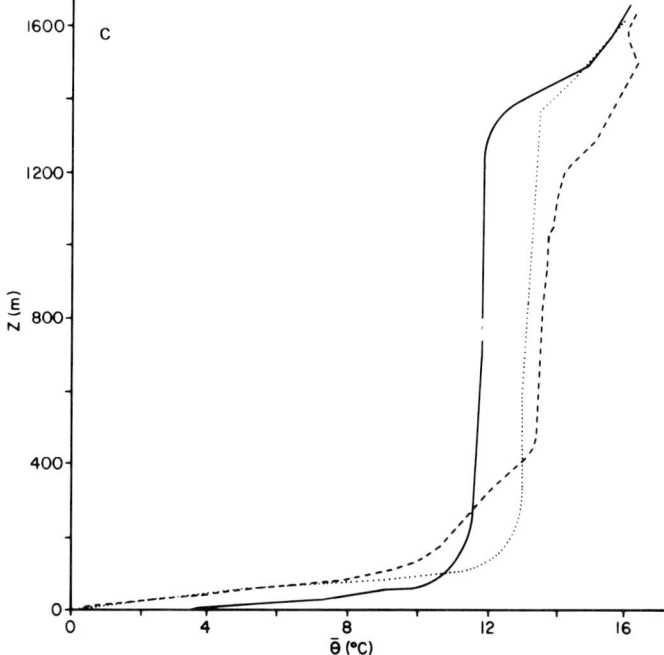

FIG. 4. Simulations and observations of the profile of virtual potential temperature for days 33 and 34 of the Wangara experiment: (1) day 33, 1200 LST; (b) day 33, 1800 LST; (c) day 34, 0300 LST. (———) From André et al., 1978 (predictions); (·····) from McNider and Pielke, 1980 (predictions); (----) from André et al., 1978 (observations).

where

- Eqs. (107)–(109) are used when $\bar{u}_* \leq 0$ (such as over land on sunny days;;
- Eq. (106) is used when $\theta_* > 0$ (such as at night over land, or on cloudy days with wet ground).

8. Radiation Parameterization

The emission, transmission, reflection, and absorption of electromagnetic radiation is a function of its wavelength and the material upon which it impinges. If an object absorbs all radiation that reaches it, it is called a *blackbody*, while if some reflection occurs but its absorption is independent of wavelength in the incident radiation, the object is referred to as a *gray body*. In a material (such as the atmosphere) it is the flux divergence of the radiation that causes heating or cooling. More radiation entering a layer in the atmosphere than leaving, for example, causes heating in that layer. In terms of the conservation equations, radiational heating enters directly through the source/sink term $\bar{S}_\theta$. In addition, chemical reactions and phase changes are influenced by radiation (e.g., photochemical smog) through the source/sink terms $\bar{S}_{q_n}$ and $\bar{S}_{\chi_m}$. Kondratyev (1969), Coulson (1975), and Haltiner and Williams (1980) provide more detailed discussion of radiation characteristics in the atmosphere as well as methods for representing radiation in mesoscale models.

Up to the present, it has been prohibitively expensive in mesoscale models to include detailed radiation calculations as a function of wavelength and still retain a three-dimensional spatial representation. Therefore, in order to include radiative effects in simulations of the mesoscale atmosphere, modelers have grouped radiation into two types:

- *short-wave* radiation
- *long-wave* radiation.

Short-wave radiation is predominantly visible electromagnetic energy that emanates from the Sun, whereas long-wave radiation is emitted from the much cooler surface of the Earth and its atmosphere. Using Planck's law (e.g., see Wallace and Hobbs, 1977) and assuming the surface temperature of the Sun to be 5780°K, while that of the Earth and its atmosphere is about 255°K, it is straightforward to show that there is little overlap between the electromagnetic spectra of the two emissions. According to Wien's displacement law, for example, which is obtained by differentiating Planck's law and solving for the maximum, the peak emission from the Sun's surface has a wavelength of 4.75×10^{-5} m (blue light), while that of the Earth is around 140×10^{-5} m (infrared).

The gases in the Earth's atmosphere respond differently to these two categories of electromagnetic energy. Except for clouds, air in the troposphere is relatively transparent to short-wave radiation with only a relatively small proportion reflected or absorbed by oxygen, carbon dioxide, water vapor, and other atmospheric gaseous and aerosol materials. The atmosphere is almost opaque to certain wavelengths in the long-wave region, however, because of the presence of such triatomic gases as water vapor, carbon dioxide, and ozone. Radiation in those wavelengths, for instance, is absorbed after being emitted from the Earth's surface and a large fraction is reradiated back downward. In addition, clouds are essentially blackbodies in the long-wave region. Wallace and Hobbs (1977, Figs. 7 and 8) illustrate the attenuation of short- and long-wave radiation as it travels through the Earth's troposphere.

8.1. Short-Wave Radiation

Short-wave radiation is composed of two components:

- direct radiation
- diffuse radiation.

8.1.1. Direct Radiation.
Direct short-wave radiation is that which reaches a point without being absorbed or scattered by the intervening atmosphere, or other objects. The image of the Sun's disk as a sharp and distinct object represents that portion of the short-wave radiation that reaches the viewer directly. Diffuse radiation, on the other hand, is the portion that reaches the observer after first being scattered. On an overcast day, for example, the Sun's disk is not visible and all of the radiation is diffuse. *Scattering* consists of one or more reflections of the original incident radiation.

The direct downward solar radiation reaching a horizontal surface of unit area in the atmosphere, R_s^D, can be expressed as

$$(110) \qquad R_s^D = \begin{cases} S \dfrac{a^2}{r^2} \cos Z \, p^{\sec Z} & \text{for } |Z| < 90° \\ 0 & \text{for } |Z| \geq 90° \end{cases}$$

where $S = 1376$ W m^{-2} (Hickey *et al.*, 1980) is the radiation from the Sun on a surface of unit area perpendicular to the direction of propagation of the Sun's energy at the average distance of the Earth from the Sun, a. The distance of the Earth from the Sun at any given time is r. The parameter Z is the *zenith angle*, while p is a measure of the clarity (also called the *turbidity*) of the air through which the Sun's rays penetrate. The zenith angle is equal to 90° when the Sun's disk bisects the horizon and is 0° when it is directly overhead.

The zenith angle is defined by

$$(111) \qquad \cos Z = \cos \Phi \cos \delta_{sun} \cos h_r + \sin \delta_{sun} \sin \Phi$$

where Φ is latitude, δ_{sun} is the declination of the Sun (which ranges between $+23.5°$ on June 21 to $-23.5°$ on December 22), and h_r is the hour angle (0° = noon). Using Eq. (111), sunrise and sunset occur when $Z = \pm 90°$ and can be obtained from

$$h_r = \cos^{-1}\{-\tan \delta_{sun} \tan \Phi\}$$

(when $\tan \delta_{sun} \tan \Phi < -1$ night occurs the entire time, while for $\tan \delta_{sun} \tan \Phi > 1$ the Sun is up for the entire 24-hr period).

The parameter p, called the *transmission coefficient*, becomes less as the clarity of the air decreases. In a vacuum, such as above the Moon's surface, $p = 1$. The value of p depends on the amount of air, including water vapor, cloudiness, and presence of other aerosols and gases above the level of interest. List (1971), for example, provides values of daily solar radiation reaching the ground using various values of p, while Threlkeld and Jordan (1958) present graphs for estimating direct solar radiation as a function of such factors as density of dust particles and precipitable water.

8.1.2. Diffuse Radiation. Using Eq. (110), the amount of radiation absorbed and scattered is given by

$$\Delta R_s = S \frac{a^2}{r^2} \cos Z (1 - p^{\sec Z}), \qquad |Z| < 90°$$

which can be further decomposed into

(112) $\qquad \Delta R_s^a + \Delta R_s^s = (1 - \tilde{\alpha}) \Delta R_s + \tilde{\alpha} \Delta R_s = \Delta R_s$

The amount of sunlight absorbed by the intervening atmosphere, ΔR_s^a, heats that region, while the amount scattered, ΔR_s^s, is divided into that reflected back out into space and reflected downward as diffuse radiation. The parameter $\tilde{\alpha}$ indicates the fraction of ΔR_s in Eq. (112) that is scattered. If $\tilde{\beta}$ represents the fraction of scattered radiation that is reflected downward, then the total short-wave radiation reaching a layer in the atmosphere from above is given by

(113) $\qquad R_s = R_s^D + \tilde{\beta} \Delta R_s^s = R_s^D + \tilde{\alpha}\tilde{\beta} \Delta R_s = S \frac{a^2}{r^2} T \cos Z$

where $T = p^{\sec Z} + \tilde{\alpha}\tilde{\beta}(1 - p^{\sec Z})$ and could be called a *net transmission*. If forward scattering predominates over back scattering, $\tilde{\beta} > 0.5$.

8.1.3. Total Solar Radiation at the Ground. Using Eq. (113), the total short-wave radiation absorbed by a horizontal ground surface can be written as

(114) $\qquad\qquad\qquad R_{s_G} = R_s(1 - A)$

where A is the *albedo* (reflectance) of the ground. The quantity $-AR_s$ is the short-wave radiation reflected back into the atmosphere and could be added to Eq. (113) (after changes by scattering and absorption between the ground and the atmospheric layer of interest); however, this effect has been neglected in mesoscale models thus far.

8.1.4. Total Solar Radiation on a Slant Surface. When the ground surface is not horizontal, Eq. (114) can be written

$$R_{S_G}^{slant} = R_{S_G} \cos i / \cos Z$$

as shown by Kondratyev (1969), where

$$\cos i = \cos \alpha_G \cos Z + \sin \alpha_G \sin Z \cos(\beta_G - \eta_G)$$

The slope angle α_G and the projections onto a horizontal surface of the solar and slope azimuths are given by

$$\alpha_G = \arctan\left[\left(\frac{\partial z_G}{\partial x}\right)^2 + \left(\frac{\partial z_G}{\partial y}\right)^2\right]^{1/2}$$

$$\beta_G = \arcsin\left[\frac{\cos \delta_{sun} \sin h_r}{\sin Z}\right]$$

$$\eta_G = \frac{\pi}{2} - \arctan\left(\frac{\partial z_G}{\partial y} \bigg/ \frac{\partial z_G}{\partial x}\right).$$

As discussed in Section 7, the terrain slope, given by α_G, must be much less than 45° if (82) is to be used as the hydrostatic equation. Even with small slopes, however, the influence on mesoscale circulations can be significant as shown, for example, by Mahrer and Pielke (1977b, Fig. 4), who found the eastern slope of a 1-km mountain (with a slope of about 2°) to be 1–2°C warmer in the morning and cooler by the same amount in the afternoon than the same location on the western slope.

8.1.5. Parameterized Forms of Short-Wave Heating. In mesoscale models, empirical functions are used to represent the albedo A and net transmission T. Atwater (1974) presents a methodology to parameterize short-wave radiation and his methodology has been modified and adopted by others (e.g., McCumber, 1980; Mahrer and Pielke, 1977a,b; and others). In cloud- and pollution-free air, the heating due to short-wave absorption by water vapor (from McDonald, 1960) has been expressed by McCumber (1980) as

(115) $$\left.\frac{\partial T}{\partial t}\right|_{sw} = \frac{S \cos Z}{C_p} \frac{\partial a_w}{\partial p} = 0.0231 \frac{S}{C_p}\left[\frac{r}{\cos Z}\right]^{-0.7} \frac{dr}{dp}$$

where

$$a_w = 0.077 \left[\frac{r}{\cos Z}\right]^{0.3}$$

and the *optical path length* of water vapor between the top of the atmosphere and the level with pressure p (with ρ_2 equal to the density of liquid water) is given as

$$r = -\frac{1}{\rho_2 g} \int_p^0 q_3 \, dp$$

The heating obtained from Eq. (115) represents the contribution to ΔR_s^a in Eq. (112) by water vapor, which in the absence of clouds and/or of significant quantities of aerosols is the dominant component of ΔR_s^a (List, 1971, p. 420).

When pollution is present, the absorption and scattering of short-wave radiation by the pollutants must be considered. Atwater (1971a,b), Bergstrom (1972), Viskanta and Daniel (1980), and others consider the modification to the radiation budget due to aerosols and gases. Wesely and Lipschultz (1976), for example, found from measurements that suspended particles increase the diffuse relative to the direct radiation, with more solar energy reflected back into space than would occur in a clean atmosphere. The magnitude of this effective increase in albedo is strongly dependent on zenith angle.

When clouds are present, their effect is primarily to reduce solar radiation transmission below the cloud, rather than to cause heating within them (Sasamori, 1972). Most of this lost short-wave radiation is reflected black into space due to the relatively high albedo of the top surface of water and ice clouds. Atwater (1974), Stephens and Webster (1980), Liou and Wittman (1979), and others discuss the influence of clouds on the short-wave flux of radiation.

The net transmissivity at the ground in a cloud- and pollution-free atmosphere has been represented by

(116) $$T = G - a_w$$

where

$$G = 1.03 - 0.08([9.49 \times 10^{-4} p \text{ (in mbar)} + 0.051]/\cos Z)^{1/2}$$

This expression for G was originally presented by Kondratyev (1969) and later modified by Atwater and Brown (1974). It accounts for forward Rayleigh scattering due to such gases as ozone, oxygen, and carbon dioxide.

In an atmosphere with clouds the reduction of R_s^D due to their reflection of visible light must be included. This effect can be quite important. Gannon (1978), for example, found that the sea breeze circulation over south Florida was terminated as the shading due to cirrus over the land markedly reduced the solar flux reaching the ground. Similarly, Welch *et al.* (1978) in a two-dimensional simulation of the effects of polluted air upon an urban–rural area found temperatures at the ground in the urban area during stagnant synoptic conditions to be reduced by 2°C because of low-level pollution sources and up to 7°C when upper level sources occur—a result that is partially due to the enhanced reflection and absorption of solar radiation by suspended aerosols.

The albedo over bare soil can be expressed as

(117) $$A = a_n + a_s$$

where, from McCumber (1980), using Idso *et al.* (1975), the variability of albedo with zenith angle is estimated by

$$a_n = [\exp(0.003286 Z^{1.5}) - 1]/100$$

while a_s is a function of the ratio of the volumetric moisture content η and porosity of the soil η_s. The determination of η and η_s requires a model of the flux of moisture into and out of the soil. Detailed relationships for computing the temporal fluctuations in soil moisture content are available (e.g., Philip, 1957) and have been used in atmospheric models such as those of Sasamori (1970) and Garrett (1978). McCumber (1980) has introduced a somewhat simpler form to obtain soil moisture flux, maintaining that detailed soil data has been unavailable over a mesoscale-sized area. If vegetation, ice, or snow cover the ground, a_s is replaced by their respective albedos. Typical albedos range upward to 0.90 for fresh, deep snow to as low as 0.03 for coniferous forests (List, 1971). Otterman (1974), Berkofsky (1977), and others have argued that changes in albedo can have a profound influence on average vertical motion—with increased albedo (due to overgrazing, for example) causing subsidence and a tendency toward *desertification* in arid areas.

8.2. Long-Wave Radiation

8.2.1. Clear Air. The parameterization of long-wave radiation is somewhat more involved than for short-wave radiation since the atmosphere is such an effective absorber of portions of the infrared electromagnetic spectrum. At any level in the atmosphere, the net flux of long-wave radiation is equal to the difference between the fluxes above and below. These two flux terms can be written as

$$R_L \downarrow = \sigma \int_p^0 \epsilon \frac{\partial T^4}{\partial \xi} d\xi; \quad R_L \uparrow = \sigma \int_{p \text{ at } z_G}^p \epsilon \frac{\partial T^4}{\partial \xi} d\xi$$

from McCumber (1980), where σ is the Stefan–Boltzmann constant ($\sigma = 8.314 \times 10^3$ J deg^{-1} kmol^{-1}). The emissivity ϵ is a function of the path length of the long-wave radiation as given below. A contribution to the heating or cooling of a layer in the atmosphere results from the vertical flux divergence of the net long radiation.

Sasamori (1972) provides a simplified and useful algorithm for computing the temperature change due to long-wave radiational flux divergence, in which the atmosphere is assumed to be isothermal at the temperature of the level of interest. As given by McCumber (1980) for use in a model, the temperature change is written as

$$\partial T/\partial t = (g\sigma/C_p)\{(T_N^4 - T_G^4)[\epsilon(p_{N+1}, p_G) - \epsilon(p_N, p_G)] \\ + (T_{\text{top}}^4 - T_N^4)[(\epsilon(p_{N+1}, p_{\text{top}}) \\ - \epsilon(p_N, p_{\text{top}})]\}/(p(N + 1) - p(N))$$

where the subscripts N, top, and G refer to the grid level of interest, the top of the model, and the ground surface, respectively. The emissivities are evaluated between the two levels given within the parentheses.

In the troposphere in clean cloudless air, carbon dioxide and water vapor are the two most important absorbers and emitters of long-wave radiation. Kuo (1979) reports that for clear air with a temperature and moisture profile approximating that of the U.S. standard atmosphere, the infrared cooling rate is about 1.2°C/day with a maximum at the surface and 7.5 km and a minimum at 2 km and the tropopause. He concludes that for this atmospheric profile, water vapor is the main absorber/emitter of long-wave radiation.

Atwater (1974), using data from Kuhn (1963), gives one representation of the emissivity of water vapor as

$$\epsilon_r(p, p_i) = \begin{cases} 0.113 \log_{10}(1 + 12.6\bar{r}) & \text{for} & \log_{10}\bar{r} \leq -4 \\ 0.104 \log_{10}\bar{r} + 0.440 & \text{for} & -4 < \log_{10}\bar{r} \leq -3 \\ 0.121 \log_{10}\bar{r} + 0.491 & \text{for} & -3 < \log_{10}\bar{r} \leq -1.5 \\ 0.146 \log_{10}\bar{r} + 0.527 & \text{for} & -1.5 < \log_{10}\bar{r} \leq -1.0 \\ 0.161 \log_{10}\bar{r} + 0.542 & \text{for} & -1.0 < \log_{10}\bar{r} \leq 0 \\ 0.136 \log_{10}\bar{r} + 0.542 & \text{for} & \log_{10}\bar{r} > 0 \end{cases}$$

while the emissivity of carbon dioxide, utilizing Kondratyev (1969), is given by

$$\epsilon_{CO_2}(p, p_i) = 0.185[1 - \exp(-0.39\bar{c}^{0.4})]$$

The terms $\bar{r}$ and $\bar{c}$ are the optical path lengths for water vapor and carbon dioxide between levels p and p_i. In these expressions

$$r = -\frac{10^3}{\rho_2 g} \int_{p_0}^{p} q_3 \, dp; \qquad \bar{r} = r(p) - r(p_i)$$

and

$$c = -0.415(p - p_0); \qquad \bar{c} = c(p) - c(p_i)$$

where p is in millibars, p_0 is the surface pressure (~ 1014 mbar), and r and

c are in centimeters. The expression for c assumes a uniform vertical distribution of 320 ppm of carbon dioxide. The total emissivity in clear, clean air, neglecting the small overlap in the water vapor and carbon dioxide absorption spectrum, is given by

$$\epsilon(p, p_i) = \epsilon_r(p, p_i) + \epsilon_{CO_2}(p, p_i)$$

8.2.2. Polluted Air. When significant concentrations of pollution are present, it is necessary to consider the absorption and reemission due to these materials. Viskanta *et al.* (1976) and Venkatram and Viskanta (1976) use one- and two-dimensional models to investigate the influence of pollutants in both the short- and long-wave radiative fluxes. For naturally occuring aerosols, Carlson and Benjamin (1980) found typical aerosol heating rates from the combined short- and long-wave spectrum to be in excess of 1°C for most of the atmosphere below 500 mbar in a region of suspended Saharan desert dust.

8.2.3. Cloudy Air. When clouds are present, they can be treated as black bodies to long-wave radiation if they are thicker than about 100 m. For this situation, clouds radiate heat from their top quite rapidly, as well as substantially reduce heat loss from below their base. Roach and Slingo (1979), for example, found a cooling rate of 8.7°C/hr from the 1 mbar layer at the top of nocturnal stratocumulus over England. Therefore, in the formulations for $R_L \uparrow$ and $R_L \downarrow$, downward fluxes from above the cloud base are ignored if clouds lie above the level of interest, and upward fluxes below the cloud top are neglected if clouds occur below. Longwave radiation is assumed to radiate from the clouds as σT_c^4, where T_c refers either to the top or base of a cloud layer. If the sky is only partly covered with clouds, it is possible to proportionally weight the upward and downward long-wave radiation by the fractional coverage of clouds as suggested by Atwater (1974).

Stephens and Webster (1981) have shown that vertical temperature structure (and therefore mesoscale dynamics) is highly sensitive to cloud height, although only sensitive to water path for optically thin clouds. They contend that high thin clouds at low and middle latitudes and all clouds at high latitudes tend to warm the surface compared to a clear sky, while all other clouds cool. Liou and Wittman (1979) give a detailed discussion of the parameterization of the radiative properties of clouds. Among their results, they conclude that cirrus clouds, in contrast to other cloud types, cannot be treated as blackbodies.

In three-dimensional mesoscale models, radiation physics has not yet received the attention that planetary boundary-layer dynamics has. The importance of long-wave radiative flux divergence to nocturnal mesoscale flows, and of both short- and long-wave fluxes in polluted atmospheres during the day suggests that the accurate parameterization of radiation in mesoscale models is a fertile area for future research.

9. Moist Thermodynamics

In many mesoscale systems such as the sea breeze and squall line, phase changes of water occur as mesoscale and/or subgrid-scale circulations lift air above the condensation level, and as precipitation falls back out of clouds and begins to evaporate. The presence of water as solid and liquid, as well as vapor, necessitates that the complete form of the conservation equations for water substance be included in a mesoscale model, as well as the proper representation of the source/sink term $\bar{S}_\theta$ in the equation for potential temperature. In Eqs. (22) and (24), or (83) and (84), for example, $\bar{S}_\theta = (L/\bar{\pi})\bar{C}^*$ and $\bar{S}_{q_3} = -\bar{C}^*$ can be used, where $\bar{C}^*$ is the total condensate (or sublimate) rate in a grid volume, and L is the latent heat of condensation ($L = 2.5 \times 10^6$ J kg^{-1} at 0°C). In addition, as mentioned in the last section, radiative flux divergence is influenced by the presence of clouds.

In order to discuss the representation of these effects in mesoscale models, it is useful to catalog moist mesoscale systems into those that occur in two distinct atmospheres above the condensation level:

- a *convectively stable atmosphere* ($\partial\bar{\theta}_E/\partial z > 0$)
- a *convectively unstable atmosphere* ($\partial\bar{\theta}_E/\partial z \leq 0$).

The variable $\bar{\theta}_E$ is the grid volume-averaged equivalent potential temperature, and is represented approximately by

$$\bar{\theta}_E = \bar{\theta} \exp\left(\frac{L}{C_p}\frac{\bar{q}}{\bar{T}}\right) = \bar{\theta} \exp\left(\frac{L\bar{q}}{\bar{\pi}\bar{\theta}}\right)$$

where L is the latent heat of vaporization [see R. H. Simpson (1978) for a precise derivation of θ_E].

9.1. Convectively Stable Atmosphere

In a convectively stable atmosphere on the resolvable scale the following applies:

- Only stratiform clouds will develop if $\partial\theta_E''/\partial z > 0$ everywhere within a grid volume.

- Some cumuliform clouds can develop if $\partial\theta''_E/\partial z \leq 0$ locally within a grid volume. The number and vigor of those that develop depend on the magnitude and distribution of the regions of subgrid scale convective instability.

In an atmosphere that is convectively stable at all points within a grid volume, the conversion of water between its phases can be represented straightforwardly using formulations such as those developed for cloud models. The degree of sophistication can range from the detailed simulation of the stochastic growth of distributions of water droplets and ice crystal sizes [such as used by Clark (1973) for a cumulus model with detailed warm cloud microphysics] to simple representations of total water conversion based only on the grid-volume vertical velocity and the saturation specific humidity [Colton (1976) used this formulation to predict precipitation caused by airflow over a mountain barrier].

When the atmosphere has subgrid regions that are potentially unstable *and* in which condensation or sublimation occur, the representations for phase change mentioned above may be unsatisfactory. Moreover, since the grid-volume variables are convectively stable, the methodology to determine when local regions have $\partial\theta''_E/\partial z < 0$ must be governed by such variables as the depth of the planetary boundary layer, lifting condensation level, and intensity of subgrid-scale mixing. This problem has not yet been addressed by mesoscale modelers.

9.2. *Convectively Unstable Atmosphere*

In a convectively unstable region on the resolvable scale the following applies:

- Only cumuliform clouds will form if $\partial\theta''_E/\partial z \leq 0$ everywhere within a grid volume.
- Some layer-form clouds will form if $\partial\theta''_E/\partial z > 0$ locally within a grid volume. The extent and levels of such clouds depend on the distribution of regions which are convectively stable.

In contrast to stratiform cloud systems, cumulus clouds generally have smaller spatial dimensions and more irregular patterns of updrafts and downdrafts. Except for cumulonimbus-size systems, individual cumulus clouds have horizontal dimensions that are smaller than can be resolved by a mesoscale model grid. Moreover, the depth that a cumulus cloud attains may be more dependent on the magnitude of convective instability than on the intensity of the mesoscale ascent once condensation is attained. Along with such effects as precipitation, downdrafts, ground shadowing, and cumulus-induced subsidence, the accurate representation of cumulus cloud influences on the mesoscale has been and will remain one of the more difficult problems in meteorology.

The ability to accurately represent cumulus clouds in a mesoscale model requires that the mesoscale dynamic and thermodynamic structure control the regions of initiation and development of this moist convective activity. There is evidence that this condition is valid. Ulanski and Garstang (1978) found that, over land, cumulus clouds are often forced by the boundary-layer convergence/divergence patterns with the surface inflow preceding rain showers by as much as 90 min. Pielke (1974a) also found qualitative agreement between predicted sea breeze convergence and the subsequent actual development of cumulonimbus activity, while Simpson *et al.* (1980) obtained a high correlation between merged thunderstorm complexes and sea breeze convergence for three case study days over south Florida. For a particular summer day over south Florida, Pielke and Mahrer (1978) obtained a 4-hr lag between this predicted mesoscale convergence and thunderstorm activity. Most of the rainfall in these sea breeze events occurs in large cumulonimbus complexes (Simpson *et al.*, 1980). In the 1974 GARP Atlantic Tropical Experiment (GATE) Ogura *et al.* (1979) found low-level convergence to be present or enhanced prior to the development of organized convective systems in all cases considered.

9.3. Types of Parameterizations

In considering methodologies to represent cumulus clouds it is useful to group them into four classes:

- convective adjustment
- use of one-dimensional cloud models
- use of a cumulus field model, or set of equivalent observations
- explicit representation of moist thermodynamics.

In the first method such as discussed by Krishnamurti *et al.* (1980), Kurihara (1973), and others, the lapse rate is forced to be moist adiabatic over all or part of the model grid when saturation occurs. In the second class, Kuo (1965, 1974), Krishnamurti and Moxim (1971), Anthes (1977), Johnson (1977), Fritsch and Chappell (1980a,b), Kreitzberg and Perkey (1976, 1977), Yenai (1975), Arakawa and Schubert (1974), Ooyama (1971), and others discuss or use one-dimensional cloud models to represent the feedback of cumulus scales to the larger scale. In using one-dimensional models, it is assumed that in deep cumulonimbus systems, the vertical distribution of heating within the clouds is essentially the same as within the model grid volumes (Anthes, 1977), although such an assumption is not true for shallow cumulus. Yenai (1975) gives a review of these types of cumulus representations.

Both of the above approaches have shortcomings, however. With the convective adjustment approach, the regions of potential instability are removed too rapidly, while the one-dimensional representations require arbitrary inputs and assumptions, such as cloud radius and a quasi-equilibrium with the large-scale environment, for them to work. If the mesoscale has a time scale on the same order as the cumulus cloud (as suggested by W. Frank, 1980), The clouds cannot be assumed to be in equilibrium. Moreover, as shown by Cotton and Tripoli (1978), one-dimensional cloud models cannot simultaneously accurately predict both cloud top height and liquid water content even for shallow cumulus clouds. Therefore, it should not be expected that the influence of deeper convective systems on three-dimensional mesoscale systems would be better represented.

The third technique listed above would utilize three-dimensional cumulus field model simulations or sets of observations in order to determine the temporal and spatial response of cumulus clouds to a particular set of mesoscale dependent variables, and their subsequent feedback to the mesoscale. Examples of possible models are the two-dimensional model of Hill (1974) and the three-dimensional simulations of Cotton and Tripoli (1978), Klemp and Wilhelmson (1978a,b), Schlesinger (1980), Miller and Pearce (1974), and Clark (1979). Chang and Oville (1973), Chen and Orville (1980), and Cotton *et al.* (1976) have begun to examine the response of cloud models to mesoscale convergence. Although it is expensive to perform the experiments needed to determine all possible interactions, it may eventually be possible to develop predictive relations that specify the interrelation between the cumulus scale and mesoscale. This procedure is discussed in more detail by Golden and Sartor (1978). Orville (1978), Simpson (1976), and Cotton (1975) provide reviews of cumulus dynamics and their numerical representations.

Recently, Rosenthal (1979a) has suggested that in tropical mesoscale models cumulus clouds should be represented explicitly in the same fashion as performed for stratiform clouds. He maintains that the successful implementation of cumulus parameterization schemes requires a strong coupling between the cloud and larger scale. Thus, while tropical cyclogenesis can be well represented with such an approach, tropical squall lines, such as reported by Zipser (1971), cannot. As discussed by Rosenthal (1979a), these squall lines have a distribution of moist and dry downdrafts such that convection is diminished near the center of the larger scale system with subsequent deep cumulus clouds forced to develop away from the region of larger scale vertical ascent. Zipser (1977) gives an example of such destructive interference between the cumulus scale and the larger scale.

Because of this limitation, Rosenthal (1978) has replaced one-dimensional cumulus parameterization with an explicit treatment of moist thermodynamics on a grid with a 20-km horizontal mesh interval. With this approach he was able to simulate tropical storms that represent constructive reinforcement between the cumulus scale and mesoscale, as well as tropical squall lines in which the larger-scale fields of dependent variables have little effect after the initiation of the system. The type of system that develops depends on the magnitude of the vertical shear of the horizontal wind and the dryness of the middle atmosphere. Studies by Yamasaki (1977), Jones (1980), and Rosenthal (1978) have shown that realistic hurricane simulations can be obtained when latent heat is released on the resolvable grid scale in such convectively unstable atmospheres. Rosenthal (1979b) concludes that the further use of cumulus parameterization schemes in hurricane simulations "seems to be of dubious value." He contends that an "experienced numerical experimenter can pick and choose closures that will provide almost any desired result."

The use of an explicit representation of moist thermodynamics in other types of mesoscale simulations would greatly simplify the representation of the conservation of heat, water substance, and other gaseous and aerosol materials if such a formulation can be accurately performed. Bhumralkar (1972), for example, obtained realistic results using an explicit formulation to model the airflow over Grand Bahama Island during the day. This important modeling question needs continued study.

A justification for the neglect of cumulus parameterization may occur when

$$\left| \bar{w} \frac{\partial \bar{\theta}_E}{\partial z} \right| \gg \left| w'' \frac{\partial \theta_E''}{\partial z} \right|$$

so that the vertical transport of heat is predominantly on the resolvable scale rather than on the subgrid scale. The need to satisfy this inequality may also determine the grid sizes used in a mesoscale model. Since at least four grid intervals in each spatial direction are needed to properly represent mesoscale variables in a numerical model (as discussed in Section 10), moist processes such as condensation and sublimation should be realized at least over this scale.

When regions of convective stability occur within the grid volume, layered clouds can form. If the areal extent of such clouds is large enough, short- and long-wave radiative fluxes will be affected, thereby altering subsequent cumulus cloud activity and the mesoscale response. The influence of such layered clouds on the subgrid-scale fluxes and source/sink terms has not been treated in mesoscale models.

10. Methods of Solution

As discussed in Section 5, sets of simultaneous nonlinear partial differential equations such as (21), (22), (24), (25), (28) and (32) or (79)–(85) cannot be solved using known analytic methods, but require numeric computational methods in which the equations are descretized and solved on a lattice. This lattice corresponds to the grid-volume average defined by (20) or (86), depending on whether a terrain-following representation is used or not.

There are several broad classes of solution techniques available to represent derivatives in these differential equations, including the following:

- *finite difference* schemes, which utilize a form of a truncated Taylor series expansion;
- *spectral* techniques in which dependent variables are transformed to wavenumber space using Fourier transforms (or other types of transformations with global basis functions);
- the *pseudospectral* method, which uses a truncated spectral series to approximate derivatives;
- the *finite element* schemes, which seek to minimize the error between the actual and approximate solutions using local basis functions;
- *interpolation schemes* in which polynomials, or other formulas, are used to approximate advection terms.

In mesoscale models only the finite difference, finite element, and interpolation schemes have generally been used. The spectral method has been shown to be highly accurate (e.g., Fox and Deardorff, 1972; Orszag, 1971) and eliminates the fictitious feedback (*aliasing*) of small-scale energy to the larger scales; however, the mathematical expressions that result from the Fourier transformation are cumbersome to handle and have required periodic boundary conditions to make it work effectively. This approach has not found acceptance among mesoscale modelers for these reasons. The pseudospectral technique has been introduced by Fox and

Orszag (1973), and has been contrasted with the spectral technique and conventional finite difference methods by Christensen and Prahm (1976). Eliassen (1980) summarizes the uses of these and other representations in air pollution transport modeling. Although the pseudospectral technique appears to be a viable tool for use in mesoscale models, it also has not been adopted. The major reason may be the recent interest in the finite element methodology.

In examining the accuracy of solution techniques in a mesoscale model, it is customary to investigate the representations of such components as advection, diffusion, and the Coriolis terms separately. The assumption is made that if the individual terms are independently well approximated, they will be accurate representations in the general equations. Also, in order to quantitatively analyze the relations, the equations can be linearized and solution techniques such as those introduced in the subsection on linear models in Section 5 can be used to evaluate them. In this case the dependent variables are given by

$$(118) \qquad \phi(x, t) = \tilde{\phi}(k, \omega)e^{i(kx+\omega t)} = \tilde{\phi}(k, \omega)e^{i(kj\,\Delta x+\omega\tau\,\Delta t)}$$

for a representation with one spatial direction, where τ is the index of the time step and j refers to the particular grid location. Substituting this representation into any linearized differential equation, the amplitude and phase speed of the solution for a given time step can be determined and are given by

$$\psi^1 = e^{i\omega\,\Delta t} = e^{-\omega_i\,\Delta t}(\cos \omega_R\,\Delta t + i \sin \omega_R\,\Delta t)$$
$$= \lambda(\cos \omega_R\,\Delta t + i \sin \omega_R\,\Delta t)*$$

where $\omega = \omega_R + i\omega_i$. The parameter λ represents the amplitude change after one time step with $c_R = -\omega_R/k$ being the numerically calculated phase speed.

The accuracy of the linear numerical solution depends on how well the calculated values of λ and c_R approximate the exact solutions of the differential equation, λ_{exact} and $c_{R_{\text{exact}}}$. If $\lambda > 1$ for any possible wavelength, the solution technique is *linearly unstable*. If it is not linearly unstable but $\lambda/\lambda_{\text{exact}} < 1$ for any wavelength, the scheme is *damping*, while if λ is identically equal to λ_{exact}, the technique is exact. When $c_R \neq c_{R_{\text{exact}}}$, the approximation representation is *erroneously dispersive* (the exact solution, of course, is *dispersive* if $c_{R_{\text{exact}}}$ is a function of k).

---

* The notational shorthand to perform the stability analysis was suggested by A. Mizzi (personal communication, 1979) and is discussed in detail in Pielke (1982).

10.1. Advection Terms

The exact solution to the linear advection equation

$$\frac{\partial \phi}{\partial t} = -U \frac{\partial \phi}{\partial x}$$

is $\lambda_{\text{exact}} = 1$ and $c_{R_{\text{exact}}} = U$. Table II lists five representations of the approximate nonlinear form of this equation, while Table III gives representative values of λ and c_R/U as a function of wavelength for these relations when u_j^τ is set equal to a constant speed U. The parameter $C = |u_j^\tau| \, \Delta t / \Delta x$ is called the *Courant number*.

Schemes IV and V in Table II are called *implicit* when $\beta_{\tau+1} \neq 0$, while with $\beta_{\tau+1} = 0$ they are *explicit*. Implicit schemes can be derived for which the solution is linearly stable regardless of the size of the time step, whereas its explicit counterpart becomes unstable if the Courant number exceeds unity. As shown by Grotjahn and O'Brien (1976), however, the stability of the implicit scheme is achieved by erroneously slowing the propagation speed of the predicted wave (e.g., see Table III, Schemes IV and V).

In Scheme II, two solutions occur. One corresponds to the physical solution and the other is an artifact of the numerical approximation, called the *computational mode*. The computational mode arises because the leapfrog scheme is a second-order difference equation. Both the slopes of the spline, N_j, in Scheme III, and the values of ϕ at the new time level in Schemes IV and V are solved by inverting a tridiagonal matrix. When $p = 0$, Scheme V reduces to Scheme IV.

When a more complete linear representation of the conservation relations is used [e.g., Eqs. (41)–(46)], the Courant number is defined as the phase speed of the fastest wave that is a solution. If the anelastic approximation is used (i.e., $\lambda_2 = 0$), the fastest wave is an internal gravity wave (see Section 5). The Courant number of physical significance in a table such as Table III, however, depends on the speed of the physically significant information in the model, as suggested by Lucero (1976). If acoustic waves are present but not considered an important component of the solution, then the Dendy method could be used with a larger time step (i.e., $C > 1$), since the phase characteristics of the sound waves are unimportant so long as the waves do not amplify and are of a small magnitude (so that nonlinear interactions with other waves are minimized).

In Table III, it is evident that the shorter the wavelength, the poorer the resolution of phase speed for each of the schemes, with larger damping occurring in Schemes I and III. The $2 \, \Delta x$ wavelength has the worst representation.

TABLE II. APPROXIMATE FORMULATIONS FOR ONE-DIMENSIONAL ADVECTION

I. Forward-in-time, upstream (e.g., Lavoie, 1972; Pielke, 1974a)

$$\frac{\phi_j^{\tau+1} - \phi_j^{\tau}}{\Delta t} = \begin{cases} -u_j^{\tau} \dfrac{\phi_{j+1}^{\tau} - \phi_j^{\tau}}{\Delta x} & \text{for } u_j^{\tau} \leq 0 \\ -u_j^{\tau} \dfrac{\phi_j^{\tau} - \phi_{j-1}^{\tau}}{\Delta x} & \text{for } u_j^{\tau} > 0 \end{cases}$$

II. Centered-in-time, centered-in-space: 2nd-order leapfrog (e.g., Haltiner and Williams, 1980)

$$\frac{\phi_j^{\tau+1} - \phi_j^{\tau-1}}{\Delta t} = -u_j^{\tau} \frac{\phi_{j+1}^{\tau} - \phi_{j-1}^{\tau}}{\Delta x}$$

III. Forward-in-time, upstream spline interpolation (e.g., Mahrer and Pielke, 1978b; Long and Pepper, 1976; Purnell, 1976)

$$\phi_j^{\tau+1} = \begin{cases} S(x_j - C\,\Delta x) = \phi_j^{\tau} - CN_j\,\Delta x + C^2[N_{j-1}\,\Delta x + 2N_j\,\Delta x + 3(\phi_{j-1}^{\tau} - \phi_j^{\tau})] \\ \quad - C^3[\Delta x N_{j-1} + \Delta x N_j + 2(\phi_{j-1} - \phi_j^{\tau})] \text{ for } u_j^{\tau} \geq 0 \\ S(x_j + C\,\Delta x) = \phi_j^{\tau} + CN_j\,\Delta x - C^2[N_{j+1}\,\Delta x + 2N_j\,\Delta x + 3(\phi_j^{\tau} - \phi_{j+1}^{\tau})] \\ \quad + C^3[\Delta x N_j + \Delta x N_{j+1} + 2(\phi_j^{\tau} - \phi_{j+1}^{\tau})] \text{ for } u_j^{\tau} < 0 \end{cases}$$

where S is the spline interpolation and N_{j+1}, N_j, and N_{j-1} are the derivatives of the spline at grid points $j+1$, j, and $j-1$.

IV. Finite element, chapeau basis function, implicit advection representation (e.g., Long and Pepper, 1976)

$$\phi_{j-1}^{\tau+1} + 4\phi_j^{\tau+1} + \phi_{j+1}^{\tau+1} = \phi_{j-1}^{\tau} + 4\phi_j^{\tau} + \phi_{j+1}^{\tau} - 3u_j^{\tau}\frac{\Delta t}{\Delta x}[\beta_{\tau+1}(\phi_{j+1}^{\tau+1} - \phi_{j-1}^{\tau+1}) + \beta_{\tau}(\phi_{j+1}^{\tau} - \phi_{j-1}^{\tau})]$$

where $\beta_{\tau+1} + \beta_{\tau} = 1$ and the values of ϕ at the new time level are obtained from this expression by inverting a tridiagonal matrix.

V. Finite element, chapeau function with damping: Dendy method (e.g., Raymond and Garder, 1976)

$$\left(1 + \frac{3p}{\Delta x}\right)\phi_{j-1}^{\tau+1} + 4\phi_j^{\tau+1} + \left(1 - \frac{3p}{\Delta x}\right)\phi_{j+1}^{\tau+1} = \left(1 + \frac{3p}{\Delta x}\right)\phi_{j-1}^{\tau} + 4\phi_j^{\tau} + \left(1 - \frac{3p}{\Delta x}\right)\phi_{j+1}^{\tau}$$
$$- \frac{3u_j^{\tau}}{\Delta x}\Delta t\left\{\beta_{\tau+1}\left[\left(1 - \frac{2p}{\Delta x}\right)\phi_{j+1}^{\tau+1} - \frac{4p}{\Delta x}\phi_j^{\tau+1} - \left(1 + \frac{2p}{\Delta x}\right)\phi_{j-1}^{\tau+1}\right]\right.$$
$$\left. + \beta_{\tau}\left[\left(1 - \frac{2p}{\Delta x}\right)\phi_{j+1}^{\tau} - \frac{4p}{\Delta x}\phi_j^{\tau} - \left(1 + \frac{2p}{\Delta x}\right)\phi_{j-1}^{\tau}\right]\right\}$$

where $\beta_{\tau+1} + \beta_{\tau} = 1$.

In selecting an approximation scheme to represent the nonlinear advection terms it is desirable to utilize a representation that, in the linear formulation, either has both accurate phase and amplitude characteristics, or if phase speed is poorly treated, the scheme is damping so that the erroneous dispersion of the technique does not remain a part of the solution. If the dispersive solutions remain, nonlinear aliasing can occur, as explained later in this section, causing the solutions to degrade into computational noise. Thus, the amplitude-preserving characteristic of schemes such as the leapfrog and the finite element schemes does not by itself guarantee meaningful simulations. Baer and Simons (1970), for example, have reported that in approximating nonlinear advection terms, individual energy components may have large errors when the total energy has essentially none. They further conclude that neither conservation of integral properties nor satisfactory prediction of amplitude is sufficient to justify confidence in the results—one must also assure the accurate calculation of the phase speeds.

The spline representation and the finite element method with damping ($p = 0.01$) have a maximum damping per time step for a 4 Δx wave of 2.8% and 0.5%, respectively, while the leapfrog and finite element techniques ($p = 0$) have no damping. The maximum erroneous dispersion per time step of the spline and damped finite element schemes applied to a 4 Δx wave is 4.5% and 18.1%, respectively, contrasted to 36.3% and 18.1% for the leapfrog and finite element ($p = 0$) techniques. Grotjahn (1977) concludes that such fictitious dispersion resulting from a poor phase representation can lead to even more serious errors in the computation of the group velocity. Because the advection speed of 4 Δx waves is handled poorly by the latter two schemes yet no amplitude damping occurs, it is necessary to apply an external filter in order to minimize the effect of these erroneously traveling waves on the meteorological important solutions.

Figure 5 (reproduced from Long and Pepper, 1976) contrasts the use of Schemes I, III, and IV to simulate the rotation of a cone about a constant grid. As evident in the figure, the poor treatment of phase for short waves in the finite element scheme generates a poorer solution than does the spline, while the large computational damping of Scheme I has almost eliminated the cone.

The solution to the advection terms in the complete conservation relations using the spline interpolation or finite element methods is usually performed separately for each spatial direction. Such a procedure for calculating individual terms in the conservation relations separately is called *splitting* (e.g., Long and Hicks, 1975; Mesinger and Arakawa, 1976) and greatly simplifies the numerical computations. Otherwise either two- or

TABLE III. Values of λ and C_R/U as a Function of Wavelength and Courant Number[a]

Scheme		Wave length	Courant number												
			0.001	0.01	0.1	0.2	0.3	0.4	0.5	0.6	0.7	0.8	0.9	1.0	1.5	2
I. Forward-in-time linear-interpolation upstream	λ	$2\Delta x$	0.998	0.980	0.800	0.600	0.400	0.200	0.000	0.200	0.400	0.600	0.800	1.000		
		$4\Delta x$	0.999	0.990	0.906	0.825	0.762	0.721	0.707	0.721	0.762	0.825	0.906	1.000		
		$10\Delta x$	1.000	0.998	0.983	0.969	0.959	0.959	0.951	0.953	0.959	0.969	0.983	1.000		($\lambda > 1$)
		$20\Delta x$	1.000	1.000	0.996	0.992	0.990	0.988	0.988	0.988	0.990	0.992	0.996	1.000		
	C_R/U	$2\Delta x$	0.000	0.000	0.000	0.000	0.000	0.000	1.000	1.667	1.429	1.250	1.111	1.000		
		$4\Delta x$	0.637	0.643	0.704	0.780	0.859	0.936	1.000	1.043	1.060	1.055	1.033	1.000		
		$10\Delta x$	0.936	0.937	0.953	0.968	0.981	0.992	1.000	1.005	1.008	1.008	1.005	1.000		
		$20\Delta x$	0.984	0.984	0.988	0.992	0.995	0.998	1.000	1.001	1.002	1.002	1.001	1.000		
II. Centered-time, centered-in-space (leapfrog)	λ	$2\Delta x$	1.000	1.000	1.000	1.000	1.000	1.000	1.000	1.000	1.000	1.000	1.000	1.000		
		$4\Delta x$	1.000	1.000	1.000	1.000	1.000	1.000	1.000	1.000	1.000	1.000	1.000	1.000		
		$10\Delta x$	1.000	1.000	1.000	1.000	1.000	1.000	1.000	1.000	1.000	1.000	1.000	1.000		($\lambda > 1$)
		$20\Delta x$	1.000	1.000	1.000	1.000	1.000	1.000	1.000	1.000	1.000	1.000	1.000	1.000		
	$C_R/U_{\text{physical mode}}$	$2\Delta x$	0.0	0.0	0.0	0.0	0.0	0.0	0.0	0.0	0.0	0.0	0.0	0.0		
		$4\Delta x$	0.637	0.637	0.638	0.641	0.647	0.655	0.667	0.683	0.705	0.738	0.762	1.000		
		$10\Delta x$	0.935	0.935	0.936	0.938	0.940	0.944	0.950	0.956	0.964	0.974	0.986	1.000		
		$20\Delta x$	0.984	0.984	0.984	0.984	0.985	0.986	0.988	0.989	0.991	0.994	0.997	1.000		
	$C_R/U_{\text{computational mode}}$	$2\Delta x$	>100	100.0	10.00	5.000	3.333	2.500	2.000	1.667	1.429	1.250	1.111	1.000		
		$4\Delta x$	>100	199.4	19.36	9.359	6.020	4.345	3.333	2.651	2.152	1.762	1.430	1.000		
		$10\Delta x$	>100	499.6	49.06	24.06	15.73	11.56	9.051	7.377	6.179	5.276	4.570	4.000		
		$20\Delta x$	>100	999.0	99.02	49.02	32.35	24.01	19.01	15.68	13.29	11.51	10.11	9.000		

													($\lambda > 1$)
III. Forward-in-time, upstream spline interpolation	λ	2 Δx	1.000	0.999	0.944	0.762	0.568	0.296	0.000	0.296	0.568	0.762	0.944	1.000
		4 Δx	1.000	1.000	0.997	0.989	0.981	0.975	0.972	0.975	0.981	0.989	0.997	1.000	0.888
		10 Δx	1.000	1.000	1.000	1.000	1.000	1.000	1.000	1.000	1.000	1.000	1.000	1.000	0.996
		20 Δx	1.000	1.000	1.000	1.000	1.000	1.000	1.000	1.000	1.000	1.000	1.000	1.000	1.000
	C_R/U	2 Δx	0.000	0.000	0.000	0.000	0.000	0.000	1.000	1.667	1.429	1.250	1.111	0.000	0.000
		4 Δx	0.955	0.955	0.958	0.967	0.979	0.980	1.000	1.007	1.009	1.008	1.005	1.000	1.042
		10 Δx	0.999	0.999	0.989	0.999	1.000	1.000	1.000	1.000	1.000	1.000	1.000	1.000	1.001
		20 Δx	1.000	1.000	1.000	1.000	1.000	1.000	1.000	1.000	1.000	1.000	1.000	1.000	1.000
IV. Finite element, chapeau basic function, implicit advection representation (equivalent to Scheme V with $p = 0.0$)	λ	2 Δx	1.000	1.000	1.000	1.000	1.000	1.000	1.000	1.000	1.000	1.000	1.000	1.000	1.000
		4 Δx	1.000	1.000	1.000	1.000	1.000	1.000	1.000	1.000	1.000	1.000	1.000	1.000	1.000
		10 Δx	1.000	1.000	1.000	1.000	1.000	1.000	1.000	1.000	1.000	1.000	1.000	1.000	1.000
		20 Δx	1.000	1.000	1.000	1.000	1.000	1.000	1.000	1.000	1.000	1.000	1.000	1.000	1.000
	C_R/U	2 Δx	0.000	0.000	0.000	0.000	0.000	0.000	0.000	0.000	0.000	0.000	0.000	0.000	0.000
		4 Δx	0.955	0.955	0.953	0.948	0.939	0.928	0.914	0.897	0.879	0.860	0.840	0.819	0.717
		10 Δx	0.999	0.999	0.999	0.998	0.996	0.994	0.991	0.988	0.983	0.979	0.974	0.968	0.934
		20 Δx	1.000	1.000	1.000	1.000	0.999	0.999	0.998	0.997	0.996	0.995	0.993	0.992	0.982
V. Finite element, chapeau basis functions with damping, implicit advection representation $p = 0.01$	λ	2 Δx	1.000	0.999	0.988	0.976	0.965	0.953	0.942	0.931	0.919	0.908	0.898	0.887	0.835
		4 Δx	1.000	1.000	0.999	0.999	0.998	0.997	0.997	0.996	0.996	0.996	0.995	0.995	0.995
		10 Δx	1.000	1.000	1.000	1.000	1.000	1.000	1.000	1.000	1.000	1.000	1.000	1.000	1.000
		20 Δx	1.000	1.000	1.000	1.000	1.000	1.000	1.000	1.000	1.000	1.000	1.000	1.000	1.000
	C_R/U	2 Δx	0.000	0.000	0.000	0.000	0.000	0.000	0.000	0.000	0.000	0.000	0.000	0.000	0.000
		4 Δx	0.955	0.955	0.953	0.948	0.939	0.928	0.914	0.897	0.879	0.860	0.840	0.819	0.717
		10 Δx	0.999	0.999	0.999	0.998	0.996	0.994	0.991	0.988	0.983	0.979	0.974	0.968	0.934
		20 Δx	1.000	1.000	1.000	1.000	0.999	0.999	0.998	0.997	0.996	0.995	0.993	0.992	0.982

[a] Calculated by C. Martin.

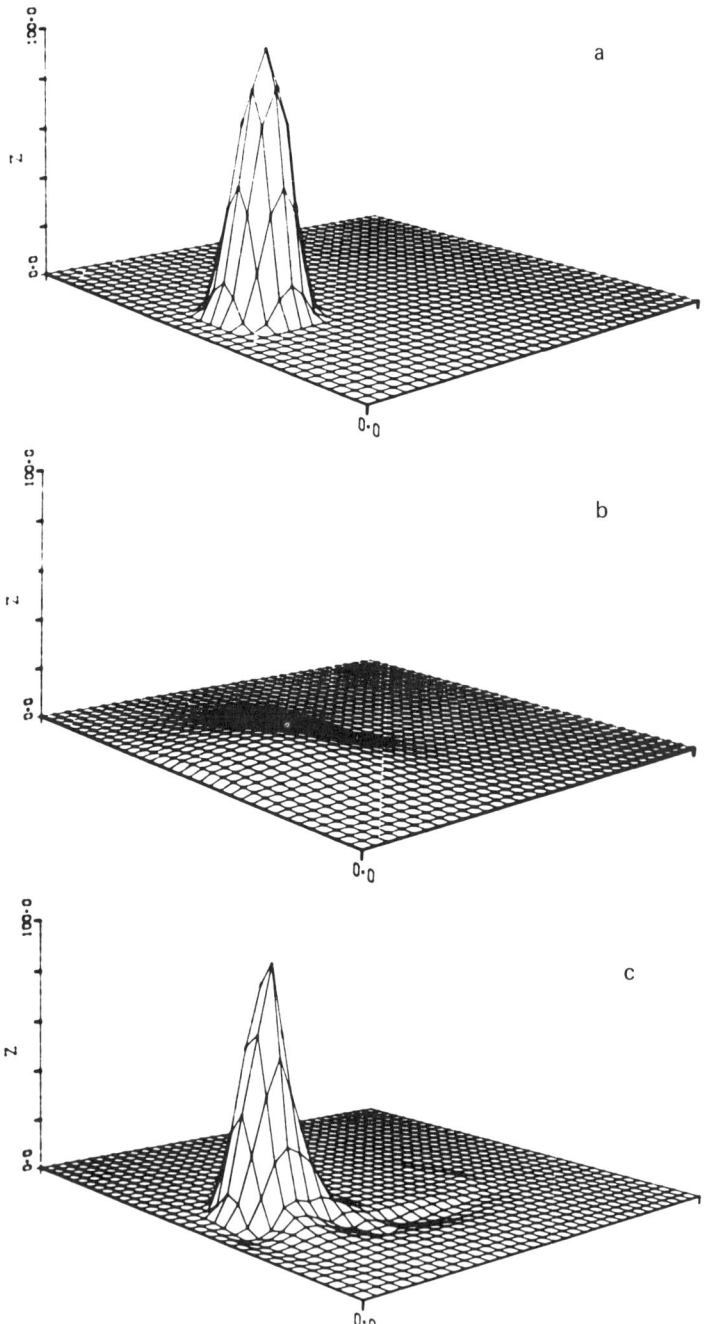

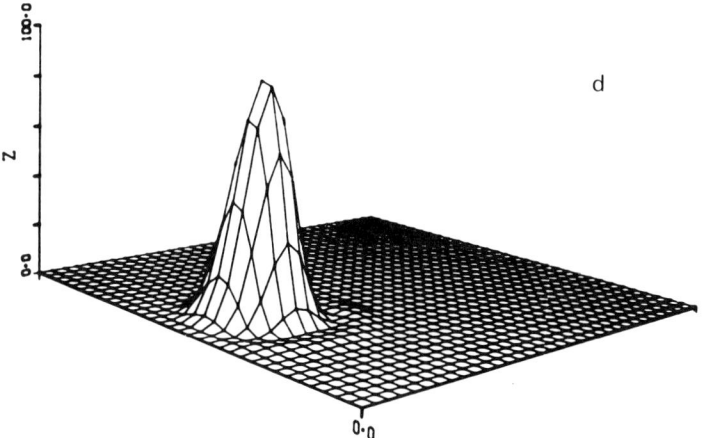

FIG. 5. Use of Schemes I, III, and IV (b, d, and c, respectively) to represent the advection of a cone (given at the initial time by (a)) one revolution around a constant grid. (From Long and Pepper, 1976.)

three-dimensional interpolation formulas (in the case of the spline) or basis functions (for the finite element method) would have to be used. Splitting is also referred to as the *Marchuk method* (Mesinger and Arakawa, 1976).

Useful discussions of formulations for the advection terms, including comparisons between the results using different forms of the finite element and finite difference techniques, are given in papers by Long and Pepper (1976), Lee *et al.* (1976), Long and Hicks (1975), Gresho *et al.* (1976), Cullen (1976), and Wang *et al.* (1972). Gresho *et al.* (1979), Lee and Gresho (1977), and other reports and publications prepared by their group at the Lawrence Livermore Laboratory discuss the finite element technique in detail as it is applied to the primitive equations. Richtmyer and Morton (1967), Kreiss and Oliger (1973), and Mesinger and Arakawa (1976) give rather extensive discussions of initial-value problems along with an outline of principal finite difference methods in use at the time of their writing.

10.2. Diffusion Terms

Other terms in the conservation relations can be similarly investigated. The vertical subgrid correlation terms, for instance, can be represented using the one-dimensional linear diffusion equation as

$$\frac{\partial \phi}{\partial t} = K \frac{\partial^2 \phi}{\partial z^2}$$

TABLE IV. APPROXIMATE FORMULATIONS FOR ONE-DIMENSIONAL DIFFUSION

I. Forward-in-time, centered-in-space, explicit

$$\frac{\phi_j^{\tau+1} - \phi_j^{\tau}}{\Delta t} = K_{j+1/2}\frac{\phi_{j+1}^{\tau} - \phi_j^{\tau}}{(\Delta z)^2} - K_{j-1/2}\frac{\phi_j^{\tau} - \phi_{j-1}^{\tau}}{(\Delta z)^2}$$

II. Forward-in-time, centered-in-space, implicit (e.g., Paegle et al., 1976)

$$\frac{\phi_{j+1}^{\tau} - \phi_j^{\tau}}{\Delta t} = \frac{K_{j+1/2}}{(\Delta z)^2}[\beta_\tau(\phi_{j+1}^{\tau} - \phi_j^{\tau}) + \beta_{\tau+1}(\phi_{j+1}^{\tau+1} - \phi_j^{\tau})]$$

$$- \frac{K_{j-1/2}}{(\Delta z)^2}[\beta_\tau(\phi_j^{\tau} - \phi_{j-1}^{\tau}) + \beta_{\tau+1}(\phi_j^{\tau+1} - \phi_{j-1}^{\tau+1})]$$

which has the solution $\phi(z, t) = \tilde{\phi}(k, \omega) \exp(-Kk^2 t + ikz)$. The value of λ_{exact} (the exact damping per time step) is therefore equal to $\exp[-\gamma(2\pi)^2/n^2]$, where $\gamma = K \Delta t/(\Delta z)^2$ is called the *Fourier number* and n is the number of grid points per wave. The linear coefficient K used here represents the turbulent exchange coefficient discussed in Section 7. Two numerical representations of this expression used in mesoscale models, with Δz constant but K variable, are given in Table IV, while values of $\lambda/\lambda_{\text{exact}}$ for K equal to a constant are given in Table V for an explicit and implicit representation.

In Table V, for a given value of γ the 2 Δx wave is most poorly represented. In addition, the 2 Δx wave is always insufficiently damped and often the value of λ is negative, yielding a wave whose amplitude reverses sign (flip-flops) each time step. The explicit approximations also become more accurate as γ becomes smaller, while the implicit form, with partial weighting of the old and new values, gives reasonable values even when the explicit form is linearly unstable.

When $\beta_\tau = \beta_{\tau+1}$ the representation is called the Crank–Nicholson scheme. As evident in Table V, a longer time step is permitted when an implicit representation of the diffusion is used—a result reported by Paegle et al. (1976), and reproduced by Mahrer and Pielke (1978b). In the latter paper, the simulation of the growth of the heated planetary boundary layer for day 33 of the Wangara experiment ran 17 times faster than the explicit formulation ($\beta_\tau = 1$) using the implicit formulation with $\beta_\tau = 0.25$ and a longer time step, as suggested by Paegle et al. (1976). The results were essentially identical.

10.3. Other Terms

The remaining terms in the conservation relations need to be similarly evaluated. It can be shown, for example, that an implicit representation

TABLE V. VALUES OF λ/λ_{exact} AS A FUNCTION OF WAVELENGTH AND FOURIER NUMBER[a]

Scheme	Wavelength	γ									
		0.1	0.2	0.3	0.4	0.5	0.6	0.7	0.8	0.9	1.0
Forward-in-time, centered-in-space diffusion											
Explicit $\beta_\tau = 1$	2 Δx	1.610	1.440	−3.863	−31.094	<−100					
	4 Δx	1.024	0.983	0.839	0.537	0.0	($\lambda > 1$ for a 2 Δx wave)				
	10 Δx	1.001	0.999	0.997	0.992	0.986					
	20 Δx	1.000	1.000	1.000	1.000	0.999					
Implicit $\beta_\tau = 0.7$	2 Δx	1.725	2.554	2.272	−4.202	−34.761	<−100	<−100	<−100	<−100	<−100
	4 Δx	1.038	1.053	1.030	0.952	0.792	0.517	0.079	−0.584	−1.555	−2.948
	10 Δx	1.001	1.001	1.001	1.000	0.998	0.996	0.992	0.988	0.982	0.975
	20 Δx	1.000	1.000	1.000	1.000	1.000	1.000	1.000	0.999	0.999	0.999
$\beta_\tau = 0.5$	2 Δx	1.789	3.085	4.829	5.758	0.00	−33.91	<−100	<−100	<−100	<−100
	4 Δx	1.047	1.092	1.129	1.150	1.145	1.099	0.993	0.800	0.485	0.00
	10 Δx	1.001	1.003	1.004	1.005	1.006	1.007	1.007	1.008	1.008	1.008
	20 Δx	1.000	1.000	1.000	1.000	1.000	1.000	1.000	1.001	1.001	1.001
$\beta_\tau = 0.3$	2 Δx	1.845	3.507	6.718	12.711	23.17	38.97	54.11	33.16	<100	<100
	4 Δx	1.055	1.126	1.211	1.307	1.414	1.529	1.648	1.766	1.875	1.965
	10 Δx	1.002	1.004	1.006	1.009	1.013	1.017	1.021	1.026	1.031	1.037
	20 Δx	1.000	1.000	1.000	1.001	1.001	1.001	1.001	1.002	1.002	1.003
$\beta_\tau = 0.1$	2 Δx	1.916	3.999	8.779	19.93	46.35	>100	>100	>100	>100	>100
	4 Δx	1.067	1.170	1.310	1.491	1.717	1.998	2.344	2.769	3.290	3.931
	10 Δx	1.002	1.005	1.010	1.016	1.023	1.031	1.040	1.050	1.062	1.074
	20 Δx	1.000	1.000	1.001	1.001	1.002	1.002	1.003	1.004	1.004	1.005

[a] Calculated by C. Martin.

for the Coriolis terms given by

$$\frac{u_j^{\tau+1} - u_j^\tau}{\Delta t} = f v_j^\tau; \qquad \frac{v_j^{\tau+1} - v_j^\tau}{\Delta t} = -f u_j^{\tau+1}$$

is neutrally stable, whereas replacing $u_j^{\tau+1}$ by u_j^τ in the right-hand expression results in a linearly unstable solution. The linear stability of approximate solution techniques is not intuitively obvious.

When internal gravity waves are an important component of the solutions to the conservation relations, it is necessary to consider how well the numerical representation approximates their speed and amplitude.

Recently, Sun (1980) has investigated the linear stability of finite difference approximations to equations of the form given by (41) and (43)–(45) with $f = 0$, $\rho'/\rho_0 = \theta'/\theta_0$, and $\lambda_2 = 0$. As shown in Section 5, this system of equations has internal gravity waves as the solution. Using approximate solution techniques, Sun showed that when the hydrostatic assumption is used ($\lambda_1 = 0$), the dependent variables are staggered in space, updated velocities are used in the potential temperature equation, and the pressure gradient is approximated by a centered-in-space scheme, a computationally stable solution results. This sequence of calculations discussed by Sun is called the forward-backward time integration scheme and has been adopted in a number of mesoscale models (e.g., Pielke, 1974a,b; Bhumralkar, 1972; Jones, 1973). Bhumralkar (1972) similarly found that unless updated values of velocity were used in the computation of potential temperature, the results would be linearly unstable. Also, as he and others have concluded, the time step must be less than or equal to the time it takes a disturbance to propagate between grid points, otherwise the solutions will be unstable.

10.4. Other Calculations

In a hydrostatic model, pressure is determined by vertical integration of Eqs. (26), (28), (35), (36), or (82), usually starting at the top of the domain. Vertical velocity is similarly computed in such models by the upward integration of Eq. (21) or (79) from the bottom.

In a nonhydrostatic model the computation of pressure is not so straightforward. When the anelastic assumption is used (e.g., $\lambda_2 = 0$), modelers obtain a diagnostic equation for pressure by taking the three-dimensional divergence of the conservation of motion relation [e.g., Eqs. (31) and (33)], using Eq. (21) to eliminate the local tendency term.

Recently, some investigators have retained sound waves as part of their solutions (by retaining $\partial \bar{p}/\partial t$) and have decomposed the conservation equations into those terms that significantly influence sound wave propagation and those that do not. The terms involving the propagation of sound waves in the model are then evaluated with very short time steps, while the remainder of the physical processes in the model, such as advection and internal gravity wave propagation, are computed with longer time steps. The short time steps are used because the computational approximation of the sound waves will become linearly unstable using the solution techniques typically applied unless the time step is less than the time it takes a wave to travel between grid points. Tapp and White (1976) use this formulation, called *time splitting,* in their mesoscale simulation of the airflow over south Florida, while Klemp and Wilhelmson (1978a,b) use it to represent three-dimensional cumulus cloud dynamics. Daley (1980) discusses the splitting of fast gravity wave computations from slower speed gravity and Rossby waves using a procedure called model normal mode expansion for the fast waves.

10.5. Nonlinear Effects

In a turbulent medium such as the atmosphere, kinetic energy produced on large scales cascades to smaller and smaller sizes until the energy is dissipated by molecular friction into heat. This transfer of energy is caused by the interaction of different scales of motion and is represented mathematically by the nonlinear advection terms. On the mesoscale, kinetic energy is produced by such forcings as horizontal temperature gradients near the ground, and is eventually removed by molecular interactions on a scale of 1 cm or less. Somewhere in between these scales, motion is not directly influenced by either molecular processes or kinetic energy generation so that the relation between kinetic energy per unit wavenumber per unit mass, $E(k)$, wavenumber, and the rate at which energy is being removed at the much smaller scales, ϵ, is given by

$$E(k)dk = a\epsilon^{2/3}k^{-5/3}$$

The region of wavenumber space in which kinetic energy is transferred to smaller scales in this fashion is called the *inertial subrange* [e.g., see Lumley and Panofsky (1964) for a detailed discussion]. Gage (1979) discusses the observed occurrence of such a $k^{-5/3}$ relation in mesoscale and larger atmospheric features.

In a numerical mesoscale model, however, this cascade of energy to smaller scales cannot occur because the smallest feature that can be resolved has a wavelength of twice the grid spacing. As discussed in Section 4, the grid intervals in meteorological models cannot be small enough to represent both molecular processes, and yet still stimulate mesoscale atmospheric phenomena.

To illustrate this problem let

$$\phi_1 = \tilde{\phi} \cos k_1 \Delta x; \qquad \phi_2 = \tilde{\phi} \cos k_2 \Delta x$$

represent two waves of the same amplitude but different horizontal wavenumbers. If the waves interact, such as through the advection terms in the prognostic conservation equations, the result can be expressed as

$$\phi_1\phi_2 = \tilde{\phi}^2 \cos k_1 \Delta x \cos k_2 \Delta x$$
$$= \tfrac{1}{2}\tilde{\phi}^2[\cos (k_1 + k_2) \Delta x + \cos (k_1 - k_2) \Delta x]$$

which represents two waves with wavenumbers $k_1 + k_2$ and $k_1 - k_2$. If the lengths of the original waves were $2 \Delta x$ and $4 \Delta x$, in terms of the model grid, then the results should be a $1.33 \Delta x$ and a $4 \Delta x$ wave. Although the latter feature can be resolved, the $1.33 \Delta x$ wave cannot and instead will appear to be the first integer multiple of $1.33 \Delta x$ equal to $n \Delta x$, where n is an integer greater or equal to 2. In this case an erroneous $4 \Delta x$ wave will appear. Waves that result because of the lack of grid resolution are said to have *aliased* or *folded* to longer wavelengths.

Listed as follows are several wave–wave interactions that will produce aliased waves:

Interacting wavelengths	Should produce	Will produce
$2 \Delta x$ and $2 \Delta x$	$1 \Delta x$	Add a constant to entire model
$2 \Delta x$ and $4 \Delta x$	$1.33 \Delta x$	$4 \Delta x$
$2 \Delta x$ and $6 \Delta x$	$1.5 \Delta x$	$3 \Delta x$
$2 \Delta x$ and $8 \Delta x$	$1.6 \Delta x$	$8 \Delta x$
$2 \Delta x$ and $10 \Delta x$	$1.67 \Delta x$	$5 \Delta x$

Even if no $2 \Delta x$ waves are present initially, they will be created since the interaction of two $4 \Delta x$ waves will generate a $2 \Delta x$ wave, while longer wave–wave interactions will produce $4 \Delta x$ waves. Kinetic energy input to the model at the larger scales, therefore, cascades to smaller scales but, because dissipation is not properly represented, it will not be removed from the model but will accumulate fictitiously. Thus, even if the computational scheme is linearly stable, the solutions can degrade into physically meaningless computational noise. Indeed, this accumulation of energy can cause the model dependent variables to increase in magnitude without bound—an error that is referred to as *nonlinear instability*.

As seen in the discussions of both linear and nonlinear stability, features with wavelengths less than $4 \Delta x$ can cause serious computational errors. Therefore, since such features are poorly resolved and should not be a crucial component in the physical interpretation of the model results, it is desirable to remove these smaller waves. This can be accomplished by the following:

- the accurate parameterization of the subgrid-scale fluxes (e.g., $\overline{u_i'' u_j''}$, $\overline{\theta'' u_j''}$) so that energy is extracted from the averaged equations in a manner consistent with reality *or*
- the use of a spatial smoother that removes the shortest waves but leaves the longer wavelengths unaffected.

The first method is the most attractive, of course, since it is based on fundamental principles; however, little is known about horizontal subgrid-scale fluxes on the mesoscale. Only with the vertical subgrid fluxes, as discussed in Section 7, is there enough physical insight to develop a realistic parameterization. For this reason, the first method, at least in the planetary boundary layer, is usually used to eliminate short waves in the vertical, while smoothers are used in the horizontal.

The form of the horizontal smoother often takes the appearance of a diffusion equation patterned after the formulation for the vertical exchange coefficient. Tag *et al.* (1979) provides a useful discussion of several forms of these variable eddy coefficient formulations.

The alternative to an explicit diffusion equation is to formally apply a filter such as discussed by Shapiro (1970). Cullen (1976) compares solutions in a simplified synoptic scale model using several types of explicit diffusion representations and filters, and shows that selective filters can be used more effectively in finite element representations than with finite difference techniques because of the greater accuracy of the first method. The use of optimal filters is also discussed recently by Storch (1978), while Jones (1977b) outlines a smoothing technique to control computational noise as information is transferred between coarse and fine grids in a hurricane model. One particularly useful selective low-pass filter was developed by P. E. Long (at the Techniques Development Lab of the NWS) and is given by

$$(119) \quad (1 - \delta)\phi_{i+1}^* + 2(1 + \delta)\phi_i^* + (1 - \delta)\phi_{i-1}^* = \phi_{i+1} + 2\phi_i + \phi_{i-1}$$

where ϕ_{i-1}^*, ϕ_i^*, and ϕ_{i+1}^* are the smoothed values and δ is an arbitrary constant. Using the transformation given by Eq. (118), this filter results in a dependent variable with the amplitude as a function of wavenumber after one application of the filter (assumed here to be one time step):

$$\lambda = \frac{\cos k \, \Delta x + 1}{(1 - \delta) \cos k \, \Delta x + 1 + \delta}$$

A $2 \, \Delta x$ wave is removed by this formulation regardless of the value of δ, while very long waves are essentially unaffected. The degree of damping of intermediate waves depends on the value of δ. For a 4 Δx and 8 Δx wave, for example, λ = 0.9901 and 0.9983 for δ = 0.01; and λ = 0.9091 and 0.9831 for δ = 0.1. Mahrer and Pielke (1978b) provide a table with values of λ. As shown by P. E. Long (personal communication, 1979) this filter is equivalent to a diffusion coefficient given by

$$K = -\frac{1}{k^2 \, \Delta t} \ln \lambda$$

when it is applied at each time step.

All mesoscale models employ some sort of horizontal filtering to control nonlinear aliasing. Either explicit smoothers, such as that given by Eq. (119), are used or implicit computational diffusion inherent to the numerical approximation, such as with Scheme I or V, provide the necessary removal of the shortest wavelengths.

In this section approximate solution techniques to the conservation relations have been discussed. Among the major conclusions are the following:

- When advection is considered an important component of the mesoscale circulation, the advection terms should be approximated with formulations that produce accurate predictions of phase *and* amplitude for wavelengths 4 Δx and larger. The interpolated spline and finite element techniques are examples of schemes that possess this attribute. When the amplitude-preserving finite element approach without damping is used, however, it is necessary to use a filter to prevent problems associated with the fictitious dispersion of the shortest waves.
- When vertical subgrid-scale mixing is considered an important component of the mesoscale circulation, it is desirable to use the implicit representation as suggested by Paegle *et al.* (1976), which is reasonably accurate for wavelengths greater than 4 Δz yet computationally economical.
- In order to obtain computationally stable results in the hydrostatic form of the conservation equations, the forward–backward time integration scheme is useful. Such a formulation is needed when internal gravity waves or other physical phenomena caused by the horizontal pressure gradient force occur. Spatial staggering of the dependent variables also appears to be required (Sun, 1980).
- When the Coriolis force is an important component of a mesoscale circulation, an implicit form must be utilized in order to assure linear computational stability.

- With a nonlinear representation, aliasing causes the fictitious accumulation of energy on wavelengths shorter than 4 Δx. To eliminate this problem it is necessary to apply horizontal diffusion either through an explicit representation such as a smoother, or implicitly as part of the computational scheme.

11. Boundary and Initial Conditions

In the last section, methodologies for obtaining solutions to the conservation relations were introduced. In those discussions, it was shown that certain approximations to the differential equations produce more accurate solutions than others. Once optimal approximate forms of the equations are selected, however, it is still necessary to define the domain and grid structure [e.g., Eqs. (30) and (19)] over which the equations will be evaluated, as well as the boundary and initial conditions required to solve the set of equations.

11.1. Grid and Domain Structure

The selection of the model domain size and grid increments in a mesoscale model are dictated by the following constraints:
- the dimensionality of the forcing
- the spatial scales of the forcing
- the available computer resources.

Three-dimensional spatial grids are the most general, but until about 10 years ago they were difficult to achieve because of limitations of computer resources. Nonetheless, studies using two-dimensional models such as Lavoie's (1972, 1974) horizontal ($x-y$) representations of the airflow over Lake Erie and over the island of Oahu, and Estoque's (1961, 1962) $x-z$ sea breeze simulations yielded considerable new insight into mesoscale processes. In order to produce economical calculations, Lavoie averaged the dependent variables over the depth of the boundary layer so that the explicit vertical dependence in the conservation relations was removed. This approach of averaging the dependent variables through one or more vertical layers, also used by Lee (1973), is called a *layered representation*. On the other hand, Estoque and others used a vertical cross section with values of the dependent variables defined at specific heights to provide vertical as well as horizontal resolution of mesoscale structure. This use of multiple vertical levels is called a *discrete-level representation*.

Although three-dimensional representations are required for most actual mesoscale circulations (Pielke, 1974b), two-dimensional cross-section simulations represent a relatively inexpensive but effective technique for examining the physics of mesoscale features.

The spatial scales of the mesoscale forcing determine the permissible domain and grid sizes. To properly represent mesoscale systems, the following are necessary:

- The meteorologically significant variations in the dependent variables caused by the mesoscale forcing must be contained within the region of model spatial integration.
- The averaging volume used to define the model grid [e.g., Eq. (20) or (86)] must be sufficiently small so that the mesoscale forcing is accurately resolved.

The first criterion listed above implies that D_x and D_y in Eq. (30) must be large enough such that $\phi' \cong 0$ at the lateral boundaries of the model. The second requirement specifies that the significant portion of the generation of energy by the mesoscale feature of interest must have a scale equal to or larger than four times the grid spacing. As examples, the model domain size for the Alps would be different than the size required for the island of Barbados, while the grid increments needed to properly represent the Blue Ridge Mountains of Virginia are smaller than those that are acceptable for over south Florida. In representative cross sections through the Blue Ridge, Pielke and Kennedy (1980) found that about 80% of the terrain irregularities occur on scales larger than about 6 km, so that a 1.5-km horizontal grid (i.e., one-fourth the scale length) is required to adequately resolve this portion of the topographic relief, while over south Florida Pielke (1974a) and Pielke and Mahrer (1978) found that an 11-km grid interval gave accurate results. The use of Fourier analysis to represent geographic forcing such as topography, z_G, or other surface characteristics (e.g., water, type of soil and vegetation) in terms of horizontal wavelengths is an effective tool to determine the necessary spatial grid resolution as determined in the preliminary study by Pielke and Kennedy (1980).

The amount of grid resolution required can also be determined by numerical experiment, where the grid intervals are decreased until the change in the solution by increasing the resolution is small. Pielke (1974b) used this approach to show that for uniform land and sea surfaces, seven vertical levels with a model top of 3.6 km was sufficient to resolve the sea breeze circulation.

To increase the domain size of mesoscale models, but without adding extra grid points, many modelers use a *stretch-grid* formulation in which the separation between grid points expands near the boundaries of the model. In the vertical a stretched grid is used to provide enhanced resolution within the planetary boundary layer (e.g., Orlanski et al., 1974), while in the horizontal such an approach is used to minimize the influence of the lateral boundaries (e.g., Lee, 1973; Anthes, 1970). Although a mathematically formal transformation can be applied to the conservation equations in order to take account of the stretching, simply expanding the grid intervals produces the same result.

Numerical results can be degraded with a stretched grid, however. Brown and Pandolfo (1979), for example, examine the computational stability associated with nonuniform grid intervals and show that instability can occur even when the analysis for a constant grid indicates linear stability.

A second approach to expand the domain size is to use *grid meshing* (also called *nested grids*), where a grid with fine resolution is embedded within one or more coarser grids. Perturbations can be permitted to enter and to leave the fine-grid mesh (*two-way interaction*). Each separate grid has a constant but different (fixed-grid) interval. Mathur (1974) used this approach to provide detailed resolution near a simulated hurricane as well as to obtain information concerning the larger-scale synoptic environment. Elsberry (1978) provides a short review of nested grid procedures.

One problem with the stretched-grid and grid-mesh representations is that, as shown in Section 10, short-wave features in a numerical grid travel at different speeds depending on the grid size (e.g., see Table III, C_R/U values) so that a wave that could be well resolved in the region of fine resolution will only be poorly handled in the region of larger grid increments. Also, near the regions where the grid interval size is changing, erroneous reflection and refraction of waves occur in much the same fashion as when light travels from a material of one index of refraction to another. Perkey and Kreitzberg (1976) have attempted to minimize this problem by permitting perturbations from the coarse grid to enter the fine mesh, but using a filter to prevent short-wave length features in the fine mesh from entering the larger-interval grid. They call this a *parasitic grid representation* since the interaction is one-way. Perkey (1976) used such an approach in modeling squall lines and other cumulus convective phenomena generated by midlatitude cyclones; however, Fritsch and Maddox (1980) have recently shown that there is strong feedback of convection to the synoptic scale. Two-way interactive grids, such as that suggested by Jones (1977b), therefore, seem essential for this mesoscale problem.

Lagrangian grids, which move relative to the Earth's surface, can also be useful. Schlesinger (1973) utilized a movable grid in order to prevent a simulated thunderstorm from existing his domain, while Jones (1977a) used three meshed grids with intervals of 10, 30, and 90 km to simulate the dynamics of a moving hurricane, with the two smaller grids moving with the storm. Ookochi (1978) and Kurihara *et al.* (1979) also report on a system of moving nested grids for tropical storm simulations.

11.2. Grid Staggering

As reported in Section 10, Sun (1980) stated that a staggered grid is required in order to generate computationally stable solutions using the forward-backward time integration with a hydrostatic model. Even if linear stability is obtained, however, the conservation relations may be better represented if certain types of staggered grids are used. Lilly (1961), for example, presents an example of a staggered grid that helps preserve such properties as total kinetic energy in a model. Grotjahn (1977) suggests that staggering can sometimes significantly reduce phase and group velocity errors present in finite differencing schemes.

Use of a staggered grid is also motivated by the differential nature of the conservation relations since the effective resolution is increased by a factor of two when derivatives are defined across one grid interval, rather than two. Sun (1980) and Pielke (1974a) present examples of staggered grids used to take advantage of this available added resolution, which results in no increase in computation time.

The solutions of a set of conservation relations given, for example, by Eqs. (21), (22), (24), (25), (28), and (32); by Eqs. (21), (22), (24)–(26), (31), and (6)–(8) using $\bar{T}_v$, $\bar{\theta}$, $\bar{\rho}$, $\bar{p}$, and $\bar{q}_3$; and by Eqs. (79)–(85), in Sections 4 and 6 represent an *initial–boundary*-value problem. In general, the linear form of these relations is *hyperbolic* (see Haltiner and Williams, 1980, p. 443). Therefore, in order to commence the integration of the model equations, initial values of the dependent variables are needed. The process to obtain these values is called *initialization*. Also, throughout the remainder of the model integration, values of the dependent variables are needed on all sides of the model domain. These values are called *boundary* conditions.

11.3. Initialization

In terms of the relationship between the wind and potential temperature fields, the mesoscale and synoptic scales are quite different, as pointed out by Hoke and Anthes (1976). On the larger scale the distribution of temperature (which through the hydrostatic equation also reflects the distribution of mass) dominates the response of the wind field so that observed winds are usually closely approximated by the gradient wind assumption [see Smith and Lin (1978) for a discussion of various methods of determining vertical motion on the synoptic scale]. On the mesoscale in the midtroposphere, as shown by Hoke and Anthes using a two-dimensional simulation of a jet core, the reverse is true—the wind field dominates the mass field. In the lower troposphere, Daniel Keyser (personal communication, Penn State University, 1980) has suggested, however, that the mass still dominates velocity changes because of the presence of the rigid bottom ground boundary.

The methodology for initializing a mesoscale model can be grouped into two categories:

- objective analysis
- dynamic initialization.

11.3.1. Objective Analysis. With objective analysis two basic procedures can be used. Available observational data can be extrapolated to grid points using weighting functions (e.g., Goodin *et al.*, 1979), where the initial dependent variables are dependent on distance from the observations and then the mass divergence is reduced to a specified value (e.g., Goodin *et al.*, 1980). Measured data can also be interpolated throughout an analysis region by utilizing *variational analysis* such as suggested by Sasaki (1970a,b,c), Sasaki and Lewis (1970), and O'Brien (1970b), where, using variational calculus, one or more of the conservation relations along with one or more specified constraints are applied in order to minimize the variance of the difference between the observed and analyzed fields. Sasaki's approach is used by Sherman (1978), Dickerson (1978), and others to obtain estimates of wind fields over rough terrain. The use of such analysis tools to obtain wind fields is called *kinematic* or *diagnostic modeling*.

Satellite imagery may also be used to initialize mesoscale models. Rodgers *et al.* (1979) and Grody *et al.* (1979), for example, have illustrated how satellite imagery can be used to estimate winds in the vicinity of hurricanes, while Wilson and Houghton (1979) performed such an analysis for a severe storm outbreak in northern Texas. Purdom (1976), Parmenter (1974), Gurka (1974), Parmenter and Anderson (1974), and others have illustrated the important benefits of using such satellite data to derive mesoscale information, while Kreitzberg (1976) has summarized how satellite observations can be used with mesoscale numerical models to improve weather forecasts.

11.3.2. Dynamic Initialization. The dynamic initialization technique involves the integration of the approximate form of the conservation equations for a period of time such that short-wavelength features in the observed fields, which cannot be properly represented in the model grid, are removed. Such small-scale features are eliminated during the integration through filtering and their propagation as internal gravity waves through the lateral boundaries of the model as shown by Hoke and Anthes (1976). The actual model simulation begins after this initialization integration has generated model fields that are in approximate equilibrium. Hoke and Anthes (1976) perform this initialization by adding extra terms to the prognostic conservation equations, of the form

$$\frac{\partial \bar{u}}{\partial t} = -\bar{u}_j \frac{\partial \bar{u}_i}{\partial x_j} - \frac{1}{\bar{\rho}} \frac{\partial}{\partial x_j} \overline{\rho u_j'' u''} - \frac{1}{\rho_0} \frac{\partial \bar{p}}{\partial x_i} + f\tilde{v} - \hat{f}\tilde{w} + G_u(u_{obs} - \bar{u})$$

[derived from Eq. (31) with $i = 1$], where G_u is referred to as the *nudging coefficient*. Similar terms can appear in the other prognostic equations. The nudging coefficients are presumed to be a function of observation accuracy, distance between the observation and the grid point, the variable nudged, and the typical magnitudes of the other terms in the predictive equation. The major disadvantage of this approach is the cost in computer resources, since the initialization could require up to 12 hr or so of simulated time in a mesoscale calculation. Other examples of a dynamic initialization include those of Kurihara and Tuleya (1978, as updated in Kurihara and Bender, 1979), Anthes (1974a), Temperton (1973), and Hoke and Anthes (1977).

11.4. Initialization with Sparse Data

The objective analysis and dynamic initialization routines are both effective tools when observational data are available. On the mesoscale, however, extensive measurements, particularly of vertical structure, are available only during intensive field programs. Usually one or two radiosonde sites with soundings taken twice a day (at 12 GMT and 00 GMT) are available in a domain-sized region. Barnes and Lilly (1975), for example, suggest that sufficient variance exists on the mesoscale that upper air stations must be spaced about 100 km apart in order to detect important severe thunderstorm triggering mechanisms. Moreover, the initialization problem is complicated even further over rough terrain since one radiosonde ascent is not likely to be representative of the domain area.

The following initialization procedure is therefore suggested when available observation data are sparse:

(1) Use the available radiosonde measurements to predict the winds within the planetary boundary layer at each observation site using one-

dimensional forms of the horizontal conservation of motion relation such as

$$\frac{\partial u_R}{\partial t} = -\frac{\partial}{\partial z}\overline{u''w''}* + f(v_R - v_{g_R})$$

$$\frac{\partial v_R}{\partial t} = -\frac{\partial}{\partial z}\overline{v''w''}* - f(u_R - u_{g_R})$$

where u_R, v_R and u_{g_R}, v_{g_R} are the components of velocity and of geostrophic velocity at the radiosonde site, respectively. The overbar with an asterisk over the subgrid-scale term is used to indicate that it is defined for a grid-sized domain centered at the radiosonde site. To obtain values for u_R and v_R, the equations are integrated (invoking the assumption that the geostrophic wind is independent of time) until a balance is achieved [about six inertial periods (Mahrer and Pielke, 1976)]. The subgrid-scale fluxes (which include the influence of vertical thermodynamic stratification) are computed using formulations such as those given in Section 7.

(2) A simple objective analysis routine such as the inverse-squared distance interpolation formula used by Segal and Pielke (1980) for two radiosonde sites, given as

$$(120) \qquad \phi_p = \left(\phi_1 \frac{1}{r_{1p}^2} + \phi_2 \frac{1}{r_{2p}^2}\right)\left(\frac{1}{r_{1p}^2} + \frac{1}{r_{2p}^2}\right)^{-1}$$

can be used to estimate the dependent variables throughout the model domain. In this formula r_{1p} and r_{2p} are the distances between observations at locations 1 and 2, and point p, which corresponds to the model grid location. ϕ_p is the interpolated value, while ϕ_1 and ϕ_2 are the values of the dependent variables at the observation sites. As long as the measurements are widely spaced, small-scale data inconsistencies will not be introduced into the model by this interpolation. As shown by Sasaki (1971), when prognostic constraints are used in the variational analysis, a low-pass filter is necessary because of the sparseness of data. This filter is required so as not to violate the uniqueness of the solution of the prognostic equation used in his formulation. His conclusion suggests that simple objective analyses, as given above, may be just as satisfactory as the variational approach when observational data resolution is poor, since Eq. (120), with sparse data should produce analyses similar to those obtained using Sasahi's approach with a low-pass filter.

(3) Using the analyzed fields, the model equations are then integrated forward for a short time in order to assure that the dependent variables obey the conservation relations. Segal and Pielke (1980), for example, using winds from the Dulles and Wallops Island rawinsonde soundings to

simulate sea breezes over the Chesapeake Bay region, found maximum vertical velocities on the order of 0.5 cm sec^{-1} after 3 hr of integration. After this period of dynamic initialization, the model simulations for a case study day can commence.

(4) When nonflat terrain occurs in the model, it can be present initially as required by Mahrer and Pielke (1977b), or it can be permitted to grow gradually during a dynamic initialization such as performed by Mahrer and Pielke (1975, 1977a), Klemp and Lilly (1978), Deaven (1976), and Yamada (1978). This latter procedure has been called *diastrophism* (because of the analogy to the original geological application of this term to mountain building). Mahrer and Pielke (1977b) eliminated the need for diastrophism by assuring that the numerical approximation of the hydrostatic equation and horizontal pressure gradient terms are defined consistently with one another.

(5) A quasi-equilibrium is required in the initial field only if the large-scale conditions in the real world have persisted for the same time as required to achieve such an equilibrium in the model simulations. Otherwise, a sudden change of synoptic conditions, such as the passage of a strong cold front over mountainous terrain, would generate a mesoscale response for some period thereafter that is quite unsteady. A steady-state requirement in the initial field, in this case, would be unrealistic. Simulated mesoscale circulations generated over a mountain ridge, for example, take about 10 hr to achieve a steady-state after a steady gradient synoptic wind is imposed (Klemp and Lilly, 1978). In the atmosphere, the actual synoptic wind is seldom constant for this period of time.

11.5. Spatial Boundaries

In discussing boundary conditions in mesoscale models, only the bottom surface represents a physical interface. The top and lateral sides are incorporated only for computational reasons because mesoscale models are necessarily *limited-area models*.

Models that have the correct number of boundary conditions are said to be *well posed,* while if a greater number are used than required, the model equations are said to be *overspecified*. Overspecified systems of the conservation equations can generate erroneous solutions at the boundaries that, as shown by Oliger and Sundström (1976), propagate across the model grid at the fastest wave speed permitted in the model. Oliger and Sundström (1976) further claim that the hydrostatic representation is ill posed for any choice of a local lateral boundary condition that, if substantiated in later studies, further argues for the need to minimize the influence of the side walls on the interior solution. Chen (1973) also reports that overspecification of boundary conditions can excite computational modes; however, smoothing at points next to the boundaries suppresses these erroneous perturbations.

Mathematical operations undertaken to make computations easier can also cause errors in the solutions. Neumann and Mahrer (1971), for example, use the conservation of mass relation to show that differentiation of Eq. (21) in order to permit a model to have a rigid top can lead to the fictitious creation or destruction of mass.

11.5.1. Lateral Boundary Conditions. Anthes and Warner (1978) have demonstrated the serious errors that result in mesoscale model simulations if lateral boundary conditions are incorrectly specified. An error of 1 mbar per 400 km in the specification of the pressure between the two boundaries, for instance, can lead to an erroneous acceleration of about 1 msec^{-1} hr^{-1} across the entire model, thereby critically influencing the results in the interior after a few hours of integration.

Examples of lateral boundary conditions applied in mesoscale models are the following:

- constant inflow, gradient outflow
- radiative boundary conditions
- specified from observations or a larger-scale model representation.

The first approach assumes that the air advecting into the domain is unperturbed by downstream conditions, while information one grid point in from the outflow boundary is instantaneously transmitted downstream. When waves are present, however, information can propagate upstream so that this this boundary condition is not entirely satisfactory.

To attempt to remedy this problem, the *radiative boundary condition* is applied in order to permit disturbances propagating and advecting outward from the model to exit the domain without being reflected back in. Pearson (1974), Orlanski (1976), Chen and Orville (1980), Klemp and Lilly (1978), Klemp and Wilhelmson (1978a), and Clark (1979) apply this technique using a prognostic equation at the lateral boundaries to minimize reflection, while Deaven (1976), Pielke (1974a), and Perkey and Kreitzberg (1976) perform the same task using smoothers that increase near the lateral wall. This latter approach is also called the *sponge boundary condition*. Specifying conditions at the boundaries has been a technique applied to the interface between course- and fine-grid regions of meshed models such as done by Mathur (1974), Jones (1976), and Perkey (1976), as discussed in Section 11.1.

Based on existing studies of lateral boundary conditions, the following approach is appropriate for use in mesoscale models:

(1) Remove the lateral boundaries as far from the region of interest as possible. Use a radiative boundary condition approach to minimize reflection of outward propagating perturbations back into the model.

(2) Permit the lateral boundaries to include larger scale tendencies as specified by a larger scale model, or derived from a set of observations. Carpenter (1979) has been very successful at changing boundary conditions in order to represent a varying synoptic situation in his simulation of the sea breezes over England for a case study day. He found his simulated sea breeze fronts to be very sensitive to the position of synoptic-scale features. Ballentine (1980) has been equally successful applying synoptic tendencies to a simulation of New England coastal frontogenesis.

11.5.2. Top Boundary Condition. The top of the mesoscale model, as with the lateral boundaries, should be removed as far as possible from the region of significant mesoscale disturbance. Ideally this would place the top sufficiently high such that the density of air approaches zero. In mesoscale studies, however, it is not necessary to go this high because of the deep layer of stable thermodynamic stratification that always exists in the stratosphere and much of the upper troposphere. Such layers, which are almost always stable even to the lifting of saturated air, inhibit vertical advection and tend to generate mesoscale circulations that have horizontal scales larger than vertical.

Top boundary conditions have been placed as follows:

- deep within the stratosphere (e.g., Peltier and Clark, 1979; Klemp and Lilly, 1978)
- at the tropopause (e.g., Mahrer and Pielke, 1975)
- in the middle or upper troposphere (e.g., Pielke, 1974a,b), or lower troposphere (e.g., Estoque, 1961, 1962)

for different mesoscale problems. Peltier and Clark, and Klemp and Lilly placed their model tops high into the stratosphere since it is known from observations (e.g., clear air turbulence in aircraft) that airflow over mountains causes the propagation of kinetic energy up into that layer, while Defant's (1950) linear study has suggested that the sea breeze over flat terrain is a shallow phenomenon.

A model top boundary can be rigid (e.g., Estoque, 1961), an impervious material surface (e.g., Pielke, 1974a), a pervious lid (e.g., Lavoie, 1972), or an absorbing layer (e.g., Klemp and Lilly, 1978). Rigid tops require that the vertical velocity be zero at that level while material surfaces move with the ascending and descending motions at the top. Pielke (1974a) assigns the material surface to be an isentropic surface. The pervious lid permits mass transport through the top and is analogous to the top of the planetary boundary layer formulation given by Eq. (108). The absorbing layer—a region at the top of the model in which smoothing increases with height in order to damp out upward-propagating perturbations and thereby prevent erroneous reflections back downward (Klemp and Lilly, 1978)—is analogous to the sponge condition as applied to the lateral boundary.

Based on existing mesoscale simulations, the following criteria should be used in determining what form of top boundary condition to use:

- if the vertical propagation of internal gravity wave energy is an important component in a mesoscale simulation, then an absorbing layer is required;
- otherwise, a material surface top coincident with an isentropic surface, well removed from the region of significant mesoscale influence, it sufficient.

Linear theory is a useful guide to determine which form of top boundary condition to apply. Klemp and Lilly's (1975) linear results, for example, indicate an absorbing layer is needed in the simulation of strong airflow over mountains, while Defant's (1950) sea breeze calculations indicate such a layer is unnecessary.

11.5.3. Bottom Boundary Condition. In a mesoscale atmospheric model, the bottom is the only well-defined physical interface. It is the differential gradient of dependent variables along this surface that generates many types of mesoscale systems, and it has a pronounced influence on the remainder. Because of the crucial importance of this boundary on the msesoscale, it is imperative that it be represented as accurately as possible.

11.5.3.a. Water Surfaces. In contrast to land, water is translucent to solar radiation and readily mixes vertically. To properly represent its influence on mesoscale circulations, it is necessary to utilize coupled atmospheric–oceanic models in which interactions between the two systems are permitted. Clancy *et al.* (1979) attempted such a simulation using a sea breeze model and an ocean upwelling model and found that the interaction between the two systems is weak. In Clancy *et al.*'s work, the temperature gradient created by the upwelling was used as the bottom boundary condition in the sea breeze model, while the sea breeze wind at the sea surface was utilized to influence the intensity of upwelling. Another interactive model is that of Jacobs and Brown (1974), who have performed preliminary three-dimensional simulations of the air–lake interactions over Lake Ontario as part of their contribution to the International Field Year over the Great Lakes (IFYGL) program.

The influence of water bodies on mesoscale atmospheric circulations is expected to be most important through the following:

- the temperature distribution at the water surface and its influence on the overlying wind and thermodynamic stratification [Jacobs (1978), for example, found that changes in the 10 m wind speed appear to be correlated with the movements of small-scale oceanic salinity and temperature fronts];
- its aerodynamic roughness [Clarke (1970), based on Charnock's (1955) work, suggests using a roughness parameterization over water of $z_0 = 0.032 u_*^2 / g$ with the requirement that $z_0 \geq 0.0015$ cm. Laykhtman and Snopkov (1970) suggest a similar form except that the constant of proportionality is 0.074 and no limitation on z_0 is given. The latter investigators report values of z_0 ranging from 0.002 to 0.2 m for wind speeds of 2 to 14 m sec^{-1}];
- effects on visibility, and hence radiative transfer, as water evaporates and aerosols are ejected from the water;
- influences on the radiational flux divergence and the heat budget at the adjoining land surface if rainfall and cloudiness develop over land as the marine air advects onshore, or if the vapor immediately condenses into haze, fog, or stratus and moves inland.

Effects of the mesoscale atmospheric circulation on the ocean and lake dynamics principally occur through wind stress and include the following:

- an increase in wind speed that produces a deepening of the ocean mixed layer (e.g., Marchuk *et al.*, 1977; Elsberry and Randy, 1978; Chang and Anthes, 1978; Kondo *et al.*, 1979);
- an increase in wind speed that produces small-scale wave breaking. This changeover from an aerodynamically smooth to a rough water surface occurs at a rather precise friction velocity ($\sim u_* \cong 23$ cm sec^{-1}) (e.g., Melville, 1977);
- spatial and temporal variations in wind velocity cause currents in coastal waters (e.g., Emery and Csanady, 1973; Blackford, 1978; Svendsen and Thompson, 1978; Sheng *et al.*, 1978);
- changes in wind speed and direction along a coastline alter the upwelling/downwelling pattern (e.g., Csanady, 1975; Knowles and Singer, 1977; Allender, 1979; Hamilton and Rattray, 1978);
- changes in wind speed alter the circulations in estuaries and harbors

through mixing and the resultant creation of horizontal gradients in the water (e.g., Long, 1977; Hachey, 1934; Wang, 1979; Weisberg, 1976);
- wind energy absorbed by coastal waters is a function of the spectral energy of the wind (e.g., Lazier and Sandstrom, 1978);
- wind velocity affects the drift of coastal pack ice (e.g., McPhee, 1979);
- the wind velocity influences the movement of pollutants in the water (e.g., Pickett and Dossett, 1979);
- the wind causes the formation of helical circulations in the water with the resultant accumulation of surface debris in lines parallel to the surface wind direction (e.g., Gross, 1977).

Pielke (1981) summarizes these and other expected interactions between the ocean, lakes, and atmosphere along coastal regions.

11.5.3.b. *Land Surfaces.* To accurately model the influence of land on mesoscale circulations, it is necessary to model the flow of heat, moisture, and other gaseous and aerosol materials into and out of bare soil, vegetation, and man-made structures. Realistic boundary conditions on velocity, called *no-slip* conditions, are specified by

$$\bar{u}(z_0) = \bar{v}(z_0) = \bar{w}(z_0) = 0$$

while the vertical subgrid-scale fluxes at z_0 are discussed in Section 7. The same conditions apply over the ocean.

Potential temperature $\bar{\theta}$ has been prescribed at the ground surface using sinusoidal heating functions (e.g., Pielke, 1974a; Neumann and Mahrer, 1971; Mahrer and Pielke, 1976), or diagnosed over bare soil using a heat budget (e.g., Physik, 1976; Mahrer and Pielke, 1977a,b; Estoque *et al.*, 1976). Specific humidity $\bar{q}_3$ at the surface over bare soil has been ignored, treated as a diurnal wave (Burk, 1977), or fractionally weighted by a surface saturation value and a value at the first model grid above the ground (Dieterle, 1976; Mahrer and Pielke, 1977a,b; Pandolfo and Jacobs, 1973; Hsu, 1979). Garrett (1978) and McCumber (1980), utilizing Deardorff's (1978) work, have recently included heat and moisture budgets for vegetation, along with a soil moisture model, in order to compute the influence of moisture on the surface values of potential temperature and specific humidity. Deardorff showed that neglecting vegetation could result in errors in the computation of the subgrid-scale moisture and heat flux as large as a factor of two, while Garrett found that the presence of plants affected the timing and intensity of convection predicted by his three-dimensional model.

Figures 6 and 7 (reproduced from McCumber, 1980) illustrate the pre-

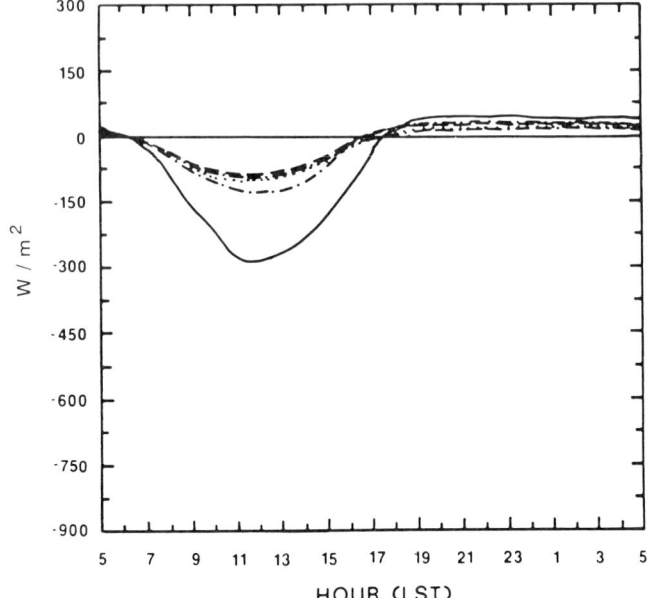

FIG. 6. Predicted surface sensible heat flux density as a function of soil type for a typical clear summer day in south Florida. Fluxes directed toward the atmosphere are negative. (——) Sand; (----) sandy loam; (-·-·-) sandy day; (····) peat; (-··-··) marsh. (From McCumber, 1980.)

dicted diurnal variations of surface sensible and latent heat fluxes over five bare soil types for a typical summer day over south Florida. The greater partitioning of the subgrid-scale heat flux into sensible heat for sand is primarily a result of its lower conductivity of heat.

Figure 8 (also reproduced from McCumber, 1980) illustrates the changes when vegetation is added. Among McCumber's results is the prediction that the temperature profile is stably stratified under a forest canopy during the day, while it is close to neutral at night ($\partial \bar{\theta}/\partial z \cong 0$), a result that is supported by the studies of Raynor (1971) and others. Also, when trees cover a significant percentage of the surface, a deeper and warmer boundary layer is predicted than when just bare soil is present. Such experiments provide useful insight into inadvertent climate modification resulting from changes in ground characteristics.

Otterman (1975) has suggested, using an observational study, that changes in vegetation, and therefore albedo, due to overgrazing in the Negev–Sinai desert area, can reduce rainfall. Mahrer and Pielke (1978a) have used mesoscale model results to examine Otterman's hypothesis and found that sea breeze convergence can be decreased if overgrazing reduces the temperature differential between the land and water; however, the results are very sensitive to the moisture content in the soil. In a different geographic area, Lord *et al.* (1972) have examined the effects of tundra vegetation on the land–air interface using a one-dimensional model. They concluded that a three-dimensional representation is necessary in order to account for the horizontal heterogeneity caused by the presence of thaw lakes on top of the permafrost. In cities, Nunez and Oke (1977) have investigated the complex energy balance below roof height—a region they call an *urban canyon*.

When either bare soil or vegetation is present a heat budget formulation, such as that used by Physik (1976), Bergstrom and Viskanta (1973a), Mahrer and Pielke (1977a,b), and others, is applied in order to obtain the equilibrium interface temperature. With this procedure the ground and surface of vegetation are considered to have no heat storage.

As an illustration, the surface heat budget for bare soil can be written as

$$(121) \quad R_{S_G} + R_L \downarrow - R_L \uparrow + Q_H + Q_E + Q_G + Q_M = F(T_G)$$

where R_{S_G} is the net solar radiation [defined by an expression such as Eq. (114)]; $R_L \downarrow$ is the downward long-wave flux immediately above the ground [defined by $R_L \downarrow = \sigma \int_{p_{z_G}}^{0} \epsilon(\partial T^4/\partial \xi) \, d\xi$ as given in Section 8.2.1];

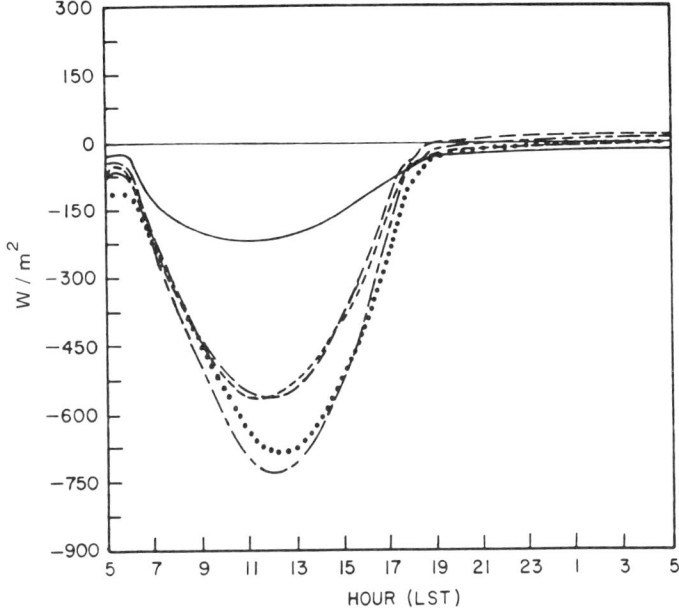

FIG. 7. Same as Fig. 6 except for surface latent heat flux. (———) Sand; (— — —) sandy loam; (—·—·—) peat; (----) sandy clay; (····) marsh. (From McCumber, 1980.)

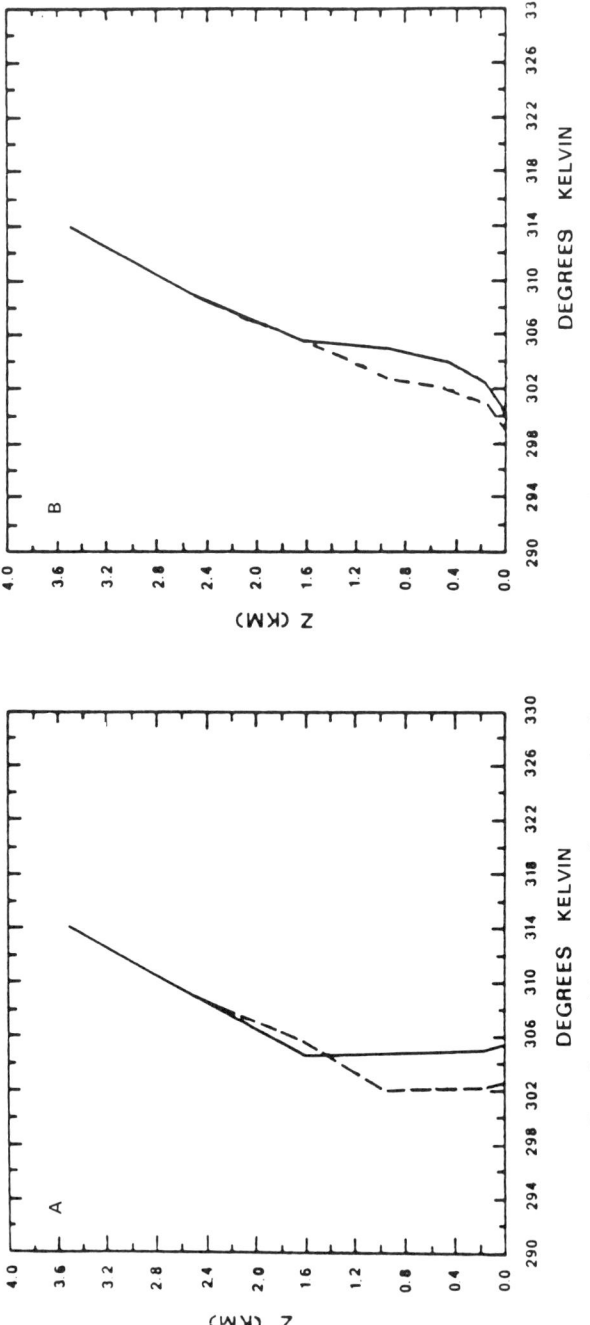

FIG. 8. Vertical profiles of potential temperature for (a) early afternoon (1315 LST) and (b) at sunrise the following day (0515 LST): over a bare sandy loam soil (dashed line) and over the same soil but with a forest canopy which shades 90% of the ground (solid line). Representative of a typical summer day over south Florida. (From McCumber, 1980.)

$R_L \uparrow$ is upward long-wave flux [given by $\epsilon_G \sigma T_G^4 + (1 - \epsilon_G)R \downarrow$, where ϵ_G is the emissivity of the ground]; Q_H and Q_E are the sensible and latent turbulent heat fluxes [proportional to $\overline{w''\theta''}$ and $\overline{w''q_3''}$ as given in Eq. (100)]; Q_G is the heat flux into or out of the ground (defined by $\rho_s C_s \partial T_G/\partial z$, where ρ_s and C_s are the density and heat capacity of the soil); and Q_M is the heat added by anthropogenic activities. $F(T_G)$ should be equal to zero if the terms are in balance. In general, however, a truncated Taylor series expansion given by

$$T_G^{n+1} = T_G^n - F(T_G)/F'(T_G)$$

is used to obtain successive estimates of temperature (the superscript n and $n + 1$ refer to the original and new estimates). This procedure, called the *Newton–Raphson procedure*, continues as an iterative calculation until $F(T_G)$ becomes arbitrarily small. $F'(T_G)$ is obtained by differentiating Eq. (121) with respect to the ground surface temperature T_G, such as performed by Mahrer and Pielke (1977a,b), Physik (1976), and others. Jacobs and Brown (1973) offer a modification to the Newton–Raphson method discussed above using what they called the method of false position. They claim this latter method guarantees a solution to an equilibrium temperature—which does not always occur when the conventional Newton–Raphson technique is used.

Deardorff (1978) investigated several simplified heat budget formulations, including that of Blackadar (1976), and concludes that the latter is the more accurate *slab* approach (i.e., treats the soil as a single layer). When a soil moisture budget is included, however, resolution of soil levels, such as performed by McCumber (1980), is essential.

For the most accurate representation of the bottom boundary in a mesoscale model it must

- interact with the overlying atmosphere
- respond to the flow of heat, moisture, and other gaseous and aerosol materials from below the surface
- properly represent the effects of such features as vegetation and buildings, as well as bare soil on the overlying atmosphere.

The use of conceptually simple soil heat and water flux models with vertical resolution in the soil, as well as a budget representation at the bottom interface [e.g., Eq. (121)], appears to be the most appropriate representation with available computer resources.

12. Model Evaluation

In utilizing a mesoscale numerical model there are six basic requirements that must be met before the credibility of model simulations will be accepted by the scientific community. In reading papers in the published literature, the same criteria must be considered and satisfied before the results and conclusions of these papers are given credence. The requirements are as follows:

- The model must be compared with known analytic solutions from linear theory. To perform these experiments, the mesoscale model is forced by very small perturbations so that linearized results are accurate.
- Nonlinear simulations with the model must be compared with results from other models, which have been developed independently by other investigators. The differences and similarities between the results should be discussed.
- The mass and kinetic energy budget of the model must be computed in order to determine the conservation of these important physical quantities.
- The model predictions must be quantitatively compared with observations.
- The computer logic of the model must be available upon request so that the flow structure of the code can be examined.
- The published version of the model must have been subjected to peer review. For this reason, model results presented in recognized professional journals should carry more weight than those distributed in report formats.

12.1. Kinetic Energy Evaluation

To illustrate the use of a kinetic energy budget in a mesoscale model, multiply Eq. (80) by $\bar{\bar{u}}_1 \bar{\rho}(s - z_\mathrm{G})/s$ and Eq. (81) by $\bar{\bar{u}}_2 \bar{\rho}(s - z_\mathrm{G})/s$, assume small slopes, add, and use Eq. (79), after multiplying by $\bar{k}$, to yield

$$(122) \quad \bar{\rho}\frac{s - z_\mathrm{G}}{s}\frac{\partial \bar{k}}{\partial t} \cong -\frac{\partial}{\partial \tilde{x}^j}\bar{\rho}\frac{s - z_\mathrm{G}}{s}\bar{\bar{u}}^j \bar{k} - \bar{\bar{u}}_1 \bar{\rho}\frac{s - z_\mathrm{G}}{s}\overline{\tilde{u}^{j''}\frac{\partial \tilde{u}^{1''}}{\partial \tilde{x}^j}}$$

$$- \bar{\bar{u}}_2 \bar{\rho}\frac{s - z_\mathrm{G}}{s}\overline{\tilde{u}^{j''}\frac{\partial \tilde{u}^{2''}}{\partial \tilde{x}^j}}$$

$$- \bar{\bar{u}}_1 \bar{\rho}\frac{s - z_\mathrm{G}}{s}\left\{\bar{\theta}\frac{\partial \bar{\pi}}{\partial \tilde{x}^1} - g\frac{z^* - s}{s}\frac{\partial z_\mathrm{G}}{\partial x}\right\}$$

$$- \bar{\bar{u}}_2 \bar{\rho}\frac{s - z_\mathrm{G}}{s}\left\{\bar{\theta}\frac{\partial \bar{\pi}}{\partial \tilde{x}^2} - g\frac{z^* - s}{s}\frac{\partial z_\mathrm{G}}{\partial y}\right\}$$

$$+ \bar{\bar{u}}_1 \bar{\rho}\frac{s - z_\mathrm{G}}{s}\{\hat{f}\bar{\bar{u}}^3 - f\bar{\bar{u}}^2\} + \bar{\bar{u}}_2 \bar{\rho}\frac{s - z_\mathrm{G}}{s}f\bar{\bar{u}}^1$$

where $\bar{k} = \frac{1}{2}(\bar{\bar{u}}_1 \bar{u}^1 + \bar{\bar{u}}_2 \bar{u}^2)$, and $\bar{\bar{u}}_1, \bar{\bar{u}}_2$ are two of the covariant velocity components in the terrain-following coordinate system. In terms of the original rectangular coordinate system

$$(123) \qquad \bar{\bar{u}}_1 = u + w\frac{s - z^*}{s}\frac{\partial z_G}{\partial x}; \qquad \bar{\bar{u}}_2 = v + w\frac{s - z^*}{s}\frac{\partial z_G}{\partial y}$$

[In the rectangular coordinate system the covariant and contravariant components are identical—see Dutton (1976) for a detailed explanation.]

Equation (122) can then be integrated over the model domain giving

$$(124) \quad \frac{\partial K}{\partial t} \cong -\int_{z_G}^{s}\left\{\int_{0}^{D_y}\bar{\rho}\frac{s-z_G}{s}\bar{u}^1\bar{k}\Big|_{0}^{D_x}d\bar{x}^2 + \int_{0}^{D_x}\bar{\rho}\frac{s-z_G}{s}\bar{u}^2\bar{k}\Big|_{0}^{D_y}d\bar{x}^1\right.$$
$$+ \int_{0}^{D_y}\int_{0}^{D_x}\bar{\rho}\frac{s-z_G}{s}\left[\overline{\bar{u}_1\bar{u}^{j\prime\prime}\frac{\partial\bar{u}^{1\prime\prime}}{\partial\bar{x}^j}} + \overline{\bar{u}_2\bar{u}^{j\prime\prime}\frac{\partial\bar{u}^{2\prime\prime}}{\partial\bar{x}^j}}\right]d\bar{x}^1\,d\bar{x}^2$$
$$+ \int_{0}^{D_y}\int_{0}^{D_x}\bar{\rho}\frac{s-z_G}{s}\left[\bar{\bar{u}}_1\bar{\theta}\frac{\partial\bar{\pi}}{\partial\bar{x}^1} + \bar{\bar{u}}_2\bar{\theta}\frac{\partial\bar{\pi}}{\partial\bar{x}^2}\right]d\bar{x}^1\,d\bar{x}^2$$
$$- \int_{0}^{D_y}\int_{0}^{D_x}\bar{\rho}\frac{s-z_G}{s}\left[\bar{\bar{u}}_1 g\frac{z^*-s}{s}\frac{\partial z_G}{\partial x} + \bar{\bar{u}}_2 g\frac{z^*-s}{s}\frac{\partial z_G}{\partial y}\right]d\bar{x}^1\,d\bar{x}^2$$
$$\left. - \int_{0}^{D_y}\int_{0}^{D_x}\bar{\rho}\frac{s-z_G}{s}[\bar{\bar{u}}_1(\hat{f}\bar{\bar{u}}^3 - f\bar{\bar{u}}^2) + \bar{\bar{u}}_2 f\bar{\bar{u}}^1]\,d\bar{x}^1\,d\bar{x}^2\right\}d\bar{x}^3,$$

where

$$K = \int_{z_G}^{s}\int_{0}^{D_x}\int_{0}^{D_y}\left(\bar{\rho}\frac{s-z_G}{s}\bar{k}\right)d\bar{x}^2\,d\bar{x}^1\,d\bar{x}^3$$

In defining K for use in Eq. (124), the quantity $((s - z_G)/s)\bar{k}\,\partial\bar{\rho}/\partial t$ is assumed to be much less than $((s - z_G)/s)\bar{\rho}\,\partial\bar{k}/\partial t$, while (124) is derived with the condition that $\bar{\bar{u}}^3 = 0$ at z_G and s.

The first two terms on the right-hand side of Eq. (124) are proportional to the net flow of kinetic energy through the sides of the model domain, while the next term represents the change in kinetic energy due to subgrid-scale effects. The terms involving $\partial\bar{\pi}/\partial\bar{x}^1$ and $\partial\bar{\pi}/\partial\bar{x}^2$ are proportional to the conversion of potential to kinetic energy by cross-isobaric flow, while the expressions containing the gradients of terrain represent the conversion of potential to kinetic energy through upslope and downslope flow. The last term in Eq. (124) (with f and $\hat{f}$) is approximately zero [see Eq. (123)] since $\partial z_G/\partial x$ and $\partial z_G/\partial y$ must be much less than unity, as discussed in Section 6, and would be identically equal to zero if the complete three-dimensional form of the conservation relation is used.

In using Eq. (124) to determine the total kinetic energy changes, it is imperative that the approximation technique used to evaluate the individual terms in that expression be the same as that used in the original approximate form of the conservation relation from which (124) was derived.

The time rate of change of kinetic energy can also be evaluated directly at each individual grid point and then summed; i.e.,

$$\frac{\partial \hat{K}}{\partial t} = \int_{z_G}^{s} \int_0^{D_x} \int_0^{D_y} \bar{\rho} \, \frac{s - z_G}{s} \frac{\partial \bar{k}}{\partial t} \, d\tilde{x}^2 \, d\tilde{x}^1 \, d\tilde{x}^3$$

is used to obtain an estimate of the total kinetic energy change instead of Eq. (124). If the kinetic energy changes computed by this expression and Eq. (124) closely agree, the modeler can be certain that mistakes, such as coding errors, are not causing significant sources of unexplained changes of kinetic energy.

Anthes and Warner (1978) discuss the use of kinetic energy budgets in mesoscale models in further detail. Among their results they have shown that the flux of kinetic energy through the side walls of a mesoscale model curcially affect the solutions in the interior. They also conclude that, because of the extreme sensitivity of mesoscale model results to domain size and the form of lateral boundary conditions, studies of the energetics of real-world mesoscale systems will be very difficult to perform and quite sensitive to errors and small-scale variations of wind, potential temperature, and pressure at the model boundaries.

12.2. Observational Validation

Observational verification is also a necessary tool of model validation. Keyser and Anthes (1977) utilize a useful technique where if (*i*) ϕ_i and $\phi_{i_{obs}}$ are individual predictions and observations at the same grid point; (*ii*) ϕ_0 and $\phi_{0_{obs}}$ are the average values of ϕ_i and $\phi_{i_{obs}}$ at a level; and (*iii*) #N is the number of observations, then

$$E = \left\{ \sum_{i=1}^{\#N} (\phi_i - \phi_{i_{obs}})^2 / \#N \right\}^{1/2}$$

$$E_{UB} = \left\{ \sum_{i=1}^{\#N} [(\phi_i - \phi_0) - (\phi_{i_{obs}} - \phi_{0_{obs}})]^2 / \#N \right\}^{1/2}$$

$$\sigma_{obs} = \left\{ \sum_{i=1}^{\#N} (\phi_{i_{obs}} - \phi_{0_{obs}})^2 / \#N \right\}^{1/2}$$

$$\sigma = \left\{ \sum_{i=1}^{\#N} (\phi_i - \phi_0)^2 / \#N \right\}^{1/2}$$

can be used to determine the skill of the model results. The parameter E is the *root-mean-square* (rms) *error*, E_{UB} is the rms error after a constant bias is removed, while σ and σ_{obs} are the standard deviations of the predictions and the observations, respectively. Keyser and Anthes found that the rms error can be reduced significantly when a constant bias is removed. Such a bias, they suggested, could be due to an incorrect specification of the initial and/or bottom and lateral boundary conditions.

Skill is demonstrated when

- $\sigma \cong \sigma_{obs}$
- $E < \sigma_{obs}$
- $E_{UB} < \sigma_{obs}$.

Pielke and Mahrer (1978) applied these criteria to their simulation of the sea breezes over south Florida in order to show that the model had skill in predicting wind velocity and temperature at 3 m. Temperature predictions over the entire daylight period, for example, had a ratio of $E_{UB}/\sigma_{obs} = 0.6$. Segal and Pielke (1981) have recently applied this analysis tool over the Chesapeake Bay region in order to evaluate the skill of a mesoscale model prediction of biological heat load.

Concepts of set theory can also be used to validate model skill. Pielke and Mahrer (1978) applied this technique to determine the degree of correspondence between predicted convergence zones and locations of radar echoes over south Florida. To illustrate their procedure, let (i) A_{xy} be the area of the model domain; (ii) M be the area of the model domain covered by predicted convergence of a given magnitude or larger; (iii) R be the area of the model domain covered by radar echoes of a specified intensity and greater. Thus (iv) $F_E = (M \cap R)/R$ (the symbol $\cap$ is an intersection in set theory symbolism) and (v) $F_M = M/A_{xy}$, where F_E is the fraction of echoes in convergence zones with values equal to or greater than a certain value of convergence, while F_M is the fraction of the model domain covered by this value of convergence or larger. Skill is demonstrated if $F_E/F_M > 1$ since the ratio would be expected to be unity by random chance. Pielke and Mahrer (1978), for example, using this analysis, found that the ratio was greater than unity in 26 out of 30 categories examined for a sea breeze simulation over south Florida, with a ratio larger than 2.0 in 20 of them. Additional details of this analysis procedure are given in the article by Pielke and Mahrer (1978).

12.3. Model Intercomparisons

Model intercomparisons have been performed; however, the degree of similarity has been studied only qualitatively. Mahrer and Pielke (1977a), for example, used the terrain data of Anthes and Warner (1974) in order to recreate their simulation of the airflow over the mountains in the White Sands Missile Range. Tapp and White (1976) and Hsu (1979) performed simulations over south Florida in order to contrast their sea breeze results against those obtained by Pielke (1974a). Using two-dimensional models, Kessler and Pielke (1980), Mahrer and Pielke (1978b), and Peltier and Clark (1979) have simulated the airflow over rough terrain and compared their results against those of Klemp and Lilly (1978).

A limited number of comparisons of nonlinear mesoscale models have also been made with linear models. Mahrer and Pielke (1978b), for instance, recreated Klemp and Lilly's (1978) linear results by inputting a very small hill (10-m maximum height) in their nonlinear calculations. Kessler and Pielke (1981) are using Klemp and Lilly's linear results to examine the relative importance of individual terms in the conservation of motion relation as the amplitude of the mountain perturbation increases.

12.4. Model Logic

The availability of mesoscale model codes is not generally known since the standard meteorological texts will not publish such material. Rather, it is necessary to publish it in a technical report format such as performed by McCumber *et al.* (1978). The intensive study of the computer logic of a model is as important as the investigation of its physics and mathematical formulation.

13. EXAMPLES OF MESOSCALE MODELS

Mesoscale atmospheric systems can be divided into two groups:

- those that are primarily forced by surface inhomogeneities (*terrain-induced mesoscale systems*)
- those that are primarily forced by instabilities in traveling synoptic disturbances (*synoptically induced mesoscale systems*).

In the first category are features such as sea and land breezes, mountain–valley winds, urban circulations, and forced airflow over rough terrain, while examples of the second include squall lines, hurricanes, and traveling mesoscale cloud clusters. The first group is the least difficult to simulate because the sources of these mesoscale circulations are geographically fixed with time scales of 12 hr or so, and they recur frequently. These mesoscale systems do not generally move far from their point of origin. Mesoscale disturbances initiated by some type of atmospheric instability [e.g., CISK (Conditional Instability of the Second Kind); see

Wallace and Hobbs, 1977, p. 442], however, usually occur less frequently at a given location and, because they are not forced by well-defined geographic features, the data requirements needed to initialize mesoscale simulations of these phenomena are more formidable. In this article, examples of the first type of mesoscale feature are emphasized.

13.1. Terrain-Induced Mesoscale Systems

13.1.1. Sea and Land Breeze Models over Flat Terrain. Of all the mesoscale phenomena, the sea and land breezes over flat terrain appear to have been the most studied, both observationally and theoretically. This is undoubtedly a result of the geographically fixed nature of the phenomenon (the location of land–water boundaries), as well as the repetitive nature of the event. The sea breeze is defined to occur when the wind is onshore, while the land breeze occurs when the opposite flow exists.

During the case of nonexistent large-scale winds, it is comparatively easy to describe the diurnal variations of the coastal wind circulations. Defant (1951) presents an excellent qualitative description for this condition. For this case, the idealized sequence of events is as follows:

- At some time in the early morning the pressure surfaces become flat and no winds occur (e.g., 0800 LST—perhaps an hour after sunrise).
- Later in the morning, mass is mixed upward over land by turbulent mixing in the unstably stratified boundary layer, creating an offshore pressure gradient at some distance above the ground. Over water, its translucent character and ability to mix prevent significant heating of the surface (e.g., 1100 LST).
- The resultant offshore flow of air above the ground near the coast creates a low-pressure region at the ground, and onshore winds (the sea breeze) develop (e.g., 1300 LST).
- The onshore winds transport cooler marine air over the land, thereby advecting the horizontal temperature gradient and, hence, the sea breeze inland. The distance the sea breeze travels inland depends most directly on the intensity of the total heat input to the air (Pearson, 1973) (e.g., 1600 LST).
- As the sun sets, long-wave radiational cooling becomes dominant over solar heating, while the local wind field removes the horizontal temperature gradient. The pressure surfaces again become horizontal (e.g., 1900 LST).
- As long-wave cooling continues, the air near the ground becomes

more dense and sinks. The resultant lowering of the pressure surfaces a short distance above the ground creates an onshore wind at that level (e.g., 2200 LST).
- In response to the loss of mass above the ground over the water, a pressure minimum develops at the ocean surface immediately off the coast. The offshore wind that then develops near the surface is called the land breeze (e.g., 0100 LST).
- The distance of offshore penetration of the land breeze depends on the amount of cooling over the land. Because the planetary boundary layer over land is stably stratified at night and, therefore, vertical mixing is weaker and closer to the ground, the land breeze is a shallower and weaker phenomenon than the daytime sea breeze.

The evolution of the sea breeze is somewhat more complicated when a prevailing synoptic flow is included. For the two distinct situations of comparatively cold water and comparatively warm water relative to land, a synoptic wind direction from the colder to the warmer surface weakens the intensity of the local wind by diminishing the horizontal temperature gradient. On the other hand, when the prevailing larger scale flow is from the warmer to the colder surface, the temperature gradient is strengthened and the subsequent local wind flow is stronger. Examples of water that is warm relative to the land include the eastern sides of continents in the tropics and midlatitudes at night and over coastal waters during a polar outbreak. Situations with water that is cold relative to the adjacent land include the eastern sides of continents in tropical and midlatitudes during sunny days, along the west side of continents in which upwelling is occuring, as well as along polar coastal areas in the summer. Fog and low stratus often form over the relatively cold water in polar and upwelling ocean areas (e.g., Noonkester, 1979; Pilié *et al.*, 1979). Noonkester discusses fog formation due to the offshore movement of warm, dry air along the coast of southern California and its subsequent movement back towards and over land in the sea breeze. Estoque (1962) has performed numerical experiments showing the influence of the prevailing synoptic flow on sea breeze convergence.

Figure 9 illustrates predicted sea breeze results for weak and strong onshore synoptic flow. With the weaker winds, the strong horizontal gradient of potential temperature (and, therefore, large horizontal gradient of pressure) results in a tight and well-defined sea breeze circulation as it moves inland. When the prevailing onshore flow is stronger, however, such a large pressure gradient cannot develop because of the rapid inland movement and greater warming of the marine air. In this and subsequent figures, the ends of the dumbbells are spaced 100 km apart in order to illus-

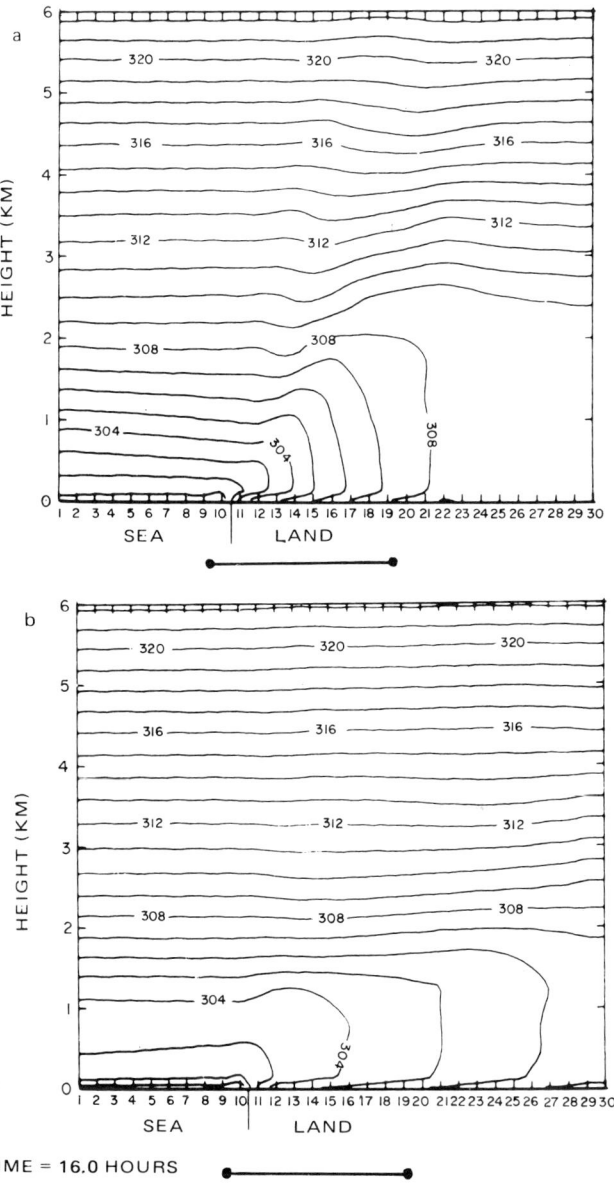

FIG. 9. The vertical cross section of potential temperature along a coastline at 1800 LST for (a) a 1 m sec^{-1} onshore synoptic wind and (b) a 6 m sec^{-1} onshore synoptic wind. Initial input was for a typical summer day over south Florida. (From Mahrer and Pielke, 1978b.)

trate the approximate resolution of the large scale, as contrasted with mesoscale grid resolution.

The magnitude of the effect of a particular horizontal temperature gradient can be estimated from existing observational and numerical studies (e.g., Hanna and Swisher, 1971; Hanna and Gifford, 1975). From these and other related works it has been found that, in the tropics and midlatitudes, a horizontal gradient of less than about 10 W m^{-2} per 30 km has only a minor influence on local wind patterns. With a gradient of 100 W m^{-2} per 30 km, however, significant effects are discernible from the statistical evaluation of observational data, whereas at 1000 W m^{-2} per 30 km the influence on local wind patterns is very pronounced in case-by-case studies.

Observational studies of significance to this phenomenon are numerous—a sampling includes those of Byers and Rodebush (1948), Gentry and Moore (1954), Day (1953), Carson (1954), Plank (1966), Frank *et al.* (1967), Randerson and Thompson (1964), Pielke and Cotton (1977), Burpee (1979), and Schwartz and Bosart (1979) for Florida; Keen and Lyons (1978) for Lake Michigan; Hsu (1969) for the Texas coast; Neumann (1951) and Skibin and Hod (1979) for Israel; and Johnson and O'Brien (1973) along the coast of Oregon.

These studies have demonstrated that land and sea breezes (and other similar mesoscale circulations) are poorly resolved in conventional weather-observing network systems. Such a lack of resolution creates serious problems in developing routine operational forecasts of these phenomena.

Lyons and Keen (1976) also concluded that studies of transport and diffusion over land are generally invalid when applied to the coastal environment. Lyons and Cole (1976), for example, discuss the accumulation of pollutants that results from the recirculation associated with the Lake Michigan sea breeze—an effect that is not considered in commonly used dispersion models. Keen *et al.* (1979) also concluded that size sorting of aerosols occurs within this lake breeze system. In a different geographic area, Carroll and Baskett (1979) concluded that the most serious degradation of air quality in Yosemite National Park occurs due to the transport of material from several hundred kilometers away by the sea breeze from the Pacific coast, as well as by the valley–mountain circulation generated by the Sierras.

Examples of early analytic studies of direct relevance to the sea breeze phenomenon include those of Defant (1950), Schmidt (1947), Smith (1955, 1957), Malkus and Stern (1953), and Stern and Malkus (1953), while more recent studies of this sort include those of Geisler and Bretherton (1969) and Kimura and Eguchi (1978). The first nonlinear numerical modeling study of this phenomenon, performed using two-dimensional models, was undertaken by Estoque (1961, 1962), followed by Fisher (1961), Neumann and Mahrer (1971), Moroz (1967), and others. More recent two-dimensional simulations, such as those of Physik (1976), Pielke (1974b), Gannon (1978), Estoque *et al.* (1976), Dalu (1978), and Neumann and Mahrer (1974, 1975), used such models to improve our understanding of the physical processes associated with the sea breeze.

In the last 10 years or so, computer capability has improved sufficiently to permit three-dimensional simulations. McPherson (1970) was the first to report such calculations of the sea breeze and was followed by the studies of Pielke (1974a), Warner *et al.* (1978), Mahrer and Pielke (1977a,b), Hsu (1979), and Carpenter (1979). In these studies, valuable new insight into the sea breeze was attained, including the conclusion that along coastlines, under undisturbed synoptic conditions during the summer in the tropics and subtropics, the sea breeze exerts a dominant influence on the sites of formation and the movement of thunderstorm complexes (Pielke, 1974a). Figure 10 presents a sea breeze model calculation during an afternoon over the Chesapeake Bay, illustrating the need for three-dimensional simulations. Figure 11 (reproduced from Carpenter, 1979) illustrates a similar complex wind field for a sea breeze over England.

13.1.2. Mountain–Valley Winds. In a region with irregular terrain, local wind patterns can develop as a result of the differential heating between the ground surface and the free atmosphere at the same elevation some distance away. A larger diurnal temperature variation usually occurs at the ground, so that during the day the higher terrain becomes an *elevated heat source,* while at night it is an *elevated heat sink.*

Two categories of mountain–valley winds are generally recognized:

- *slope flow*
- *valley winds.*

These types are easiest to recognize when the prevailing large-scale flow is light. Slope flow refers to cool, dense air flowing down elevated terrain at night, with warm, less dense air moving towards higher elevations during the day. Such air movement is often referred to, respectively, as *nocturnal drainage flow* and the *daytime upslope.* The nocturnal drainage flow is also called a *katabatic wind* (e.g., Manins and Sawford, 1979a,b). Manins and Sawford (1979b), for example, found that such drainage winds are three-dimensional phenomena and that a critical gradient Richardson number of about 0.25 is required to maintain mixing between the

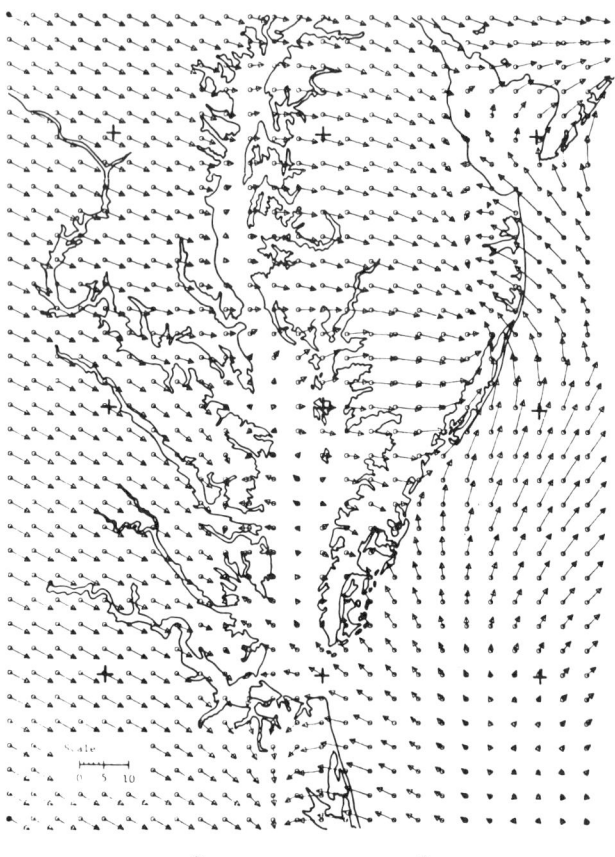

Fig. 10. The predicted winds at 4 m at about 1500 LST over the Chesapeake Bay for August 9, 1975. Model simulation performed by W. Snow using the University of Virginia Mesoscale Model. Scale bar in m sec^{-1}.

katabatic and ambient winds. Other studies of drainage flows include those of Briggs (1979a,b). The Atmospheric Studies in Complex Terrain (ASCOT) program (e.g., Dickerson and Gudiksen, 1980) is studying this type of wind field in detail in the Geysers area of California.

Valley winds are up- and down-valley circulations that develop as a result of air movement into and out of valley floors due to the slope flow. Slope flows are generally assumed to occur when topographic gradients along the slope are steeper than those found along the valley bottom, hence the former circulation tends to develop more quickly.

During sunny days, these local circulations are deeper than at night, as with the sea breeze, because the heating of the ground by the Sun is mixed upward effectively by turbulent heat fluxes. At night, radiational cooling predominates if the winds are light and the resultant perturbation flow field is more shallow. Figure 12 (reproduced from Mahrer and Pielke, 1977b) illustrates the differences in depth and strength of upslope and downslope winds in a two-dimensional simulation in the absence of a prevailing synoptic flow. These figures also illustrate that the airflow tends to form a closed circulation, so that if pollutants were released in one segment of the flow, they would tend to accumulate in a region. Such *recirculation* is ignored in the Gaussian plume models commonly used to estimate concentrations of pollutants in rough terrain.

When a large-scale flow (often including a vertical shear of the horizontal wind), variable surface characteristics, and/or three-dimensional topographic features are present, the resultant mesoscale flow field can become quite complex. Without extensive observations or an accurate three-dimensional mesoscale model, it is generally impossible to anticipate the details of the diurnal variations in the wind field. Figure 13 (re-

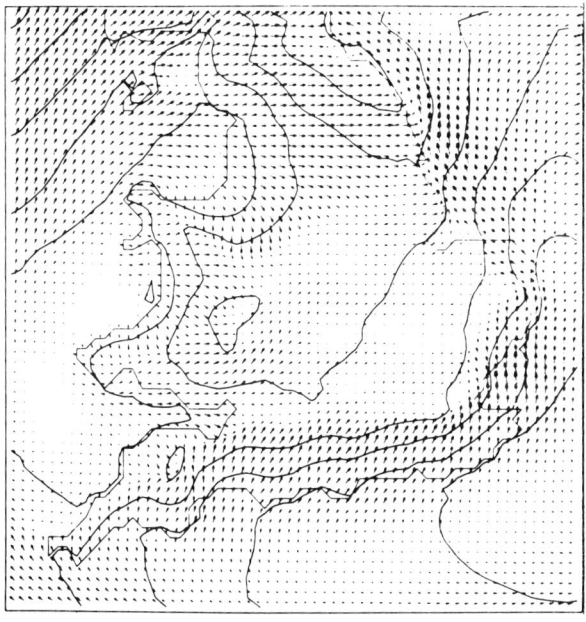

FIG. 11. The 50 m wind and sea level pressure forecast for 1800 LST on June 14, 1973. Experiment includes orographic effects, the movement of a synoptic-scale anticyclone, and differential heating between land and water. The isopleth interval is 0.5 mbar. One grid interval corresponds to 10 msec^{-1}. (From Carpenter, 1979.)

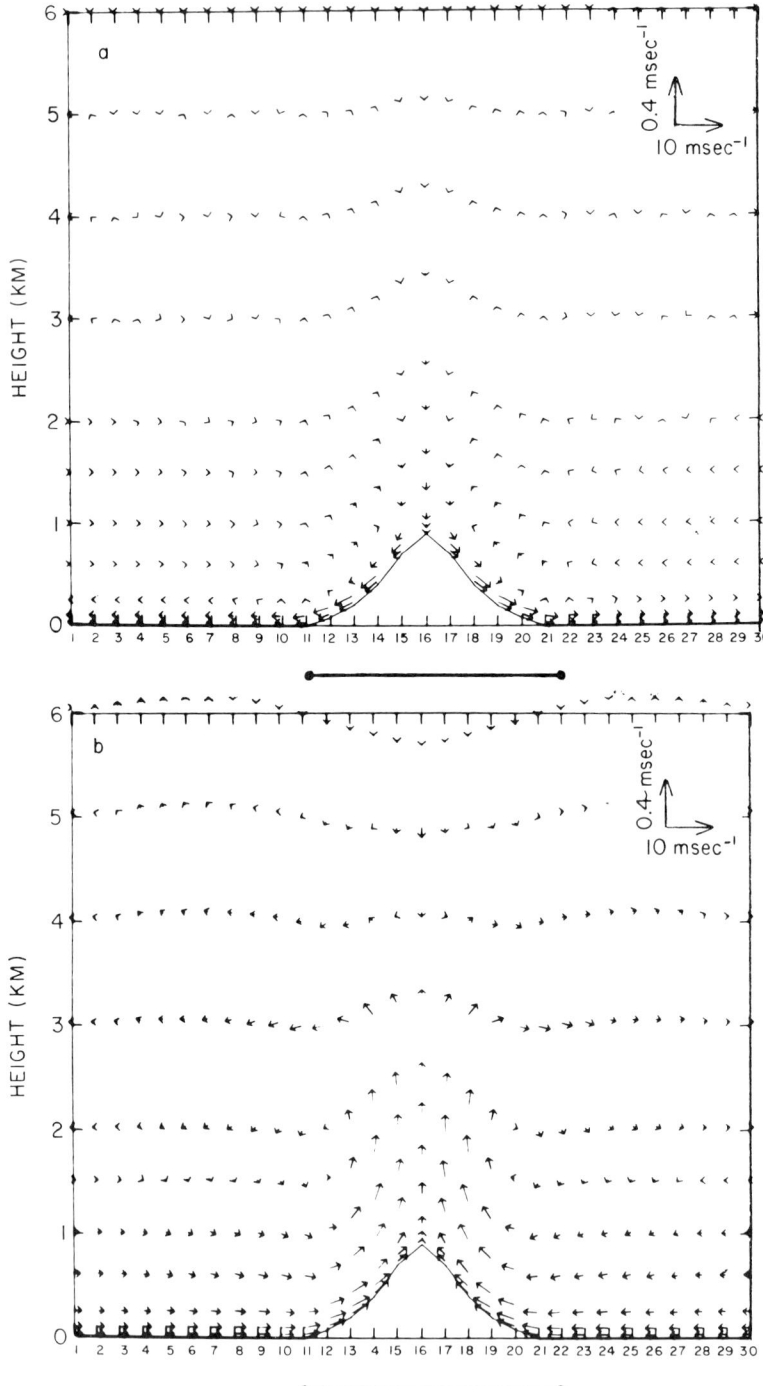

FIG. 12. Two-dimensional simulation of (a) nocturnal drainage flow and (b) upslope flow with no prevailing synoptic flow. Input conditions typical of summer in midlatitudes. (From Mahrer and Pielke, 1977b.)

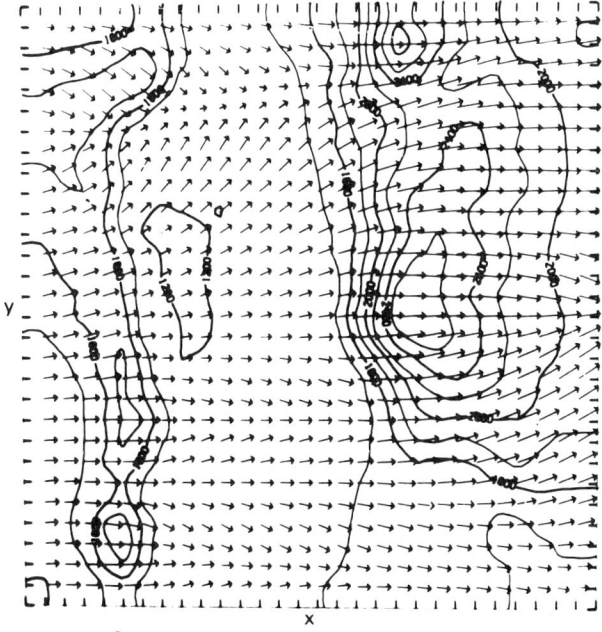

FIG. 13. The predicted surface winds at 3 m at 1300 LST over the White Sands Missile Range for an average June day. One grid interval corresponds to 6 m sec^{-1}. Terrain is contoured at intervals of 200 m. (From Mahrer and Pielke, 1977a.)

produced from Mahrer and Pielke, 1977a) illustrates the complicated wind field predicted over the Sacramento and San Andreas Mountains of New Mexico at about 1300 LST during a summer afternoon with a prevailing westerly wind of 5 m sec^{-1}. Another study of three-dimensional mountain–valley type flow patterns is that of Hughes (1978).

Observational studies of mountain–valley winds include those of MacHattie (1968) and George (1979). In Colorado, the South Park Area Cumulus Experiment (SPACE, Danielson and Cotton, 1977) was an investigation of the influence of mountain winds on cumulus cloud development. As an example of results from these studies, MacHattie found that the synoptic wind was most strongly coupled to the flow in the direction of the main valley. The diurnal perturbation was therefore reduced more in that direction than it was normal to the valley axis. Also, the diurnal variation of the wind was observed to be less well defined on days with intense solar radiation because the strong heating was effective in developing a deep planetary boundary layer, which enhanced mixing of the gradient wind down to the ground.

Sea and land breeze circulations interacting with mountain–valley systems have also been studied (e.g., Mahrer and Pielke, 1977a,b; Ookouchi *et al.*, 1978). Such interactions can be quite complex and, as shown by Segal *et al.* (1981), are not simply a superposition of the two different phenomena. Rather, mountains along coastal regions, acting as elevated heat sources, create subsidence over and just inland from the coastal waters, thereby influencing the intensity and distribution of the sea breeze. Asai and Mitsumoto (1978) investigate the influence of slope on sea and land breezes with a linear and nonlinear representation.

13.1.3. Forced Airflow over Rough Terrain. When air flows over terrain features that have horizontal scales of 25–100 km or so, another type of mesoscale system develops. This atmospheric feature is different from the sea and land breezes, and mountain–valley winds because forced ascent of air in a prevailing stably stratified air mass, rather than differential heating of the ground by the Sun, generates the mesoscale perturbation. The intensity of this mesoscale system is directly proportional to the pressure gradient generated by this forced movement of air.

Since the pressure gradient force [of the form $-\bar{\theta}\, \partial\bar{\pi}/\partial\tilde{x}^1 + g((z^* - s)/s)\, \partial z_G/\partial\tilde{x}' - f\bar{u}^2$, for example, from Eq. (80)] is of such importance in the evolution of this type of flow and because it is approximately a linear term, exact analytic wave solutions have been applied with considerable success. Early investigators who used exact solutions include Queney (1947, 1948), Scorer (1949), Eliassen and Palm (1960), and Covez (1971). Eliassen and Palm (1960) found that, depending on wavelength, 65–100% of the wave energy generated as air flow is forced over mountains could be reflected downward from layers of strong wind in the upper troposphere. More recently, Klemp and Lilly (1975) have had some success at applying a linear model to estimate the occurrence or nonoccurrence of extreme downslope winds in the lee of the Colorado Rockies, while Sangster (1977) has tested a statistical procedure to forecast these winds using observed synoptic information and parameters derived from linear theory. Klemp and Lilly found that the maximum downslope winds occur when an inversion is present near mountain top level upstream and if the temperature and wind profiles are such that the wave induced by the terrain approximately reverses phase between the surface and the tropopause. Sangster determined that temperature differences in the vertical and strong westerly winds at 700 mbar were important parameters in causing strong downslope winds, although in contrast to Klemp and Lilly's result, such information as the vertical wavelength, Scorer parameter (defined below), and the presence of an inversion were not. Extending Klemp and

Lilly's work, Hyun and Kim (1979) give another example of a linear two-dimensional model of this type of mesoscale system.

Three-dimensional linear models (e.g., Blumen and McGregor, 1976) provide guidance as to what fraction of the airstream goes around topographic barriers and how much advects over it as a function of factors such as the thermodynamic stability. When air can neither go over nor go around because it is too stable, the influence of the mountain propagates rapidly upwind—a process called *blocking*. Richwien (1978) and Baker (1970) give examples of such blocking by segments of the Appalachian Mountains in the eastern United States.

Linear theory predicts that the vertical wavelength of the wave induced by a single ridge is given by $L = 2\pi/S_0^{1/2} = 2\pi\bar{u}/((g/\bar{\theta})(\partial\bar{\theta}/\partial z))^{1/2}$ (assuming $\bar{v} = 0$), where S_0 is called the *Scorer parameter* (e.g., Anthes and Warner, 1978; Alaka, 1960). According to linear theory, for well-developed waves to develop (as listed by Anthes and Warner) the Scorer parameter must be less in the upper troposphere than at lower levels. This requires that if $\partial\bar{\theta}/\partial z$ is constant, $\bar{u}$ must increase with height, while if $\bar{u}$ is a constant, $\partial\bar{\theta}/\partial z$ must be less stable in the higher levels. According to linear theory, in the absence of the Coriolis effect two types of wave motions are induced as air flows over rough terrain—the *forced wave,* which is colocated with the underlying topography, and the *lee wave,* which propagates downstream. Only the forced wave is realistically simulated in a hydrostatic model.

The use of nonlinear models to simulate the airflow over mountains originated with Hovermale (1965), who felt that the large perturbation velocities observed in actual mountain flows violated the requirements of linear theory in which the products of perturbations must be small. Nonlinear studies have continued with the more recent work of such investigators as Furukawa (1973), Klemp and Lilly (1978), Gal-Chen and Somerville (1975b). Deaven (1976). Mahrer and Pielke (1978b), Anthes and Warner (1974), Clark (1977), Clark and Peltier (1977), and Peltier and Clark (1979). The latter authors, for example, disagree with Klemp and Lilly's (1975) explanation for strong downslope wind events, and suggest that downward reflection of energy from breaking waves in the stratosphere is the primary mechanism. Figure 14 illustrates a simulation by Klemp and Lilly (1978) for a particular windstorm in Colorado on January 11, 1972—a study day also studied by Peltier and Clark (1979) and Mahrer and Pielke (1978b). Peltier and Clark maintain that a nonhydrostatic model is necessary to properly simulate the windstorm on this day, while Klemp and Lilly, and Mahrer and Pielke claim a hydrostatic representation is adequate. The scale analysis introduced earlier in this chapter [e.g., Eq. (59) and page 215] indicates that the hydrostatic formulation is

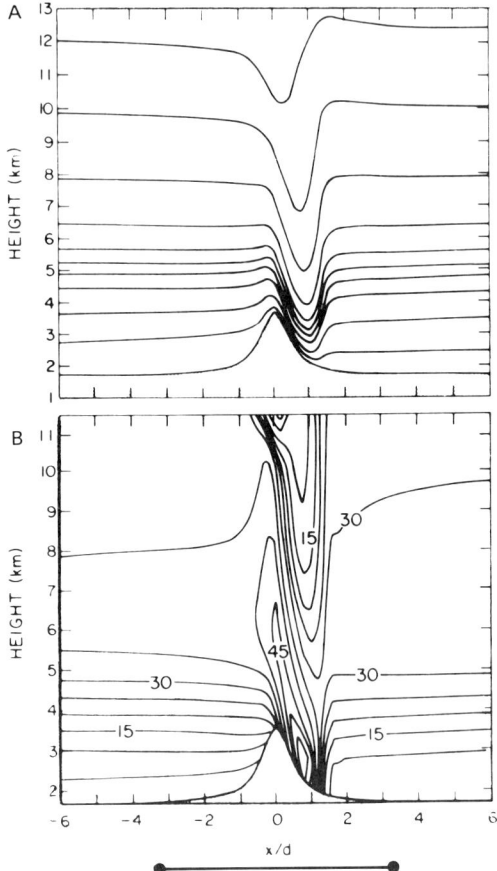

FIG. 14. The predicted potential temperature cross section (a) and horizontal wind field (b) for a simulation of the January 11, 1972 windstorm along the east slopes of the Colorado Rockies. (From Klemp and Lilly, 1978.)

adequate in representing this windstorm; however, additional quantitative experiments are needed to conclusively settle the issue.

When the Coriolis effect and boundary-layer dynamics are included, the response of the atmosphere to terrain is more complex. As shown by Kessler and Pielke (1981), air that becomes ageostrophic after passing over one ridge, does not adjust again to equilibrium for a long distance downstream. Therefore, if a second ridge is situated a short distance downstream, its upstream wind profile would be markedly different from that obtained if the Coriolis effect were not included.

Kessler and Pielke also showed that net boundary-layer warming occurs downwind of the second ridge relative to the first, even in the absence of the Coriolis effect, due to the enhanced mixing of potentially warmer air downward as it accelerates over the upstream ridge. This net warming, a contribution to the *chinook* or *föhn* effect, can occur in the absence of precipitation on the upwind side of the mountain barrier.

Observational studies of strong airflow over rough terrain include those of Lilly and Zipser (1972), Brinkmann (1974), and Lilly and Kennedy (1973). Lilly and Zipser, for example, observed wind gusts of 166–200 km hr^{-1} associated with a chinook immediately downwind of the Rockies.

Over many mountainous regions of the world, this forced lifting on the upwind slopes causes condensation and/or sublimation and precipitation, and is an important factor in the local water budget. Huge snow packs of over 800 cm, for instance, occur in the San Juan Mountains of southwestern Colorado in large part due to this effect. Because of the increase in potential temperature that results from the release of latent heat (and entrainment of potentially warmer air from above the planetary boundary layer), comparatively dry, and even arid, regions often occur in the lee of mountains, particularly when the prevailing flow is persistently from one direction. Figure 15 (reproduced from the two-dimensional results of Colton, 1976) illustrates an example of the predicted and observed orographic rainfall pattern over the Sierra Nevadas of California, with precipita-

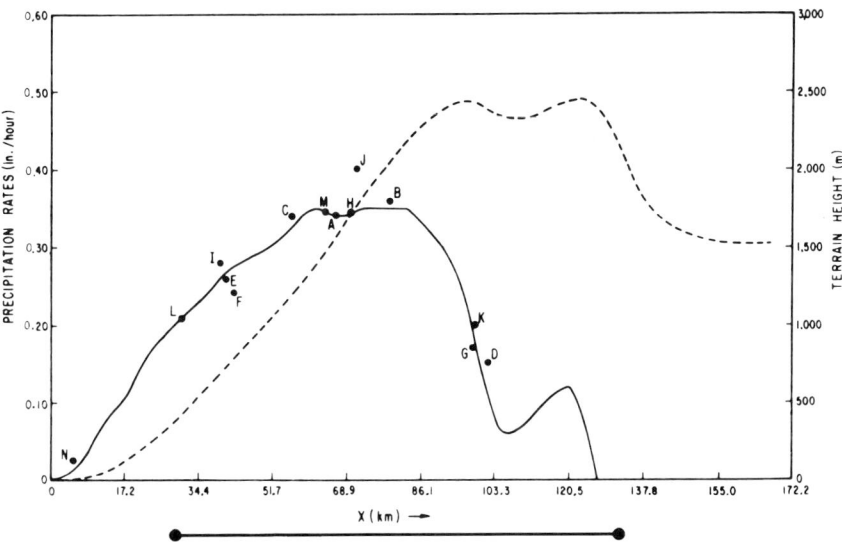

FIG. 15. The predicted precipitation (solid line) over the Sierra Nevada of central California (dashed line—profile of mountain range) for December 21–22, 1964, obtained using a two-dimensional nonlinear mesoscale model. The dots with letters refer to observations. (From Colton, 1976.)

tion confined to the windward slope. In Colton's work, precipitation rates are parameterized using a single conservation equation to represent both cloud and precipitation water. Fraser *et al.* (1973) and Hobbs *et al.* (1973) describe a diagnostic two-dimensional simulation of the airflow over the Cascades of Washington State in which a detailed description of the cloud and precipitation microphysics is included. Gocho (1978) used a two-dimensional linear steady-state model to investigate the influence of microphysical processes on rainfall over the Suzuka Mountains in Japan.

Simulations of clouds and/or precipitation over rough terrain using three-dimensional models include those of Chappell *et al.* (1978) for the San Juan Mountains, and those of Nickerson and Magaziner (1976) and Nickerson (1979) for the island of Hawaii. Lavoie (1974) presents a one-layer simulation of Oahu in Hawaii, while Chang (1970) applied Lavoie's model to the Black Hills of South Dakota. Raddatz and Khandekar (1979) also successfully applied Lavoie's model to the western plains of Canada using a 47.6-km horizontal grid. When the atmosphere is particularly moist and potential instability is released, the resultant rains over rough terrain can be heavy and can cause disastrous flash floods such as occurred over the Black Hills of South Dakota in 1972 and in the Big Thompson watershed in Colorado in 1976 (e.g., Caracena *et al.*, 1979). Mesoscale models may provide an effective tool to forecast these extreme events.

Accurate simulations of airflow over rough terrain when precipitation and cloudiness occur must not only represent properly the complex terrain but also the dynamic and thermodynamic changes caused by the phase transformations of water. Hill (1978), for example, found that circulation cells are formed over mountain areas by the precipitation itself, while Reid *et al.* (1976) determined that cloud shadowing over irregular terrain also affects the intensity of the airflow over mountains.

Simulations of wind flow over rough terrain using a form of dynamic initialization or objective analysis include those of Rhea (1977), Fosberg *et al.* (1976), Collier (1975, 1977), Dickerson (1978), Danard (1977), and Sherman (1978). Ludwig and Byrd (1980) recently outlined what they claim to be a particularly efficient procedure to compute mass-consistent flow fields from wind observations in rough terrain. Such models are called *diagnostic*, even if the conservation relations are used, because they are not used to forecast forward in time through integration of the conservation relations. Figures 16 and 17 give examples of results from two of these models.

Diagnostic models are very economical and appear to be effective mesoscale analysis tools when

FIG. 16. The predicted precipitation pattern over Colorado with a westerly synoptic wind. The darkest shading indicates the heaviest precipitation. (From Rhea, 1977.)

- the dominant forcing is the terrain
- sufficient observational data is available to input to the analysis.

Intercomparisons between diagnostic and prognostic mesoscale models are being undertaken as part of an international wind energy verification study under the coordination of the Atmospheric Sciences Section of Battelle Northwest Laboratories in order to determine how well these two tools perform in wind energy applications. Hiester and Pennell (1980) give a useful review of the application of modeling techniques to large wind turbine siting. One question being addressed is whether the prognostic models, without a dense network of observations, can produce the same fields as obtained by the diagnostic models (which require considerable input data to be accurate). It is expected that since the prognostic models are based on fundamental physical principles as discussed in Section 1 of

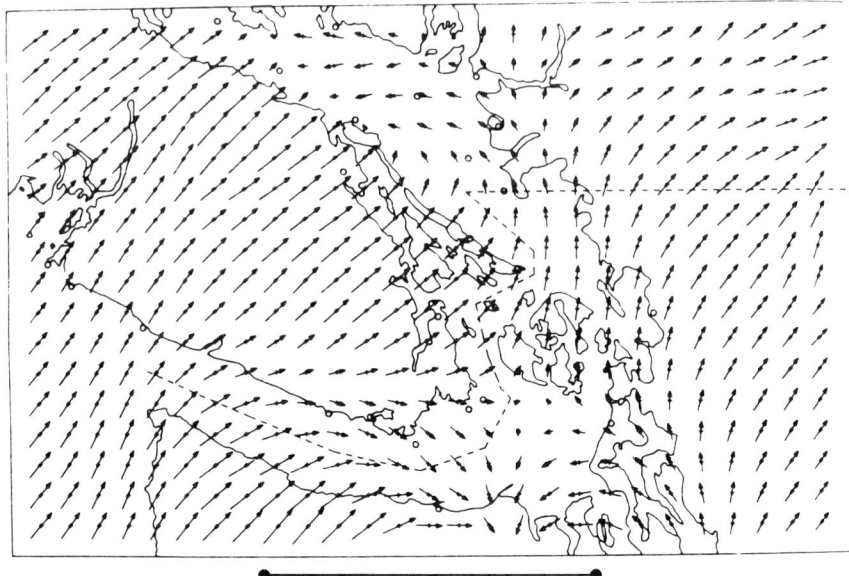

Fig. 17. The simulated winds over the region around the southern end of Vancouver Island, British Columbia. One grid distance represents 10 m sec^{-1}. (From Danard, 1977.)

this chapter, the results should be an accurate modeling of reality as long as sufficient resolution of the forcings is provided.

The report of the first planning on the GARP Mountain Sub-Programme (1978) provides a useful overview of many aspects of our current knowledge of mountain meteorology.

13.1.4. Urban Circulations. Urban circulations are similar to sea and land breezes, and mountain–valley winds in that it is the differential heating and cooling between the rural and urban areas that generates and sustains the wind system. Over the cities and suburban areas, such alterations as asphalting, buildings, and the removal of vegetation have markedly altered the surface heat budget, and therefore the intensity of heat flux to the air.

The influence of these urban areas on the local weather pattern has received increased attention in recent years as the areal extent of such regions expand, and as we begin to realize the major influence of industrial and populated areas on climate, and on human health and well-being. It is in the study of urban circulations that Eq. (5) becomes an important component in the conservation laws relevant to atmospheric flows. As reported by Pielke (1978), an estimated 15,000 deaths per year in the United

States are due to air pollution. This number exceeds the annual average number of fatalities of all the remaining weather-related hazards combined.

Health effects from poor air quality occur throughout the world. In Italy, during July 1976, for example, the accidental venting of the highly toxic organic compound called dioxin (2,3,7,8-tetrachlorodibenzo-*p*-dioxin) from a factory and its transport and dispersal by the local flow field caused death to farm animals and sickness to people, and forced the permanent evacuation of individuals from their homes (Fuller, 1978; *Science News*, 1976). In the eastern United States, the emission of sulfates from coal-burning power plants and nitrates from automobile exhausts and power plants has greatly increased the acidity of the precipitation (Likens and Bormann, 1974). J. Galloway of the University of Virginia (personal communication, 1978) has sampled precipitation in Virginia and found that on occasion it has acidities over 250 times the naturally expected value (pH as low as 3.2). The impact of anthropogenic gases and aerosol contaminants in the atmosphere demands greater complexity in mesoscale models since the number of interactions is greater. In addition, Gaussian plume simulations, as originally proposed by Pasquill (1961) and used by the Environmental Protection Agency (e.g., Turner, 1969), are inaccurate representations of pollutant distributions when such phenomena as the recirculation of the urban air occurs. Calder (1977), for example, in a serious oversimplification of urban meteorology, assumes a single wind speed and direction are representative of an entire city area for 1-hr periods in his multiple-source plume model formulation.

During the past few years, urban models have tended to evolve separately in the areas of air chemistry and meteorology. In the former case, models have treated detailed chemical interactions (e.g., Appel *et al.*, 1978; Kowalczyk *et al.*, 1978; Peterson, 1970) but have neglected to handle properly the mesoscale dynamics. Such models are often called *Box models* (e.g., Schere and Demerjian, 1978). Surveys of our knowledge of atmospheric chemistry are given by Heicklen (1976) and by McEwan and Phillips (1975).

On the other hand, the meteorological observations and simulations have concentrated on the effects of an urban area on the wind, temperature, and moisture fields, rather than chemical interactions. Loose and Bornstein (1977), for instance, investigated the influence of New York City on the synoptic flow and observed that when a heat island was well developed, synoptic fronts decelerated over the upwind half of the city, and accelerated over its downwind half in response to the higher surface roughness of the urban area. Recently, Bornstein has directed a study of the influence of New York City on the sea breeze (Anderson, 1979; Fontana, 1979; Thompson, 1979) and has further illustrated the large drag effect of the buildings in this urban area.

The St. Louis area has been studied extensively as part of the METROMEX program (see Project METROMEX, 1976, for a summary and the May 1978 issue of the *Journal of Applied Meteorology* for a series of articles with results from this program). Vukovich *et al.* (1976), for example, has found from a mesoscale model that the urban effect of St. Louis depends on wind direction, when the synoptic wind is above a certain threshold. Other mesoscale modeling studies of this urban area include those of Vukovich *et al.* (1979) and Hjelmfelt (1980).

Additional observational and theoretical urban studies include those of Sisterson and Dirks (1978), Shreffler (1979), Sawai (1978), Lee and Olfe (1974), Loose and Bornstein (1977), Oke (1976), Taylor (1974), Yap and Oke (1974), Oke and Fuggle (1972), Fuggle and Oke (1976), Nunez and Oke (1976), Anderson (1971), and Olfe and Lee (1971). Oke (1973) determined from an observational study that the heat island effect of a city on its surroundings under cloudless skies is inversely proportional to the large-scale wind speed and directly related to the logarithm of the population.

Studies of islands have also been performed in order to estimate the influence of urban areas on climate and weather (as well as to study the effects of the islands themselves, of course). Such investigations are particularly useful because pollution is not generally significant over an island, whereas it may be in the city environment. Mahrer and Pielke (1976) performed such a study, using the island of Barbados in the West Indies, and found a downwind pressure minimum that was created by the advection offshore of the heat generated by the island. Figure 18 illustrates the resultant low-level convergent zone that was produced downwind of the island. Scofield and Weiss (1976) and the principal investigators of Project METROMEX (1976) have reported on preferred regions of thunderstorm development downwind of urban areas, apparently at least partially due to this type of convergent wind field. Figure 19 (reproduced from Scofield and Weiss, 1977) illustrates the strong heating of a metropolitan area (in this case Washington, D.C., and Baltimore) relative to the rural area.

Matson *et al.* (1978) in a more recent study used satellite imagery to illustrate maximum urban–rural differences ranging from 2.6° to 6.5°C in the midwestern and northeastern United States on a particular summer day. Price (1979), using high-resolution satellite imagery, found peak rural–urban temperature differences as large as 17°C over New York City—a result that is substantially larger than surface-based observed temperatures. He suggests that this difference could be due to satellite

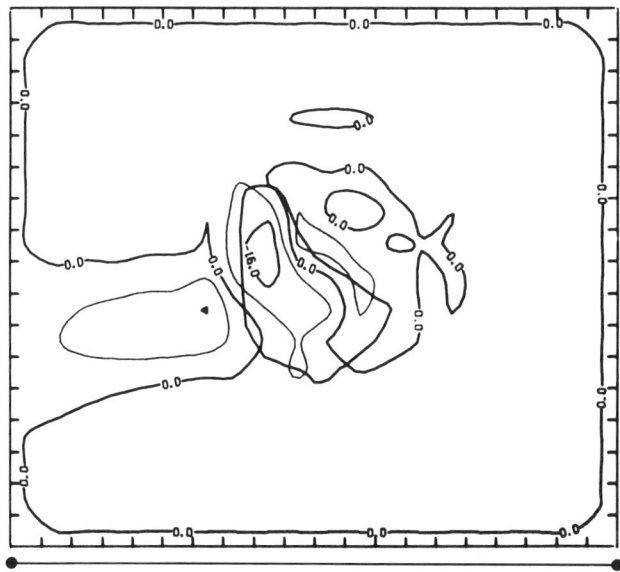

Fig. 18. The vertical velocity field at $z^* = 1$ km at 1300 LST in the vicinity of Barbados during a typical summer afternoon. The contour interval is 8 cm sec^{-1}. The synoptic geostrophic wind was 10 m sec^{-1} from the east. (From Mahrer and Pielke, 1976.)

sensing of industrial areas, rooftops, as well as the trapping of energy within urban canyons (Nunez and Oke, 1977), which are not sensed by the surface observations. Actual heat fluxes into the atmosphere must, therefore, be proportionately larger than if only surface-based observed data is used to estimate fluxes in mesoscale models of the urban circulation. The cover of the March 1980 issue of the *Bulletin of the American Meteorological Society* (Matson and Legeckis, 1980) illustrates another satellite image of urban heat islands (this example is for eastern New England), while Carlson and Augustine (1978) present an image for the Los Angeles area.

Other studies of heated islands include those of Lee (1973), Estoque and Bhumralkar (1969), Delage and Taylor (1970), Bhumralkar (1972), and Lal (1979), while Chopra (1973) summarizes these, and other aspects of the influence of islands on atmospheric flow patterns. Garstang *et al.* (1975) present a summary of heat island studies.

On a somewhat larger scale, Keyser and Anthes (1977) report on predictions of planetary boundary-layer depths over the Middle Atlantic States using a mesoscale model. The concentration of pollution, of course, is closely related to the boundary-layer depth. Sheih (1978),

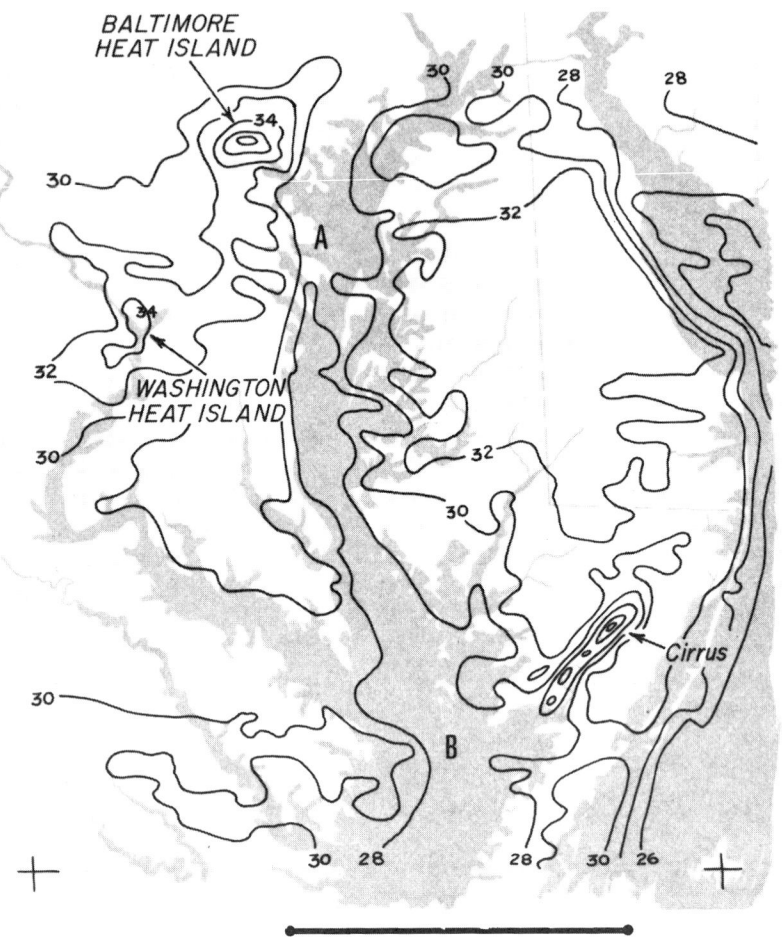

FIG. 19. The observed surface temperature over the Chesapeake Bay Region (as seen via the NOAA-4 satellite) at 0848 LST on June 26, 1976. (From Scofield and Weiss, 1977.)

Hanna (1979), McNider *et al.* (1980), and others have begun to develop more accurate representations of pollution dispersion for use in urban, and other types of mesoscale models.

Recently, efforts have commenced to couple air chemistry and meteorology; however, no three-dimensional mesoscale simulations of this interaction have yet been undertaken because of the large computer resources required. Because of the Clean Air Act Amendments of 1977 [P.L. 95–95, Clean Air Act (1977)], however, increased study of these interactions is required. Table VI lists Primary Air Quality Standards for particular gaseous and aerosol pollutants that are known to have large temporal fluctuations, and which additionally are expected to have large spatial mesoscale variations. Many urban areas do not satisfy these and other air quality standards and are said to be in *nonattainment*. Although these primary standards were chosen for health reasons, the effect on local weather could also be substantially influenced by lesser concentrations than given in Table VI. Viskanta and Weirich (1978) used a two-dimensional mesoscale model in which initial pollutant concentrations at the surface in a rural and an urban area were 20 and 100 $\mu g/m^3$, respectively, for gases and aerosols. They showed that surface temperatures are reduced about 0.3°C during midday downwind of an urban area in both winter and summer in midlatitudes, while around sunrise it was 0.8°C warmer in winter and 0.5°C warmer in summer. Even upwind of the city, changes in surface temperature were predicted due to the influence of changes in radiative flux divergence on the urban circulation. The 100 $\mu g/m^3$ concentration used by Viskanta and Weirich is well below the primary standard for sulfur oxide and suspended particulate material given in Table VI.

Other studies of the effect of pollution on urban weather include Zdunkowski *et al.* (1976), Atwater (1971a,b, 1974, 1977), Bergstrom and Viskanta (1973a,b), Viskanta and Daniel (1980). Pandolfo *et al.* (1976), Welch, *et al.* (1978), and Viskanta *et al.* (1977). This last article reported on a two-dimensional simulation of St. Louis during the summer and found a maximum heat island effect of +3°C. Viskanta *et al.* found that the interaction of radiation with air pollution acts to decrease stability near the ground at night and to increase it during the day. Welch *et al.* (1978) found significant changes in the modeled planetary boundary-layer structure

TABLE VI. SEVERAL PRIMARY AIR QUALITY STANDARDS AS MANDATED BY THE CLEAN AIR ACT AMENDMENTS OF 1977

	Primary standards ($\mu g/m^3$)
Sulfur oxides	
24-hr maximum[a]	365
Suspended particulate material	
24-hr maximum[a]	260
Carbon monoxide	
8-hr maximum[a]	10,290
1-hr maximum[a]	40,010
Hydrocarbons	
3-hr (6–9 AM) maximum[a]	160
Ozone	
1-hr maximum[b]	235

[a] Not to be exceeded more than once a year.
[b] Not to be exceeded on more than one day per year (averaged over 3 years).

over and downwind from a city due to changes in atmospheric turbidity, roughness, heating, and soil types over the urban area. Atwater (1977) investigated urban effects in desert, tropical, midlatitude, and tundra locations and concluded that the largest thermal effects were in the tundra, while the smallest were in the tropics and deserts. In contrast to Viskanta and Daniel (1980) and others, however, he concluded that, except in the tundra, pollutants are only minor factors in the formation of heat islands. Pandolfo *et al.* (1976) also have suggested that NO_2 and the particulate aerosols are the only commonly found anthropogenic pollutant constituents with significant radiative effects, and that their concentrations can be represented as a fixed fraction of CO, which is the only pollutant they predicted explicitly. Future work needs to establish the validity of this simplification as well as to clarify whether or not pollution is a significant contributor to urban heat islands.

Bornstein and Oke (1980) and Robinson (1977) review the current understanding of the influence of pollution and urbanization on urban climate. In future modeling work of these phenomena, both air chemistry and the meteorology must interact. The primary interactions are as follows:

- The rate of input, transport and dispersion, and fallout of pollutants, as well as the types and speeds of chemical reactions depend on the mesoscale meteorological dynamics and thermodynamics.
- The mesoscale circulations are influenced by alterations in radiative characteristics due to changes in the clarity of the atmosphere because of pollutant gases and aerosols.

13.1.5. Lake Effect Models. When cold air advects over warmer ocean or lake water, the sensible and latent heat fluxes to the atmosphere can be quite large, deepening the planetary boundary layer as the air continues its traverse over water. If the body of water is sufficiently broad, then marked changes in local weather occur along the windward shore, relative to conditions found on the lee coast. The shorelines of the Great Lakes and the Sea of Japan, for example, suffer from major, localized snowstorms due to the overwater advection of arctic and polar air during the winter. Lavoie (1972) and Estoque and Ninomiya (1976) successfully represented this phenomena using a one-layer mesoscale model. Boudra (1977), with a 50-km grid interval, has a somewhat larger simulation of the influence of all the Great Lakes on local circulations, while, as described most recently, Ellenton and Danard (1979) have applied a model with a 48-km grid interval to Lake Huron and vicinity. Figure 20 illustrates an example from Lavoie's work showing the vertical velocity at the top of

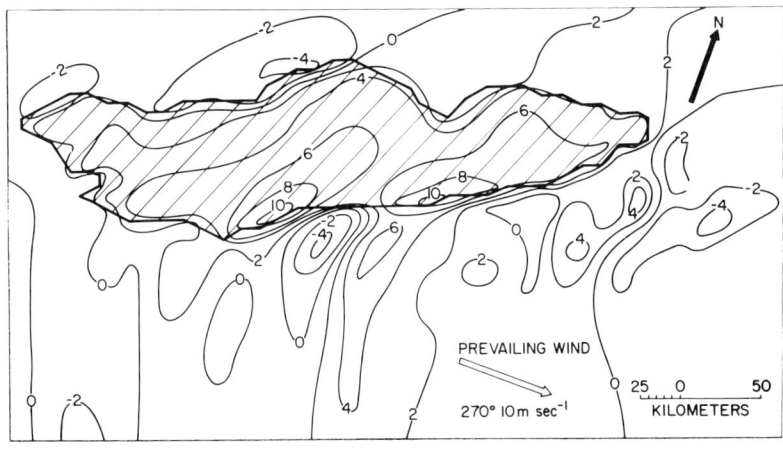

FIG. 20. The predicted vertical motion over Lake Erie with a westerly synoptic wind of 10 m sec^{-1}. (From Lavoie, 1972.)

the mixed layer as air advects across Lake Erie, where significant upward motion is associated with deep clouds and heavy precipitation. From his study he also found that upslope winds over low topographic relief enhance the precipitation. This type of phenomenon, highly localized in space, occurs over many midlatitude windward coastal regions of the world during cold air outbreaks. The weather over the entire Great Lakes region, for instance, is substantially influenced by this juxataposition of land and water. Figure 21 (reproduced from Muller, 1966) gives a climatological estimate of the annual snowfall for the Great Lakes area. The large spatial variability, evident in the figure, illustrates the strong influence of lake effect snowstorms on local weather.

13.2 Synoptically-Induced Mesoscale Systems

13.2.1. Convective Bands Imbedded in Stratiform Cloud Systems. In midlatitude cyclones and along synoptic-scale fronts, precipitation is often not uniformly distributed, but occurs in well-organized mesoscale-sized bands of heavier snow or rain (e.g., Akiyama, 1978). These smaller scale systems are generally a part of the synoptic system but are usually not resolvable with conventional meteorological observations except by satellite and radar. They occur when organized local regions of the atmosphere are convectively unstable, while the mean atmosphere is stable to moist adiabatic displacements in the vertical. Such bands can be reinforced by terrain inhomogeneities, such as the development of

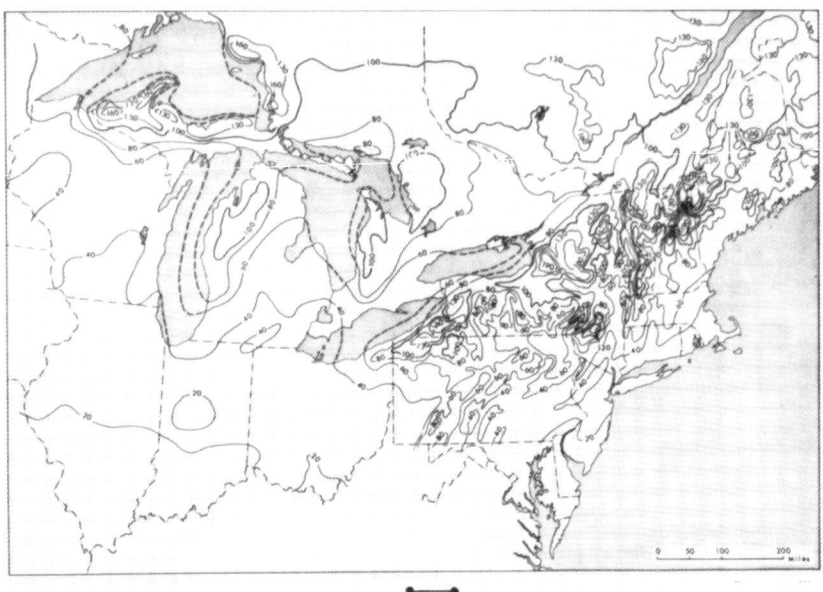

FIG. 21. The mean winter snowfall over a portion of eastern North America. U.S. data, 1951–1960; Canadian data, 1931–1960. (From Muller, 1966.)

small-scale fronts along the coast associated with the passage of extratropical storms as reported by Bosart (1975) and Marks and Austin (1979), or they can be disrupted or destroyed as they descend large terrain barriers. Over the open ocean they are well defined in satellite imagery and clearly an important component of the extratropical storm system. In the vicinity of northwest Europe, distinct subsynoptic scale disturbances of this type, apparently driven to some extent by latent heat release (e.g., Oerlemans, 1980; Rasmussen, 1979) are relatively common features during polar outbreaks. Similar features also occur in the wintertime over the north Pacific (e.g., Mullen, 1979; Reed, 1979).

Since the interactions between the meso- and synoptic scales are complex, little mesoscale modeling has been performed for this type of mesoscale system. Ballentine's (1980) work represents an initial numerical study of the quasi-stationary coastal fronts reported by Bosart, while the Cyclonic Extratropical Storms Project (CYCLES) (e.g., Matejka *et al.*, 1980; Hobbs *et al.*, 1980; Herzegh and Hobbs, 1980) represents an extensive observational program to understand these systems along the northwest Washington coast.

13.2.2. Squall Lines. Along with the convective bands imbedded in stratiform clouds, the squall line is among the most difficult of mesoscale phenomena to simulate. As mentioned earlier, although it is undoubtedly influenced by fixed geographic features such as terrain, the squall line is highly variable in space and transient in time so that accurate lateral boundary and initial conditions, essential to satisfactory predictions, are difficult and expensive to obtain. While it may be possible to build a climatology of mesoscale model forecasts for representative conditions for terrain-induced mesoscale systems, it is probably necessary to perform squall line predictions for each event.

Squall lines often form in association with synoptic weather features such as cold and warm fronts, dry lines (e.g., Ogura and Liou, 1980; Schaeffer, 1974), and tropical waves (e.g., Fortune, 1980), although they typically travel at a greater speed than these larger-scale weather phenomena. Squall lines develop in air masses that are convectively unstable and, as shown by Weiss and Purdom (1974), their appearance is strongly influenced by factors such as the occurrence of early morning cloudiness. Using linearized forms of the conservation relations, studies (e.g., Raymond, 1975) indicate that squall lines apparently propagate as waves with particularly intense convective activity occurring where two or more of these waves constructively reinforce one another. Fritsch and Maddox (1980) have recently shown that the occurrence of these squalls causes major alterations in the synoptic flow field. These areas of intense cumulus convection can be tracked for days across the United States, apparently in association with an upper tropospheric short-wave trough. Sun and Ogura (1979) have suggested that squall lines may be initiated through differential temperature gradients in the boundary layer interacting with the synoptic flow in an analogous manner to that which causes sea and land breezes. From a case study for June 8, 1966 over Oklahoma, Sun and Ogura observed a well-defined band of horizontal convergence at low levels prior to the appearance of the first radar echoes in a region of large horizontal temperature contrast. Perkey (1976), Hane (1973), Schaeffer (1974), and Ross and Orlanski (1978) give examples of nonlinear mesoscale simulations of features related to squall lines, while a review of the current understanding of squall line dynamics is presented by Lilly (1979). Brown (1979) has modeled mesoscale unsaturated downdrafts driven by rainfall evaporation from precipitating cloud features such as anvils created by squall lines. Squall lines that become stagnant over one geographic location (e.g., Johnstown, Pennsylvania, in July 1977; Hoxit *et al.,* 1978) can produce devastating floods. The Severe Environmental Storms and Mesoscale Experiment (SESAME; e.g., Alberty and Barnes, 1979; Alberty *et al.*, 1979; Lilly, 1975) is a mesoscale observational program designed to improve our understanding of the influence of the local environment on the generation of these intense cumulus convective systems over the Great Plains of the United States. The GARP Atlantic Tropical Experiment (GATE; e.g., Zipser and Gautier, 1978; Frank, 1978; Warner *et al.*, (1979) and the Venezuelan International Meteorological and Hydrological Experiment (VIMHEX; e.g., Betts *et al.*, 1976) represent similar recent observational programs in the tropical eastern Atlantic and over land in tropical South America, while PROFS (the Prototype Regional Observing and Forecast Service— Beran, 1978) proposes development of a mesoscale forecast service (initially in the Denver area) to forecast squall lines and other mesoscale phenomena.

13.2.3. Tropical Cyclones. Tropical cyclones, which are generally smaller than the extratropical cyclones that form along the polar front, represent one of the largest mesoscale phenomena. They form when the heating due to cumulonimbus activity positively reinforces the low-level convergent wind flow such that an increasingly more intense mesoscale vortex develops. Simulations of this phenomenon include two-dimensional studies such as Rosenthal (1970, 1971) and Kurihara (1975), while Anthes *et al.* (1971), Anthes (1972), Jones (1980), Kurihara and Tuleya (1974), and others have performed idealized three-dimensional calculations. Anthes *et al.* (1971) have contrasted solutions using two- and three-dimensional simulations. Ceselski (1974) and Mathur (1974, 1975) give examples of simulations for actual observed tropical storms. The effect of landfall on tropical cyclone structure has been studied by Tuleya and Kurihara (1978) and Moss and Jones (1978), while wind flow in mountainous terrain caused by such storms has been investigated by a diagnostic model and a physical model (Brand *et al.*, 1979). An overview of tropical storm modeling is given by Simpson and Pielke (1976), while a more extensive review is presented by Anthes (1974b, 1980). Krishnamurti and Kanamitsu (1973) simulate the more common tropical disturbance referred to as the nondeveloping tropical wave. The latter model is described in Krishnamurti *et al.* (1973).

Figure 22 (reproduced from the December 1969 issue of the *Virginia Climatological Summary*) illustrates the complex precipitation patterns that can occur over land due to the overland track of a tropical storm interacting with a cold front (in this case Hurricane Camille). Zhou Xiao Ping (1980) presents a similar study over central China. The ability of nonlinear numerical models to reproduce such complicated precipitation patterns has not yet been achieved, and with the severe lateral and initial

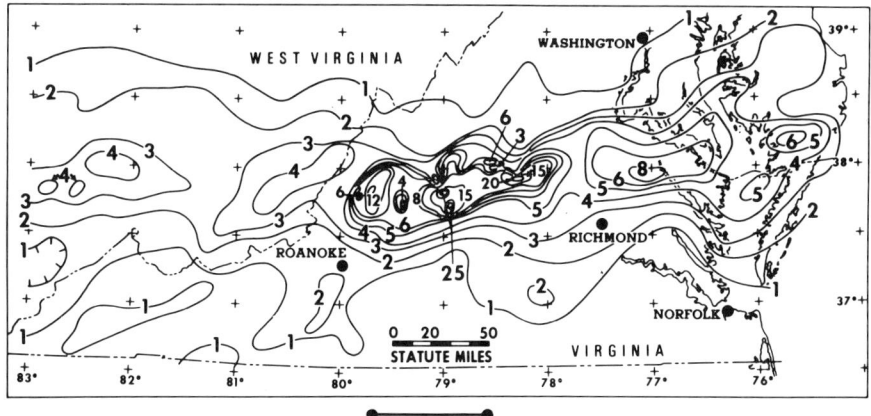

FIG. 22. The observed precipitation (in inches) from the remnants of Hurricane Camille (noon Aug. 19 to midnight Aug. 20, 1969). (From the December 1969 issue of the *Virginia Climatological Summary*, NOAA, Asheville, North Carolina.)

boundary constraints required for actual mesoscale situations (e.g., Anthes and Warner, 1978) the skill may never be attained.

14. Conclusions

In this article

- fundamental physical principles were used to derive the conservation relations applicable to tropospheric circulations in the Earth's atmosphere;
- simplifying assumptions were made to apply these conservation relations to the mesoscale, and the equations were averaged in order to permit their solution on available computers;
- physical and linear models were introduced and shown to have shortcomings in the accurate simulation of most mesoscale phenomena, although they have useful specialized applications;
- the conservation relations were also written in a terrain-following representation;
- parameterizations of planetary boundary-layer processes, and radiation and moist processes were briefly overviewed;
- appropriate solution techniques were then introduced and shown to have deficiencies when the resolution of the mesoscale feature, relative to the model grid, is small;
- initialization and spatial boundary conditions were discussed; and
- finally, examples of specific mesoscale features were introduced.

Among the major conclusions presented were the following:

- Conservation equations can be developed from fundamental physical principles. The consideration of the assumptions used in obtaining these relations is a necessary requirement for the successful application of these expressions.
- *Mesoscale* atmospheric systems are defined to be hydrostatic and with the winds significantly out of gradient wind balance above the planetary boundary layer (ageostrophic if the isobars are straight). The hydrostatic criterion requires that $L_z^2/L_x^2 \ll 1$ if internal gravity waves are an important part of the solution, while $H_\rho/L_x \lesssim 1$ is needed otherwise. To satisfy the nongradient wind requirement the Rossby number cannot be much less than unity.
- Density fluctuations can be ignored relative to the average value of density, except when they are multiplied by gravity, permitting the anelastic form of the conservation of mass relation to be used. Such an approximation eliminates sound waves as a possible solution to the conservation equations. If the mesoscale circulation is shallow compared with H_ρ, the incompressible continuity of mass equation can be used.
- In order to use the conservation relations in a mesoscale model, they must be integrated over a grid volume. Such averaging introduces subgrid-scale correlation terms that must be parameterized in terms of grid-volume-averaged quantities.
- Linear and physical models offer useful theoretical guidance and qualitative understanding of some actual mesoscale flows; however, general simulations of such phenomena require nonlinear models.
- The parameterization of the planetary boundary layer can be performed successfully using first-order closure. When the boundary layer is being heated from below, a profile formulation for the exchange coefficient is needed, while a local form should be used otherwise. Boundary-layer theory developed for horizontally homogeneous, steady-state conditions is the only information available for application in mesoscale models; however, it appears adequate as long as the horizontal grid increment in the model is sufficiently large.
- Radiation physics are parameterized in terms of short- and longwave components. Such a representation appears satisfactory for mesoscale phenomena where information concerning the radiative fluxes as a function of wavelength is not required. When chemical transformations and aerosols are included in such models, however, a detailed spectral representation may be needed.

- Moist thermodynamics in a potentially stably stratified environment can be handled by an explicit representation. When convectively unstable regions are present, it may be necessary to parameterize cumulus cloud effects; however, recent studies have given encouragement that an explicit representation may suffice for this case as well.
- The representation of chemical transformations in mesoscale meteorological models requires additional conservation equations. Up to the present no work has been performed on the interactions between atmospheric chemistry and meteorology in three-dimensional mesoscale models. But with the increased concerns regarding air pollution, interest in such studies is warranted. Among the chemical budgets in the atmosphere, the sulfur budget is perhaps the easiest to investigate.
- A terrain-following coordinate system can be used to represent the conservation relations in rough terrain. When the slope of the topography is small and the hydrostatic equation applies, the resultant equations are similar in form to those required over flat terrain. Terrain variations, however, must not be significant on scales less than about four times the horizontal grid, otherwise they will not be resolved adequately.
- Commonly used approximate solution techniques for the conservation relations do not properly represent variations in the dependent variables that have wavelengths less than four times the grid interval. These small features are erroneously dispersive even if their amplitude is preserved, and can cause aliasing. To minimize their influence, a selective low-pass filter and/or a solution technique that damps the shortest waves should be used. The cubic upstream spline interpolation scheme or the Dendy method to represent advection and the implicit diffusion scheme are examples of effective solution techniques for use in mesoscale models. The Long filter is an appropriate horizontal filter.
- Initial conditions required in nonlinear models can be obtained efficiently by using an objective analysis followed by a period of dynamic initialization. Ground characteristics, including moisture content, along with the wind field and vertical thermodynamic stratification are the most important information.
- Heat and moisture budgets, including the influence of vegetation and urbanization, if present, are necessary in order to obtain accurate moisture and temperature values at the lower boundary of the model. To complete such a budget the flow of heat and water into and out of the ground surface must be represented.

- Lateral and top boundary conditions are included for computational necessity, and are not an accurate representation of actual conditions. To the extent practical with available computer resources, they should be placed as far from the region of interest as possible.
- *Terrain-induced mesoscale systems,* which are primarily forced by the underlying geographic situation, are easier to represent than those that are *synoptically-induced.* The second type are generated and strongly influenced by synoptic features such as fronts. The major problem with properly representing synoptically-induced mesoscale phenomena is the difficulty in obtaining accurate boundary and initial conditions since the forcing is not spatially fixed.

The comments listed above summarize some of the major conclusions discussed in this article. Also included in this article are some of the accomplishments to date in mesoscale dynamical modeling as well as a discussion of some of the shortcomings that still need to be investigated. Three-dimensional synoptic dynamic modeling has progressed further than its mesoscale counterpart, and is used operationally in many parts of the world. Mesoscale modeling, with its much more limited support, lags its larger-scale counterpart. It is anticipated, however, that in the coming years the meteorological community will be increasingly exposed to this powerful theoretical tool.

Acknowledgments

I wish to acknowledge my appreciation to a number of individuals, including R. Artz, A. K. Blackadar, J. H. S. Bradley, P. T. Gannon, M. Garstang, K. P. Hoinka, R. Holle, J. B. Hovermale, R. C. Kessler, R. L. Lavoie, P. E. Long, O. Lucero, Y. Mahrer, C. L. Martin, M. C. McCumber, A. P. Mizzi, P. Pettré, M. Segal, J. M. Simpson, R. H. Simpson, J. W. Snow, J. L. Song, J. Trout, E. C. Varona, G. Warneke, V. Wiggert, W. L. Woodley, and to my students who have taken my course in the theory and practice of dynamic modeling. Each of these individuals constructively commented on either this work or on material that led to its preparation. In the individual sections, Klaus-Peter Hoinka's correspondences and discussions were instrumental in the detailed investigation by Charlie Martin and myself of the terrain-following coordinate system using tensor transformation procedures. In the section on solution techniques, Paul Long deserves the highest appreciation for his suggestions and comments to Charlie Martin and myself. Arthur Mizzi is recognized for introducing the notational procedure used in performing the stability analysis of the solution techniques discussed. Ytzhaq Mahrer is acknowledged for his major contributions to our modeling effort during his tenure at the University of Virginia. Continuing collaboration with W. Frank, M. Garstang, R. Kessler, C. Martin, M. McCumber, R. McNider, A. Mizzi, M. Segal, and W. Snow has been of considerable value in the completion of this work. R. Artz, R. Kessler, M. McCumber, M. Segal, and J. L. Song provided useful comments on the final draft of this article. I would also like to recognize Bill R. Cotton and Joanne M. Simpson, who provided me the opportunity to enter into the field of mesoscale meteorology.

Susan M. Grimstead performed the difficult job of typing and editing this material as well as managing the compilation of the figures and tables with her usual conscientious and professional dedication. Portions of the research reported in this article were supported by the Atmospheric Sciences Section of the National Science Foundation (NSF), the Environmental Protection Agency (EPA), and the Department of Energy (DOE) through Battelle Northwest Labs. Some of the computer calculations presented were performed at the National Center for Atmospheric Research, which is supported by the NSF. The original outline for this material was prepared for the European Centre for Medium Range Weather Forecasts as part of a series of seminars entitled "The Interpretation and Use of Large-Scale Numerical Forecast Products," 1978, Shinfield Park, England.

References

Akiyama, T. (1978). Mesoscale pulsation of convective rain in medium-scale disturbances developed in Bain front. *J. Meteorol. Soc. Jpn.* **56**, 267–283.

Alaka, M. A. (1960). The airflow over mountains. *WMO Tech. Rep.* **34**, 1–135.

Alberty, R. L., and Barnes, S. L. (1979). "SESAME 1979 Plans for Operations and Data Archival," NOAA rep. NOAA, ERL, Boulder, Colorado.

Alberty, R. L., Burgess, D. W., Hand, C. E., and Weaver, J. F. (1979). "SESAME 1979 Operations Summary." NOAA ERL, Boulder, Colorado.

Allender, J. H. (1979). Model and observed circulation throughout the annual temperature cycle of Lake Michigan. *J. Phys. Oceanogr.* **9**, 573–579.

Anderson, G. E. (1971). Mesoscale influences on wind fields. *J. Appl. Meteorol.* **10**, 377–386.

Anderson, S. F. (1979). "Effects of New York City on the Horizontal and Vertical Structure of Sea Breeze Fronts," Vol. II. Observations of Sea Breeze Frontal Slopes and Vertical Velocities over an Urban Area. R. D. Bornstein (P. I.). Report from Dept. of Meteorology, San Jose State University, San Jose, California.

André, J. C., DeMoor, G., Lacarcere, P., Therry, G., and du Vachat, R. (1978). Modeling the 24-hour evolution of the mean and turbulent structures of the planetary boundary layer. *J. Atmos. Sci.* **35**, 1862–1883.

Anthes, R. A. (1970). Numerical experiments with a two-dimensional horizontal variable grid. *Mon. Weather Rev.* **98**, 810–822.

Anthes, R. A. (1972). The development of asymmetries in a three-dimensional numerical model of the tropical cyclone. *Mon. Weather Rev.* **100**, 461–476.

Anthes, R. A. (1974a). Data assimilation and initialization of hurricane prediction models. *J. Atmos. Sci.* **31**, 702–719.

Anthes, R. A. (1974b). The dynamics and energetics of mature tropical cyclones. *Rev. Geophys. Space Phys.* **12**, 495–522.

Anthes, R. A. (1976). Numerical prediction of severe storms—certainty, possibility, or dream? *Bull. Am. Meteorol. Soc.* **57**, 423–430.

Anthes, R. A. (1977). A cumulus parameterization scheme utilizing a one-dimensional cloud model. *Mon. Weather Rev.* **105**, 270–286.

Anthes, R. A. (1980). "Tropical Cyclones." Monograph prepared for Unidad Multidisciplenaria de Ciencias Basicas, Zona Universitaria, Xalapa, Veracruz, Mexico. Available from R. A. Anthes, Dept. of Meteorology, Pennsylvania State University, University Park.

Anthes, R. A., and Warner, T. T. (1974). "Prediction of Mesoscale Flows Over Complex Terrain," U.S. Army Res. Dev. Tech. Rep., ECOM-5532. U.S. Army Electronics Command, White Sands Missile Range, New Mexico.

Anthes, R. A., and Warner, T. T. (1978). Development of hydrodynamic models suitable for air pollution and other mesometeorological studies. *Mon. Weather Rev.* **106**, 1045–1078.

Anthes, R. A., Trout, J. W., and Rosenthal, S. L. (1971). Comparisons of tropical cyclone simulations with and without the assumption of circular symmetry. *Mon. Weather Rev.* **99**, 759–766.

Anthes, R. A., Seaman, N. L., and Warner, T. T. (1980). Comparisons of numerical simulations of the planetary boundary layer by a mixed-layer and multi-level model. *Mon. Weather Rev.* **108**, 365–376.

Appel, B. R., Kothny, E. L., Hoffer, E. M., Hidy, G. M., and Wesolowski, J. J. (1978). Sulfate and nitrate data from the California aerosol characterization experiment (ACHEX). *Environ. Sci. Technol.* **12**, 418–425.

Arakawa, A., and Schubert, W. H. (1974). Interaction of a cumulus cloud ensemble with the large scale environment. Part I. *J. Atmos. Sci.* **31**, 674–701.

Asai, T., and Mitsumoto, S. (1978). Effects of an inclined land surface on the land and sea breeze circulation: A numerical experiment. *J. Meteorol. Soc. Jpn.* **56**, 559–570.

Atwater, M. A. (1971a). The radiation budget for polluted layers of the urban environment. *J. Appl. Meteorol.* **10**, 205–214.

Atwater, M. A. (1971b). Radiation effects of pollutants in the atmospheric boundary layer. *J. Atmos. Sci.* **23**, 1367–1373.

Atwater, M. A. (1974). "The Radiation Model," Sect. 4 in Vol. I, CEM Rep. No. 5131–4099, pp. 67–82. A description of a general three-dimensional numerical simulation model of a coupled air–water and/or air-land boundary layer. The Center for the Environment and Man, Inc., Hartford, Connecticut.

Atwater, M. A. (1977). Urbanization and pollutant effects on the thermal structure in four climatic regions. *J. Appl. Meteorol.* **16**, 888–895.

Atwater, M. A., and Brown, P., Jr. (1974). Numerical calculation of the latitudinal variation of solar radiation for an atmosphere of varying opacity. *J. Appl. Meteorol.* **13**, 289–297.

Baer, F., and Simons, T. J. (1970). Computational stability and time truncation of coupled nonlinear equations with exact solutions. *Mon. Weather Rev.* **98**, 665–679.

Baker, D. G. (1970). A study of high pressure ridges to the east of the Appalachian Mountains. Ph.D. Thesis, Massachusetts Institute of Technology, Cambridge.

Ball, F. K. (1960). Control of inversion height by surface heating. *Q. J. R. Meteorol. Soc.* **86**, 483–494.

Ballentine, R. J. (1980). A numerical investigation of New England coastal frontogenesis. *Mon. Weather Rev.* **108**, 1479–1497.

Barnes, S. L., and Lilly, D. K. (1975). Covariance analysis of severe storm environments. *Prepr. Vol., Am. Meteorol. Soc. Conf. Severe Local Storms, 9th, 1975* pp. 301–306.

Beran, D. W. (1978). "Prototype Regional Observing and Forecasting Service," NOAA Executive Summary of a Program Development Plan. NOAA, ERL, Boulder, Colorado.

Bergstrom, R. W. (1972). Predictions of the spectral absorption and extinction coefficients of an urban air pollution aerosol model. *Atmos. Environ.* **6**, 247–258.

Bergstrom, R. W., Jr., and Viskanta, R. (1973a). Modeling of the effects of gaseous and particulate pollutants in the urban atmosphere. Part I. Thermal structure. *J. Appl. Meteorol.* **12**, 901–912.

Bergstrom, R. W., Jr., and Viskanta, R. (1973b). Modeling of the effects of gaseous and particulate pollutants in the urban atmosphere. Part II. Pollutant dispersion. *J. Appl. Meteorol.* **12**, 913–918.

Berkofsky, L. (1977). The relation between surface albedo and vertical velocity in a desert. *Contrib. Atmos. Phys.* **50**, 312–320.

Betts, A. K., Grover, R. W., and Moncrieff, M. W. (1976). Structure and motion of tropical squall lines over Venezuela. *Q. J. R. Meteorol. Soc.* **102**, 395–404.
Bhumralkar, C. M. (1972). "An Observational and Theoretical Study of Atmospheric Flow over a Heated Island," NSF Final Rep., Grant No. GA-14156. Prepared by the Rosenstiel School of Marine and Atmospheric Science, University of Miami, Coral Gables, Florida.
Blackadar, A. K. (1957). Boundary layer wind maxima and their significance for the growth of nocturnal inversions. *Bull. Am. Meteorol. Soc.* **38**, 283–290.
Blackadar, A. K. (1976). Modeling the nocturnal boundary layer. *Symp. Atmos. Turbul., Diffus., Air Qual., Prepr., 3rd, 1976*, pp. 46–49.
Blackadar, A. K. (1979). High resolution models of the planetary boundary layer. *Adv. Environ. Sci. Eng.* **I**, 50–85.
Blackadar, A. K. (1980). Efficient modeling of convection in the planetary boundary layer (draft manuscript). An earlier version of this work is reported in A. K. Blackadar, *Proc. AMS Symp. Turb., Diffus., Air Poll., 4th, 1979*, pp. 443–447.
Blackadar, A. K., and Tennekes, H. (1968). Asymptotic similarity in neutral barotropic planetary boundary layers. *J. Atmos. Sci.* **25**, 1015–1020.
Blackford, B. L. (1978). Wind-driven inertial currents in the Magdalen Shallows, Gulf of St. Lawrence. *J. Phys. Oceanogr.* **8**, 655–664.
Bleck, R. (1978). On the use of hybrid vertical coordinates in numerical weather prediction models. *Mon. Weather Rev.* **106**, 1233–1244.
Blondin, C. (1978). "Un modèle de meso-echelle-conception-utilisation-développement," Note Technique de L'Etablissement d'Etudes et de Recherches Météorologiques, Direction de la Météorologie, Ministère des Transports, Paris.
Blumen, W., and McGregor, C. D. (1976). Wave drag by three-dimensional mountain leewaves in nonplanar shear flow. *Tellus* **28**, 287–298.
Bornstein, R. D., and Oke, T. (1979). Influence of pollution and urbanization on urban climates. *Adv. Environ. Sci. Eng.* **3**.
Bosart, L. F. (1975). New England coastal frontogenesis. *Q. J. R. Meteorol. Soc.* **101**, 957–978.
Boudra, D. B. (1977). A numerical study describing regional modification of the atmosphere by the Great Lakes. Ph.D. Dissertation, University of Michigan, Ann Arbor.
Brand, S., Chambers, R. P., Woo, W. J. C., Cermak, J. E., Lou, J. J., and Denard, M. (1979). "A Preliminary Analysis of Mesoscale Effects of Topography on Tropical Cyclone-associated Surface Winds," NAVENVPREDRESCHFAC Tech. Rep. TR79-04. Naval Environmental Prediction Research Facility, Monterey.
Briggs, G. A. (1979a). Analytic modelling of drainage flows. (Unpublished manuscript.)
Briggs, G. A. (1979b). Entrainment coefficient versus Ri for drainage flows and compatibility with turbulence scaling assumptions. (Unpublished manuscripts.)
Brinkmann, W. A. R. (1974). Strong downslope winds at Boulder, CO. *Mon. Weather Rev.* **102**, 592–602.
Brost, R. A., and Wyngaard, J. C. (1978). A model study of the stably stratified planetary boundary layer. *J. Atmos. Sci.* **35**, 1427–1440.
Brown, J. M. (1979). Mesoscale unsaturated downdrafts driven by rainfall evaporation: A numerical study. *J. Atmos. Sci.* **36**, 313–338.
Brown, P. S., Jr., and Pandolfo, J. P. (1979). Numerical stability of the combined advection-diffusion equation with nonuniform spatial grid. *Mon. Weather Rev.* **107**, 959–962.
Burk, S. D. (1976). Diurnal winds near the Martian polar caps. *J. Atmos. Sci.* **33**, 923–939.
Burk, S. D. (1977). The moist boundary layer with a higher order turbulence closure model. *J. Atmos. Sci.* **34**, 629–638.

Burpee, R. W. (1979). Peninsula-scale convergence in the south Florida sea breeze. *Mon. Weather Rev.* **107**, 852–860.
Bush, N. E., Chang, S. W., and Anthes, R. A. (1976). A multi-level model of the planetary boundary layer suitable for use with mesoscale dynamic models. *J. Appl. Meteorol.* **15**, 909–919.
Businger, J. A. (1973). Turbulent transfer in the atmosphere surface layer. "Workshop in Micrometeorology," Chapter 2. Am. Meteorol. Soc., Boston, Massachusetts.
Businger, J. A., Wyngaard, J. C., Izumi, Y., and Bradley, E. F. (1971). Flux-profile relationships in the atmospheric surface layer. *J. Atmos. Sci.* **28**, 181–189.
Byers, H. R., and Rodebush, H. R. (1948). Causes of thunderstorms of the Florida Peninsula. *J. Meteorol.* **5**, 275–280.
Calder, K. L. (1977). Multiple-source plume models of urban air pollution—their general structure. *Atmos. Environ.* **11**, 403–414.
Caracena, F., Maddox, R. A., Hoxit, L. R., and Chappell, C. F. (1979). Mesoanalysis of the Big Thompson storm. *Mon. Weather Rev.* **107**, 1–17.
Carlson, T. N., and Augustine, J. A. (1978). Temperature mapping of land use in urban areas using satellite data. *Earth Miner. Sci.* **47**, 41–45.
Carlson, T. N., and Benjamin, S. G. (1980). Radiative heating rates for Saharan dust. *J. Atmos. Sci.* **37**, 193–213.
Carpenter, K. (1979). An experimental forecast using a nonhydrostatic mesoscale model. *Q. J. Roy. Meteorol. Soc.* **105**, 629–655.
Carroll, J. J., and Baskett, R. L. (1979). Dependence of air quality in a remote location on local and mesoscale transports: A case study. *J. Appl. Meteorol.* **18**, 474–486.
Carson, R. B. (1954). Some objective quantitative criteria for summer showers at Miami, Florida. *Mon. Weather Rev.* **82**, 9–28.
Cermak, J. E. (1970). "Air Motion in and near Cities—Determination by Laboratory Simulation." Report available from Fluid Dynamics and Diffusion Lab. Colorado State University, Fort Collins.
Cermak, J. E. (1971). Laboratory simulation of the atmospheric boundary layer. *AIAA J.* **9**, 1746–1754.
Cermak, J. E. (1975). Applications of fluid mechanics to wind engineering—a Freeman Scholar Lecture. *J. Fluids Eng.* **97**, 9–38.
Ceselski, B. F. (1974). Cumulus convection in weak and strong tropical disturbance. *J. Atmos. Sci.* **31**, 1241–1255.
Chang, C. (1970). A mesoscale numerical model of airflow over the Black Hills. Master of Science Thesis, South Dakota School of Mines and Technology, Rapid City.
Chang, S. W. (1979). An efficient parameterization of convective and nonconvective planetary boundary layers for use in numerical models. *J. Appl. Meteorol.* **18**, 1205–1215.
Chang, S. W., and Anthes, R. A. (1978). Numerical simulations of the ocean's nonlinear, baroclinic response to translating hurricanes. *J. Phys. Oceanogr.* **8**, 468–480.
Chang, S. W., and Orville, H. D. (1973). Large-scale convergence in a numerical cloud model. *J. Atmos. Sci.* **30**, 947–950.
Chappell, C. F., Smith, D. R., and Nickerson, E. C. (1978). Numerical simulation of clouds and snowfall over mountainous terrain. *Prepr. Vol., Am. Meteorol. Soc. Conf. Cloud Phys. Atmos. Electr. 1978* pp. 259–265.
Charnock, H. (1955). Wind stress on a water surface. *Q. J. R. Meteorol. Soc.* **81**, 639–640.
Chaudhry, F. H., and Cermak, J. E. (1971). "Wind-Tunnel Modeling of Flow and Diffusion over an Urban Complex," Proj. Themis Tech. Rep. No. 17. Fluid Dynamics and Diffusion Lab., Colorado State University, Fort Collins.

Chen, C., and Orville, H. D. (1980). Effects of mesoscale convergence on cloud convection. *J. Appl. Meteorol.* **19,** 256–274.
Chen, J. H. (1973). Numerical boundary conditions and computational modes. *J. Comput. Phys.* **13,** 522–535.
Chopra, K. P. (1973). Atmospheric and oceanic flow problems introduced by islands. *Adv. Geophys.* **16,** 297–421.
Christensen, O., and Prahm, L. P. (1976). A pseudospectral model for dispersion of atmospheric pollutants. *J. Appl. Meteorol.* **15,** 1284–1294.
Churchill, R. V. (1963). "Fourier Series and Boundary Value Problems. McGraw-Hill, New York.
Clancy, R. M., Thompson, J. D., Lee, J. D., and Hurlburt, H. E. (1979). A model of mesoscale air-sea interaction in a sea breeze-coastal upwelling regime. *Mon. Weather Rev.* **107,** 1476–1505.
Clark, T. L. (1973). Numerical modeling of the dynamics and microphysics of warm cumulus convection. *J. Atmos. Sci.* **30,** 857–878.
Clark, T. L. (1979). A small-scale dynamic model using a terrain-following coordinate transformation. *J. Comput. Phys.* **24,** 186–215.
Clark, T. L. (1979). Numerical simulations with a three-dimensional cloud model: lateral boundary condition experiments and multicellular severe storm simulations. *J. Atmos. Sci.* **36,** 2191–2215.
Clark, T. L., and Peltier, W. R. (1977). On the evolution and stability of finite-amplitude mountain waves. *J. Atmos. Sci.* **34,** 1715–1730.
Clarke, R. H. (1970). Recommended methods for the treatment of the boundary layer in numerical models. *Aust. Meteorol. Mag.* **18,** 51–73.
Clarke, R. H., Dyer, A. J., Brook, P. R., Reid, D. G., and Troup, A. J. (1971). "The Wangara Experiment—Boundary Layer Data," Pap. No. 19. Division Meteorological Physics, CSIRO, Australia.
Clean Air Act (1977). Clean Air Act Amendments of 1977. P. L. 95–95. U.S. Code Congressional and Administrative News. West Publ. Co., St. Paul, Minn., 685–796.
Collier, C. G. (1975). A representation of the effects of topography on surface rainfall within moving baroclinic disturbances. *Q. J. R. Meteorol. Soc.* **101,** 407–422.
Collier, C. G. (1977). The effect of model grid length and orographic rainfall efficiency on computed surface rainfall. *Q. J. R. Meteorol. Soc.* **103,** 247–253.
Colton, D. E. (1976). Numerical simulation of the orographically induced precipitation distribution for use in hydrologic analysis. *J. Appl. Meteorol.* **15,** 1241–1251.
Cotton, W. R. (1975). Theoretical cumulus dynamics. *Rev. Geophys. Space Phys.* **13,** 419–448.
Cotton, W. R., and Tripoli, G. J. (1978). Cumulus convection in shear flow three-dimensional numerical experiments. *J. Atmos. Sci.* **35,** 1503–1521.
Cotton, W. R., Pielke, R. A., and Gannon, P. T. (1976). Numerical experiments on the influence of the mesoscale circulation on the cumulus scale. *J. Atmos. Sci.* **33,** 252–261.
Coulson, K. L. (1975). "Solar and Terrestrial Radiation." Academic Press, New York.
Covez, L. (1971). Mountain waves in a turbulent atmosphere. *Tellus* **23,** 104–109.
Csanady, G. T. (1975). Lateral momentum flux in boundary currents. *J. Phys. Oceanogr.* **5,** 705–717.
Cullen, M. J. P. (1976). On the use of artificial smoothing in Galerkin and finite difference solutions of the primitive equations. *Q. J. R. Meteorol. Soc.* **102,** 77–93.
Daley, R. (1980). The development of efficient time integration schemes using model normal modes. *Mon. Weather Rev.* **108,** 100–110.

Dalu, G. A. (1978). A parameterization of heat convection for a numerical sea breeze model. *Q. J. R. Meteorol. Soc.* **104,** 797–807.

Danard, M. (1977). A simple model for mesoscale effects of topography on surface winds. *Mon. Weather Rev.* **105,** 572–581.

Danielson, K. S., and Cotton, W. R., eds. (1977). "Space Log 1977." Dept. Atmos. Sci., Colorado State University, Fort Collins.

Day, S. (1953). Horizontal convergence and the occurrence of summer precipitation at Miami, Florida. *Mon. Weather Rev.* **81,** 155–161.

Deardorff, J. W. (1966). The contragradient heat flux in the lower atmosphere and in the laboratory. *J. Atmos. Sci.* **23,** 503–506.

Deardorff, J. W. (1972). Parameterization of the planetary boundary layer for use in general circulation models. *Mon. Weather Rev.* **100,** 93–106.

Deardorff, J. W. (1974). Three-dimensional numerical study of the height and mean structure of a heated planetary boundary layer. *Bound. Layer Meteorol.* **7,** 81–106.

Deardorff, J. W. (1978). Efficient prediction of ground surface temperature and moisture, with inclusion of a layer of vegetation. *J. Geophys. Res.* **83**(C4), 1889–1903.

Deaven, D. G. (1974). A solution for boundary problems in isentropic coordinate models. Ph.D. Dissertation, Pennsylvania State University, University Park.

Deaven, D. G. (1976). A solution for boundary problems in isentropic coordinate models. *J. Atmos. Sci.* **33,** 1702–1713.

Defant, F. (1950). Theorie der land- und seewind. *Arch. Meteorol., Geophys. Bioklimatol., Ser. A* **2,** 404–425.

Defant, F. (1951). Local winds. "Compendium of Meteorology," pp. 655–672. Am. Meteorol. Soc., Boston, Massachusetts.

Delage, Y., and Taylor, P. A. (1970). A numerical study of heat island circulations. *Bound. Layer Meteorol.* **1,** 201–226.

Dickerson, M. H. (1978). MASCON—A mass-consistent atmospheric flux model for regions with complex terrain. *J. Appl. Meteorol.* **17,** 241–253.

Dickerson, M. H., and Gudiksen, P. H. (1980). "ASCOT FY-1979 Progress Report," UCRL-52899/ascot//1. Laurence Livermore Laboratory, Livermore, California. Available from NTIS ($5.00 per copy).

Dieterle, D. (1976). "Simulation of the Urban Surface Energy Balance Including the Effects of Anthropogenic Heat Production," Rep. No. G320-3344. IBM Palo Alto Scientific Center, Palo Alto, California.

Dobosy, R. (1979). Dispersion of atmospheric pollutants in flow over the shoreline of a large body of water. *J. Appl. Meteorol.* **18,** 117–132.

Donaldson, C. du P. (1973). Construction of a dynamic model of the production of atmospheric turbulence and dispersal of atmospheric pollutants. "Workshop in Micrometeorology," pp. 313–390. Am. Meteorol. Soc. Boston, Massachusetts.

Dutton, J. A. (1976). "The Ceaseless Wind, an Introduction to the Theory of Atmospheric Motion." McGraw-Hill, New York.

Dutton, J. A., and Fichtl, G. H. (1969). Approximate equations of motion for gases and liquids. *J. Atmos. Sci.* **26,** 241–254.

Eliassen, A. (1980). A review of long-range transport modeling. *J. Appl. Meteorol.* **19,** 231–240.

Eliassen, A., and Palm, E. (1960). On the transfer of energy in stationary mountain waves. *Geophys. Norv.* **22,** 1–23.

Ellenton, G. E., and Danard, M. B. (1979). Inclusion of sensible heating in convective parameterization applied to lake-effect snow. *Mon. Weather Rev.* **107,** 551–565.

Elsberry, R. L. (1978). Prediction of atmospheric flows on nested grids. *In* "Computational

Techniques for Interface Problems" (K. C. Park and D. K. Gartling, eds.), AMD-Vol. 30, pp. 67–85. Am. Soc. Mech. Eng., New York.

Elsberry, R. L., and Randy, S. D. (1978). Sea surface temperature response to variations in atmospheric wind forcing. *J. Phys. Oceanogr.* **8**, 881–887.

Emery, K. O., and Csanady, G. T. (1973). Surface circulation of lakes and nearly landlocked seas. *Proc. Natl. Acad. Sci. U.S.A.* **70**, 93–97.

Estoque, M. A. (1961). A theoretical investigation of the sea breeze. *Q. J. R. Meteorol. Soc.* **87**, 136–146.

Estoque, M. A. (1962). The sea breeze as a function of prevailing synoptic situation. *J. Atmos. Sci.* **19**, 244–250.

Estoque, M. A., and Bhumralkar, C. M. (1969). Flow over a localized heat source. *Mon. Weather Rev.* **97**, 850–859.

Estoque, M. A., and Ninomiya, K. (1976). Numerical simulation of Japan Sea effect snowfall. *Tellus* **28**, 243–253.

Estoque, M. A., Gross, J., and Lai, H. W. (1976). A lake breeze over southern Lake Ontario. *Mon. Weather Rev.* **104**, 386–396.

Fisher, E. L. (1961). A theoretical study of the sea breeze. *J. Meteorol.* **18**, 215–233.

Fontana, P. H. (1979). "Effects of New York City on the Horizontal and Vertical Structure of Sea Breeze Fronts," Vol. I. Observations of Frictional Retardation of Sea Breeze Fronts. R. D. Bornstein (P.I.). Report from Dept. of Meteorology, San Jose State University, San Jose, California.

Fortune, M. (1980). Properties of African squall lines inferred from time-lapse satellite imagery. *Mon. Weather Rev.* **108**, 153–168.

Fosberg, M. A., Marlatt, W. E., and Krupnak, L. (1976). Estimating airflow patterns over complex terrain. *U.S., For. Serv., Rocky Mt. For. Range Exp. Stn., Res. Pap. RM-162*, pp. 1–16.

Fox, D. G., and Deardorff, J. W. (1972). Computer methods for simulation of multidimensional, nonlinear, subsonic, incompressible flow. *J. Heat Transfer, Ser. C* **94**, 337–346.

Fox, D. G., and Orszag, S. A. (1973). Pseudospectral approximation to two-dimensional turbulence. *J. Comput. Phys.* **11**, 612–619.

Frank, N. L., Moore, P. L., and Fisher, G. E. (1967). Summer shower distribution over the Florida Peninsula as deduced from digitized radar data. *J. Appl. Meteorol.* **6**, 309–316.

Frank, W. M. (1978). The life cycles of GATE convective systems. *J. Atmos. Sci.* **35**, 1256–1264.

Frank, W. M. (1980). Modulations of the net tropospheric temperature during GATE. *J. Atmos. Sci.* **37**, 1056–1064.

Fraser, A. B., Easter, R. C., and Hobbs, P. V. (1973). A theoretical study of the flow of air and fallout of solid precipitation over mountainous terrain. Part I. Airflow model. *J. Atmos. Sci.* **30**, 801–812.

Friend, A. L., Djurić, D., and Brundidge, K. C. (1977). A combination of isentropic and sigma coordinates in numerical weather prediction. *Contrib. Atmos. Phys.* **50**, 290–295.

Fritsch, J. M., and Chappell, C. F. (1980a). Numerical prediction of convectively driven mesoscale pressure systems. Part I. Convective parameterization. *J. Atmos. Sci.* **37**, 1722–1733.

Fritsch, J. M., and Chappell, C. F. (1980b). Numerical prediction of convectively driven mesoscale pressure systems. Part II. Mesoscale model. *J. Atmos. Sci.* **37**, 1734–1762.

Fritsch, J. M., and Maddox, R. A. (1980). Analyses of upper tropospheric wind perturbations associated with midlatitude mesoscale convective complexes. *Prepr. Vol., Am. Meteorol. Soc., Conf. Weather Forecasting Anal., 1980*, p. 339–345.

Fuggle, R. F., and Oke, T. R. (1976). Long-wave radiative flux divergence and nocturnal cooling of the urban atmosphere. I. Above Roof-Level. *Bound. Layer Meteorol.* **10**, 113–120.

Fuller, J. G. (1978). "The Poison that Fell from the Sky." Random House, New York (shortened version published in Reader's Digest, August 1977, pp. 191–236).

Furukawa, T. (1973). Numerical experiments of the airflow over mountains. I. Uniform current with constant static stability. *J. Meteorol. Soc. Jpn.* **51**, 400–419.

Gage, K. S. (1979). Evidence of a $k^{-5/3}$ low inertial range in mesoscale two-dimensional turbulence. *J. Atmos. Sci.* **36**, 1950–1954.

Gal-Chen, T., and Somerville, R. C. J. (1975a). On the use of a coordinate transformation for the solution of the Navier-Stokes equations. *J. Comput. Phys.* **17**, 209–228.

Gal-Chen, T., and Somerville, R. C. J. (1975b). Numerical solution of the Navier-Stokes equations with topography. *J. Comput. Phys.* **17**, 276–309.

Gambo, K. (1978). Notes on the turbulence closure model for atmospheric boundary layers. *J. Meteorol. Soc. Jpn.* **56**, 466–480.

Gannon, P. T., Sr. (1978). Influence of earth surface and cloud properties on the south Florida sea breeze. *NOAA Tech. Rep. ERL 402-NHEML2.*

GARP Mountain Sub-Programme (1978). Venice, October 24–28, 1977 (copies of report can be obtained from Joint Planning Staff for GARP c/o World Meteorological Organization. Case Postale No. 5, CH-1211 Geneva 20, Switzerland).

Garrett, A. J. (1978). "Numerical Simulations of Atmospheric Convection over the Southeastern U.S. in Undisturbed Conditions," Rep. No. 47. Atmospheric Science Group, University of Texas, Coll. Eng., Austin.

Garstang, M., Tyson, P. D., and Emmitt, G. D. (1975). The structure of heat islands. *Rev. Geophys. Space Phys.* **13**, 139–165.

Gedzelman, S. D., and Donn, W. L. (1979). Atmospheric gravity waves and coastal cyclones. *Mon. Weather Rev.* **107**, 667–681.

Geisler, J. E., and Bretherton, F. P. (1969). The sea-breeze forerunner. *J. Atmos. Sci.* **26**, 82–95.

Gentry, R. C., and Moore, P. L. (1954). Relation of local and general wind interaction near the sea coast to time and location of air-mass showers. *J. Meteorol.* **11**, 507–511.

George, R. L. (1979). "Evolution of Mesoscale Convective Systems over Mountainous Terrain," Atmos. Sci. Pap. No. 318. Dept. Atmos. Sci., Colorado State University, Fort Collins.

Gocho, Y. (1978). Numerical experiment of orographic heavy rainfall due to a stratiform cloud. *J. Meteorol. Soc. Jpn.* **56**, 405–423.

Golden, J. H., and Sartor, J. D. (1978). AMS Workshop on mesoscale interactions with cloud processes. 24–25 October 1977, Boulder, CO. *Bull. Am. Meteorol. Soc.* **59**, 720–730.

Goodin, W. R., McRae, G. J., and Seinfeld, J. H. (1979). A comparison of interpolation methods for sparse data: Application to wind concentration fields. *J. Appl. Meteorol.* **18**, 761–771.

Goodin, W. R., McRae, G. J., and Seinfeld, J. H. (1980). An objective analysis technique for constructing three-dimensional urban scale wind fields. *J. Appl. Meteorol.* **19**, 98–108.

Gresho, P. M., Lee, R. L., and Sani, R. L. (1976). Advection-dominated flows, with emphasis on the consequences of mass lumping. *Int. Symp. Finite Element Methods Flow Probl., 2nd, 1976* pp. 745–756 (preprint available from Lawrence Livermore Lab., Livermore, California).

Gresho, P. M., Lee, R. L., and Sani, R. L. (1979). On the time-dependent solution of the incompressible Navier-Stokes equations in two and three dimensions. "Recent Advances in Numerical Methods in Fluids" (C. Taylor and K. Morgan, eds.). Pineridge Press, Ltd., Swansea, United Kingdom, pp. 27–80.

Grody, N. C., Hayden, C. M., Shen, W. C. C., Rosenkranz, P. W., and Staelin, D. H. (1979). Typhoon June winds estimated from scanning microwave spectrometer measurements at 55.45 GHz. *J. Geophys. Res.* **84,** 3689–3695.
Gross, M. G. (1977). "Oceanography: A View of the Earth," 2nd ed. Prentice-Hall, Englewood Cliffs, New Jersey.
Grotjahn, R. (1977). Some consequences of finite differencing revealed by group velocity errors. *Contrib. Atmos. Sci.* **50,** 231–238.
Grotjahn, R., and O'Brien, J. J. (1976). Some inaccuracies in finite differencing hyperbolic equations. *Mon. Weather Rev.* **104,** 180–194.
Gurka, J. J. (1974). Using satellite data for forecasting fog and stratus dissipation. *Prepr. Vol., Am. Meteorol. Soc., Conf. Weather Forecasting Anal., 5th, 1974* pp. 54–57.
Gutman, L. N. (1972). "Introduction to the Nonlinear Theory of Mesoscale Meteorological Processes." Keter Press, Jerusalem, Israel.
Hachey, H. B. (1934). Movements resulting from mixing of stratified waters. *J. Biol. Board Can.* **1,** 133–143.
Haltiner, G. J. (1971). "Numerical Weather Prediction." Wiley, New York.
Haltiner, G. J., and Williams, R. T. (1980). "Numerical Prediction and Dynamic Meteorology," 2nd ed. Wiley, New York.
Hamilton, P., and Rattray, M., Jr. (1978). A numerical model of the depth-dependent, wind-driven upwelling circulation on a continental shelf. *J. Phys. Oceanogr.* **8,** 437–457.
Hane, C. E. (1973). The squall line thunderstorm: Numerical experimentation. *J. Atmos. Sci.* **30,** 1672–1690.
Hane, C. E. (1978). Scavenging of urban pollutants by thunderstorm rainfall: Numerical experimentation. *J. Appl. Meteorol.* **17,** 699–710.
Hanna, S. R. (1979). Some statistics of Lagrangian and Eulerian wind fluctuation. *J. Appl. Meteorol.* **18,** 518–531.
Hanna, S. R., and Gifford, F. A. (1975). Meteorological effects of energy dissipation at large power parks. *Bull. Am. Meteorol. Soc.* **56,** 1069–1077.
Hanna, S. R., and Swisher, S. D. (1971). Meteorological effects of the heat and moisture produced by man. *Nuc. Saf.* **12,** 114–122.
Heicklen, J. (1976). "Atmospheric Chemistry." Academic Press, New York.
Herzegh, P. H., and Hobbs, P. V. (1980). The mesoscale and microscale structure and organization of clouds and precipitation in midlatitude cyclones. II. Warm-frontal clouds. *J. Atmos. Sci.* **37,** 597–611.
Hickey, J. R., Stowe, L. L., Jacobowitz, H., Pellegrino, P., Maschkoff, R. H., House, F., and Vonder Haar, T. H. (1980). Initial solar irradiance determinations from Nimbus 7 cavity radiometer measurements. *Science* **208,** 281–283.
Hiester, T. R., and Pennell, W. T. (1980). "The Meteorological Aspects of Siting Large Wind Turbines," DOE Energy Rep.—Contract EY-76-C-06-1830. Battele Northwest Labs, Richland, Washington.
Hildebrand, F. B. (1962). "Advanced Calculus for Applications." Prentice-Hall, Englewood Cliffs, New Jersey.
Hill, G. E. (1974). Factors controlling the size of cumulus clouds as revealed by numerical experiments. *J. Atmos. Sci.* **31,** 646–673.
Hill, G. E. (1978). Observations of precipitation-forced circulations in winter orographic storms. *J. Atmos. Sci.* **35,** 1463–1472.
Hjelmfelt, M. (1980). Numerical simulation of the effects of St. Louis on boundary layer airflow and convection. Ph.D. Dissertation, University of Chicago, Chicago, Illinois.
Hobbs, P. V., Easter, R. C., and Fraser, A. B. (1973). A theoretical study of the flow of air

and fallout of solid precipitation over mountainous terrain. Part II. Microphysics. *J. Atmos. Sci.* **30**, 813–823.
Hobbs, P. V., Matejka, T. J., Herzegh, P. H., Locatelli, J. D., and Houze, R. A., Jr. (1980). The mesoscale and microscale structure and organization of clouds and precipitation in midlatitude cyclone. I. A case study of a cold front. *J. Atmos. Sci.* **37**, 568–596.
Hoke, J. E., and Anthes, R. A. (1976). The initialization of numerical models by a dynamic-initialization technique. *Mon. Weather Rev.* **104**, 1551–1556.
Hoke, J. E., and Anthes, R. A. (1977). Dynamic initialization of a three-dimensional primitive-equation model of hurricane Alma of 1962. *Mon. Weather Rev.* **105**, 1266–1280.
Holton, J. R. (1972). "An Introduction to Dynamic Meteorology." Academic Press, New York.
Hovermale, J. B. (1965). A non-linear treatment of the problem of airflow over mountains. Ph.D. Thesis, Pennsylvania State University, University Park.
Hoxit, L. R. (1975). Diurnal variations in planetary boundary layer winds over land. *Bound. Layer Meteorol.* **8**, 21–38.
Hoxit, L. R., Maddox, R. A., Chappell, C. F., Zurkerberg, F. L., Mogil, H. M., Jones, I., Greene, D. R., Saffle, R. E., and Scofield, R. A. (1978). Meteorological aspects of the Johnstown, PA, flash flood 19020 July 1977. *NOAA Tech. Rep. ERL 401-APCL 43*, 1–71.
Hsu, H.-M. (1979). Numerical simulations of mesoscale precipitation systems. Ph.D. Dissertation, Dept. Atmos. Oceanic Sci., University of Michigan, Ann Arbor.
Hsu, S. (1969). "Mesoscale Structure of the Texas Coast Sea Breeze," Rep. No. 16 of the Atmospheric Science Group. University of Texas, College of Engineering, Austin.
Hsu, S. (1973). Dynamics of the sea breeze in the atmospheric boundary layer: A case study of the free convection region. *Mon. Weather Rev.* **101**, 187–194.
Hughes, R. L. (1978). "A Numerical Simulation of Mesoscale Flow over Mountainous Terrain," Pap. No. 303, US ISSN 0067-0340, Dept. of Atmospheric Science, Colorado State University, Fort Collins.
Hunt, J. C. R., Synder, W. H., and Lawson, R. E., Jr. (1978). Flow structure and turbulent diffusion around a three-dimensional hill. Part I. Flow structure. *U.S. Environ. Prot. Agency, Off. Res. Dev. [Rep.] EPA EPA-600/4-78-041*, 1–83.
Hyun, J. M., and Kim, M. (1979). The effect of nonuniform wind shear on the intensification and reflection of mountain waves. *J. Atmos. Sci.* **36**, 2379–2384.
Idso, S., Jackson, R., Kimball, B., and Nakayama, F. (1975). The dependence of bare soil albedo on soil water content. *J. Appl. Meteorol.* **14**, 109–113.
Iribarne, J. V., and Godson, W. L. (1973). "Atmospheric Thermodynamics." Reidel Publ., Dordrecht, Netherlands.
Jacobs, C. A. (1978). Observed oceanic variability at and near the air-sea interface during Phase III of Project BOMEX. *J. Phys. Oceanogr.* **8**, 103–118.
Jacobs, C. A., and Brown, P. S., Jr. (1973). An investigation of the numerical properties of the surface heat-balance equation. *J. Appl. Meteorol.* **12**, 1069–1072.
Jacobs, C. A., and Brown, P. S., Jr. (1974). "IFYCL Final Report, Volume IV Three-dimensional Results," CEM Rep. No. 4131-509d. The Center for the Environment and Man, Inc., Hartford, Connecticut.
Janjić, Z. I. (1977). Pressure gradient force and advection scheme used for forecasting with steep and small scale topography. *Contrib. Atmos. Sci.* **50**, 186–199.
Johnson, A., Jr., and O'Brien, J. J. (1973). A study of an Oregon sea breeze event. *J. Appl. Meteorol.* **12**, 1267–1283.
Johnson, R. H. (1977). Effects of cumulus convection on the structure and growth of the mixed layer over south Florida. *Mon. Weather Rev.* **105**, 713–724.

Jones, R. W. (1973). A numerical experiment on the prediction of the northeast (winter) monsoon in southeast Asia. *NOAA Tech., Rep. ERL 272-WMPO-3,* 1–56.
Jones, R. W. (1976). Integration of a tropical cyclone model on a nested grid. *NOAA Tech. Memo. ERL-WMPO-30,* 1–37.
Jones, R. W. (1977a). A nested grid for a three-dimensional model. *J. Atmos. Sci.* **34,** 1528–1553.
Jones, R. W. (1977b). Noise control for a nested grid tropical cyclone model. *Contrib. Atmos. Phys.* **50,** 393–402.
Jones, R. W. (1980). A three-dimensional tropical cyclone model with release of latent heat by the resolvable scales. *J. Atmos. Sci.* **37,** 930–938.
Kasahara, A. (1974). Various vertical coordinate systems used for numerical weather prediction. *Mon. Weather Rev.* **102,** 509–522.
Keen, C. S., and Lyons, W. A. (1978). Lake/land breeze circulations on the western shore of Lake Michigan. *J. Appl. Meteorol.* **17,** 1843–1855.
Keen, C. S., Lyons, W. A., and Schuh, J. A. (1979). Air pollution transport studies in a coastal zone using kinematic diagnostic analysis. *J. Appl. Meteorol.* **18,** 606–615.
Kessler, R. C., and Pielke, R. A. (1981). A numerical study of airflow over adjacent ridges. *J. Atmos. Sci.* (in preparation).
Keyser, D., and Anthes, R. A. (1977). The applicability of a mixed-layer model of the planetary boundary layer to real-data forecasting. *Mon. Weather Rev.* **105,** 1351–1371.
Kimura, R., and Eguchi, T. (1978). On dynamical processes of sea and land breeze circulations. *J. Meteorol. Soc. Jpn.* **56,** 67–85.
Klemp, J. B., and Lilly, D. K. (1975). The dynamics of wave-induced downslope winds. *J. Atmos. Sci.* **32,** 320–339.
Klemp, J. B., and Lilly, D. K. (1978). Numerical simulation of hydrostatic mountain waves. *J. Atmos. Sci.* **32,** 78–107.
Klemp, J. B., and Wilhelmson, R. B. (1978a). The simulation of three-dimensional convective storm dynamics. *J. Atmos. Sci.* **35,** 1070–1096.
Klemp, J. B., and Wilhelmson, R. B. (1978b). Simulations of right- and left-moving storms produced through storm splitting. *J. Atmos. Sci.* **35,** 1097–1110.
Klöppel, M., Stilke, G., and Wamser, C. (1978). Experimental investigations into variations of ground-based inversions and comparisons with results of simple boundary-layer models. *Bound. Layer Meteorol.* **15,** 135–145.
Knowles, C. E., and Singer, J. J. (1977). Exchange through a barrier island inlet: Additional evidence of upwelling off the northeast coast of North Carolina. *J. Phys. Oceanogr.* **7,** 146–152.
Kondo, J., Sasano, Y., and Ishii, T. (1979). On wind-driven current and temperature profiles with diurnal period in the oceanic planetary boundary layer. *J. Phys. Oceanogr.* **9,** 360–372.
Kondratyev, J. (1969). "Radiation in the Atmosphere." Academic Press, New York.
Kowalczyk, G. S., Choquette, C. E., and Gordon, G. E. (1978). Chemical element balances and identification of air pollution sources in Washington, D.C. *Atmos. Environ.* **12,** 1143–1153.
Kreiss, H., and Oliger, J. (1973). Methods for the approximate solution of time dependent problems. *GARP Publ. Ser.* **10.** Available from Secretariat of the World Meteorological Organization. Case Postale No. 1, CH-1211, Geneva 20, Switzerland.
Kreitzberg, C. W. (1976). Interactive applications of satellite observations and mesoscale numerical models. *Bull. Am. Meteorol. Soc.* **57,** 679–685.
Kreitzberg, C. W., and Perkey, D. J. (1976). Release of potential instability. Part I. A sequential plume model within a hydrostatic primitive equation model. *J. Atmos. Sci.* **33,** 456–475.

Kreitzberg, C. W., and Perkey, D. J. (1977). Release of potential instability. Part II. The mechanism of convective/mesoscale interaction. *J. Atmos. Sci.* **34,** 1569–1595.
Krishnamurti, T. N., and Kanamitsu, M. (1973). A study of a coasting easterly wave. *Tellus* **25,** 568–585.
Krishnamurti, T. N., and Moxim, W. J. (1971). On parameterization of convective and nonconvective latent heat releases. *J. Appl. Meteorol.* **10,** 3–13.
Krishnamurti, T. N., Kanamitsu, M., Ceselski, B., and Mathur, M. B. (1973). Florida State University's Tropical Prediction Model. *Tellus* **25,** 523–535.
Krishnamurti, T. N., Ramanathan, Y., Pan, H.-L., Pasch, R. L., and Molinari, J. (1980). Cumulus parameterization and rainfall rates. I. *Mon. Weather Rev.* **108,** 465–472.
Kuhn, P. (1963). Radiometeorsonde observations of infrared flux emissivity of water vapor. *J. Appl. Meteorol.* **2,** 368–378.
Kuo, H. L. (1965). On formation and intensification of tropical cyclones through latent heat release by cumulus convection. *J. Atmos. Sci.* **22,** 40–63.
Kuo, H. L. (1974). Further studies of the parameterization of the influence of cumulus convection on large-scale flow. *J. Atmos. Sci.* **31,** 1232–1240.
Kuo, H. L. (1979). Infrared cooling rate in a standard atmosphere. *Contrib. Atmos. Phys.* **52,** 85–94.
Kurihara, Y. (1973). A scheme for moist convective adjustment. *Mon. Weather Rev.* **101,** 547–553.
Kurihara, Y. (1975). Budget analysis of a tropical cyclone simulated in an axisymmetric numerical model. *J. Atmos. Sci.* **32,** 25–59.
Kurihara, Y. (1976). On the development of spiral bands in a tropical cyclone. *J. Atmos. Sci.* **33,** 940–958.
Kurihara, Y., and Bender, M. A. (1979). Supplementary note on a scheme of dynamic initialization of the boundary layer in a primitive equation model. *Mon. Weather Rev.* **107,** 1219–1221.
Kurihara, Y., and Tuleya, R. E. (1974). Structure of a tropical cyclone developed in a three-dimensional numerical simulation model. *J. Atmos. Sci.* **31,** 893–919.
Kurihara, Y., and Tuleya, R. E. (1978). A scheme for dynamic initialization of the boundary layer in a primitive equation model. *Mon. Weather Rev.* **106,** 113–123.
Kurihara, Y., Tripoli, G. J., and Bender, M. A. (1979). Design of a movable nested-mesh primitive equation model. *Mon. Weather Rev.* **107,** 239–249.
Lal, M. (1979). Application of Pielke model to air quality studies. *Mausam* **30,** Journal of the Met. Dept. of India, 69–78.
Lavoie, R. L. (1972). A mesoscale numerical model of lake-effect storms. *J. Atmos. Sci.* **29,** 1025–1040.
Lavoie, R. L. (1974). A numerical model of trade wind weather over Oahu. *Mon. Weather Rev.* **102,** 630–637.
Laykhtman, D. L., and Snopkov, V. G. (1970). On the problem of the roughness of the sea surface. *Izv. Atmos. Oceanic. Phys.* **6,** 379–380.
Lazier, J., and Sandstrom, H. (1978). Migrating thermal structure in a freshwater thermocline. *J. Phys. Oceanogr.* **8,** 1070–1079.
Lee, H. N., and Kao, S. K. (1979). Finite element numerical modeling of atmospheric turbulent boundary layer. *J. Appl. Meteorol.* **18,** 1287–1295.
Lee, J. D. (1973). Numerical simulation of the planetary boundary layer over Barbados, W. I. Ph.D. Thesis, Florida State University, Tallahassee.
Lee, R. L., and Gresho, P. M. (1977). Development of a three-dimensional model of the atmospheric boundary layer using the finite element method. *Lawrence Livermore Lab.* [*Rep.*] *UCRL-52366.* Available in microfiche from NTIS.

Lee, R. L., and Olfe, D. B. (1974). Numerical calculations of temperature profiles over an urban heat island. *Bound. Layer Meteorol.* **7**, 39–52.

Lee, R. L., Gresho, P. M., and Sani, R. L. (1976). A comparative study of certain finite element and finite difference methods in advection-diffusion simulations. *Proc. Summer Comut. Simul. Conf., 1976* p. 37–42.

Likens, G. E., and Bormann, F. H. (1974). Acid rain: A serious regional environmental problem. *Science* **184**, 1176–1179.

Lilly, D. K. (1961). A proposed staggered-grid system for numerical integration of dynamic equations. *Mon. Weather Rev.* **89**, 59–65.

Lilly, D. K., ed. (1975). "Open SESAME," Proceedings of SESAME Open Meeting at Boulder, CO, Sept. 4–6, 1974. Prepared by NOAA, ERL, Boulder, Colorado.

Lilly, D. K. (1979). The dynamical structure and evolution of thunderstorms and squall lines. *Annu. Rev. Earth Plant. Sci.* **7**, 117–161.

Lilly, D. K., and Kennedy, P. J. (1973). Observations of a stationary mountain wave and its associated momentum flux and energy dissipation. *J. Atmos. Sci.* **30**, 1135–1152.

Lilly, D. K., and Zipser, E. J. (1972). The Front Range windstorm of 11 January 1972. *Weatherwise* **25**, 56–63.

Liou, K.-N., and Wittman, G. D. (1979). Parameterization of the radiation balance of clouds. *J. Atmos. Sci.* **36**, 1261–1273.

List, R. J. (1971). "Smithsonian Meteorological Tables." Smithson. Inst. Press, Washington, D.C.

Long, P. E., Jr., and Hicks, F. J. (1975). "Simple Properties of Chapeau Functions and Their Application to the Solution of the Advection Equation," NOAA NWS TDL Office Note 75-8. Techniques Development Lab, Gramex Bldg., Silver Spring, Maryland.

Long, P. E., Jr., and Pepper, D. W. (1976). A comparison of six numerical schemes for calculating the advection of atmospheric pollution. *Am. Meteorol. Soc. Symp. Atmos. Turbul., Diffus. Air Qual., Prepr., 3rd, 1976* pp. 181–187.

Long, R. L. (1977). Three-layer circulations in estuaries and harbors. *J. Phys. Oceanogr.* **7**, 415–421.

Long, R. R. (1954). "Some Aspects of the Flow of Stratified Fluids. II. Experiments with a Two-fluid System," Tech. Rep. No. 4. Johns Hopkins University, Baltimore, Maryland.

Loose, T., and Bornstein, R. D. (1977). Observations of mesoscale effects on frontal movement through an urban area. *Mon. Weather Rev.* **105**, 562–571.

Lord, N. W., Pandolfo, J. P., and Atwater, M. A. (1972). Simulations of meteorological variations over arctic coastal tundra under various physical interface conditions. *Arct. Alp. Res.* **4**, 189–209.

Lucero, O. A. (1976). "Comportamiento de esquemos de diterencias tinitas en campos cinematicos con y sin brechas de velocidades." Report of the Instituto de Hidrologia, Buenos Aires, Argentina. Available from O. A. Lucero, Facultad de Fisica, Univ. Veracruzana, Ap. postal 270, Xalapa, Veracruz, Mexico.

Ludwig, F. L., and Byrd, G. (1980). An efficient method for deriving mass-consistent flow fields from wind observations in rough terrain. *Atmos. Environ.* **14**, 585–587.

Lumley, J. L., and Khajeh-Nouri, B. (1974). Computational modeling of turbulent transport. *Adv. Geophys.* **18A**, 169–192.

Lumley, J. L., and Panofsky, H. A. (1964). The structure of atmospheric turbulence. *Intersc. Monogr. Texts Phys. Astron.*, **12**.

Lyons, W. A., and Cole, H. S. (1976). Photochemical oxidant transport: Mesoscale lake breeze and synoptic-scale aspects. *J. Appl. Meteorol.* **15**, 733–743.

Lyons, W. A., and Keen, C. S. (1976). "Particulate Transport in a Great Lakes Coastal

Environment" (reprint from Proceedings of the Second Federal Conference of the Great Lakes), pp. 222–237. Available from Mesomet, Inc., Suite 3330, 35 E. Wacker Drive, Chicago, Illinois 60601.

McCumber, M. C. (1980). A numerical simulation of the influence of heat and moisture fluxes upon mesoscale circulations. Ph.D. Dissertation, University of Virginia, Charlottesville.

McCumber, M. C., Schuh, J. A., McNider, R. T., Pielke, R. A., and Mahrer, Y. (1978). "The University of Virginia Mesoscale Model," UVA Rep. No. UVA-ENV SCI-MESO-1978-1, October. Copies available from Dept. Environ. Sci., Clark Hall, University of Virgina, Charlottesville.

McDonald, J. (1960). Direct absorption of solar radiation by atmospheric water vapor. *J. Meteorol.* **17**, 319–328.

McEwan, M. J., and Phillips, L. F. (1975). "Chemistry of the Atmosphere." Wiley, New York.

MacHattie, L. B. (1968). Kananaskis Valley winds in summer. *J. Appl. Meteorol.* **7**, 348–352.

McNider, R. T., and Pielke, R. A. (1981). Diurnal boundary layer development over sloping terrain. *J. Atmos. Sci.* (in press).

McNider, R. T., Hanna, S. R., and Pielke, R. A. (1980). Sub-grid scale plume dispersion in coarse resolution mesoscale models. *Proc. Jt. Am. Meteorol. Soc./Air Pollution Control Ass. Conf. Appl. Air Pollut. Meteorol., 2nd, 1980* pp. 424–429.

McPhee, M. G. (1979). The effect of the oceanic boundary layer on the mean drift of pack ice: Application of a simple model. *J. Phys. Oceanogr.* **9**, 388–400.

McPherson, R. D. (1970). A numerical study of the effect of a coastal irregularity on the sea breeze. *J. Appl. Meteorol.* **9**, 767–777.

Mahrer, Y., and Pielke, R. A. (1975). A numerical study of the air flow over mountains using the two-dimensional version of the University of Virginia mesoscale model. *J. Atmos. Sci.* **32**, 2144–2155.

Mahrer, Y., and Pielke, R. A. (1976). Numerical simulation of the air flow over Barbados. *Mon. Weather Rev.* **104**, 1392–1402.

Mahrer, Y., and Pielke, R. A. (1977a). A numerical study of the air flow over irregular terrain. *Contrib. Atmos. Phys.* **50**, 98–113.

Mahrer, Y., and Pielke, R. A. (1977b). The effects of topography on the sea and land breezes in a two-dimensional numerical model. *Mon. Weather Rev.* **105**, 1151–1162.

Mahrer, Y., and Pielke, R. A. (1978a). The meteorological effect of the changes in surface albedo and moisture. *Isr. Meteorol. Res. Pap.* **2**, 55–70.

Mahrer, Y., and Pielke, R. A. (1978b). A test of an upstream spline interpolation technique for the advective terms in a numerical mesoscale model. *Mon. Weather Rev.* **106**, 818–830. Corrigendum published, *Monb. Weather Rev.* **106**, 1758.

Mahrt, L. J. (1972). Some basic theoretical concepts of boundary layer flow at low latitudes. *In "Dynamics of the Tropical Atmosphere: Notes from a Colloquium, Summer 1972" (J. A. Young, coordinator), pp. 411–420*. NCAR, Boulder, Colorado.

Mahrt, L. J. (1974). Time-dependent integrated planetary boundary layer flow. *J. Atmos. Sci.* **31**, 457–464.

Mahrt, L. J. (1976). Mixed layer moisture structure. *Mon. Weather Rev.* **104**, 1403–1407.

Malkus, J. S., and Stern, M. E. (1953). The flow of a stable atmosphere over a heated island. Part I. *J. Meteorol.* **10**, 30–41.

Manins, P. C., and Sawford, B. L. (1979a). A model of katabatic winds. *J. Atmos. Sci.* **36**, 619–630.

Manins, P. C., and Sawford, B. L. (1979b). Katabatic winds: A field case study. *Q. J. R. Meteorol. Soc.* **105**, 1011–1025.

Marchuk, G. I., Kochergin, V. P., Klimok, V. I., and Sukhorukov, V. A. (1977). On the dynamics of the ocean surface mixed layer. *J. Phys. Oceanogr.* **7**, 865–875.

Marks, F. D., Jr., and Austin, P. M. (1979). Effects of the New England coastal front on the distribution of precipitation. *Mon. Weather Rev.* **107**, 53–67.

Martin, C. (1981). Numerical accuracy in a mesoscale meteorological model. M.S. Thesis, Dept. Environ. Sci., Univ. of Virginia.

Matejka, T. J., Houze, R. A., Jr., and Hobbs, P. V. (1980). Microphysics and dynamics of clouds associated with mesoscale rainbands in extratropical cyclones. *Q. J. R. Meteorol. Soc.* **106**, 29–56.

Mathur, M. B. (1974). A multiple grid primitive equation model to simulate the development of an asymmetric hurricane (Isbell 1964). *J. Atmos. Sci.* **31**, 371–393.

Mather, M. B. (1975). Development of banded structure in a numerically simulated hurricane. *J. Atmos. Sci.* **32**, 512–522.

Matson, M., and Legeckis, R. V. (1980). Urban heat islands detected by satellite. *Bull. Am. Meteorol. Soc.* **61**, 212 pp.

Matson, M., McClain, E. P., McGinnis, D. F., Jr., and Pritchard, J. A. (1978). Satellite detection of urban heat islands. *Mon. Weather Rev.* **106**, 1725–1734.

Mellor, G. L., and Yamada, T. (1974). A hierarchy of turbulence closure models for planetary boundary layers. *J. Atmos. Sci.* **31**, 1791–1806.

Melville, W. K. (1977). Wind stress and roughness length over breaking waves. *J. Phys. Oceanogr.* **7**, 702–710.

Meroney, R. N., Bowen, A. J., Lindley, B., and Pearse, J. R. (1978). Wind characteristics over complex terrain: laboratory simulation and field measurements at Rakaia Gorge, New Zealand. Final report: Part II. Fluid Mech. and Wind Eng. Program, Colorado State Univ., Fort Collins, Colorado.

Mesinger, F., and Arakawa, A. (1976). Numerical methods used in atmospheric models. *GARP Publ. Ser.* **17**, 1–64.

Miller, M. J., and Pearce, R. P. (1974). A three-dimensional primitive equation model of cumulonimbus convection. *Q. J. R. Meteorol. Soc.* **100**, 133–154.

Moroz, W. J. (1967). A lake breeze on the eastern shore of Lake Michigan: Observations and model. *J. Atmos. Sci.* **24**, 337–355.

Moss, M. S. (1978). Low-layer features of two limited area hurricane regions. *NOAA Tech. Rep. ERL 394-NHEML 1*, 1–47.

Moss, M. S., and Jones, R. W. (1978). A numerical simulation of hurrican landfall. *NOAA Tech. Memo ERL-NHEML-3*, 1–15.

Mullen, S. L. (1979). An investigation of small synoptic-scale cyclones in polar air stream. *Mon. Weather Rev.* **107**, 1636–1647.

Muller, R. A. (1966). Snowbelts of the Great Lakes. *Weatherwise* **19**, 248–255.

Murdoch, D. C. (1957). "Linear Algebra for Undergraduates." Wiley, New York.

Murray, F. W. (1970). Numerical models of a tropical cumulus cloud with bilateral and axial symmetry. *Mon. Weather Rev.* **98**, 14–28.

Neumann, J. (1951). Land breezes and nocturnal thunderstorms. *J. Meteorol.* **8**, 60–67.

Neumann, J., and Mahrer, Y. (1971). A theoretical study of the land and sea breeze circulation. *J. Atmos. Sci.* **28**, 532–542.

Neumann, J., and Mahrer, Y. (1974). A theoretical study of the sea and land breezes of circular islands. *J. Atmos. Sci.* **31**, 2027–2039.

Neumann, J., and Mahrer, Y. (1975). A theoretical study of the lake and land breezes of circular lakes. *Mon. Weather Rev.* **103**, 474–485.

Nickerson, E. C. (1979). On the numerical simulation of airflow and clouds over mountainous terrain. *Contrib. Atmos. Phys.* **52**, 161–177.

Nickerson, E. C., and Magaziner, E. L. (1976). A three-dimensional simulation of winds and non-precipitating orographic clouds over Hawaii. *NOAA Tech. Rep. ERL 377-APCL* **39**, 1–35.

Nieuwstadt, F. T. M., and Driedonks, A. G. M. (1979). The nocturnal boundary layer: A case study compared with model calculations. *J. Appl. Meteorol.* **18**, 1397–1405.

Noonkester, V. R. (1979). Coastal marine fog in southern California. *Mon. Weather Rev.* **107**, 830–851.

Nunez, M., and Oke, T. R. (1976). Long-wave radiative flux divergence and nocturnal cooling of the urban atmosphere. II. Within an urban canyon. *Bound. Layer Meteorol.* **10**, 121–135.

Nunez, M., and Oke, T. R. (1977). The energy balance of an urban canyon. *J. Appl. Meteorol.* **16**, 11–19.

O'Brien, J. J. (1970a). A note on the vertical structure of the eddy exchange coefficient in the planetary boundary layer. *J. Atmos. Sci.* **27**, 1213–1215.

O'Brien, J. J. (1970b). Alternative solutions to the classical vertical velocity problem. *J. Appl. Meteorol.* **9**, 197–203.

Oerlemans, J. (1980). A case study of a subsynoptic disturbance in a polar outbreak. *Q. J. R. Meteorol. Soc.* **106**, 313–325.

Ogura, Y. (1963). A review of numerical modeling research on small-scale convection in the atmosphere. *Meteorol. Monogr.* **5**, 65–76.

Ogura, Y., and Charney, J. G. (1961). A numerical model of thermal convection in the atmosphere. *Proc. Int. Symp. Numer. Weather Predict., 1960*, pp. 431–450.

Ogura, Y., and Liou, M.-T. (1980). The structure of a midlatitude squall line: a case study. *J. Atmos. Sci.* **37**, 553–567.

Ogura, Y., and Phillips, N. A. (1962). Scale analysis of deep and shallow convection in the atmosphere. *J. Atmos. Sci.* **19**, 173–179.

Ogura, Y., Chen, Y.-L., Russell, J., and Soong, S.-T. (1979). On the formation of organized convective systems observed over the eastern Atlantic. *Mon. Weather Rev.* **107**, 426–441.

Oke, T. R. (1973). City size and the urban heat island. *Atmos. Environ.* **7**, 769–779.

Oke, T. R. (1976). The distinction between canopy and boundary-layer urban heat islands. *Atmosphere* **14**, 268–277.

Oke, T. R. (1978). "Boundary Layer Climates." Methuen, London.

Oke, T. R., and Fuggle, R. F. (1972). Comparison of urban/rural counter and net radiation at night. *Bound. Layer Meteorol.* **2**, 290–308.

Olfe, D. B., and Lee, R. L. (1971). Linearized calculations of urban heat island convection effects. *J. Atmos. Sci.* **28**, 1374–1388.

Oliger, J., and Sundström, A. (1976). "Theoretical and Practical Aspects of Some Initial-boundary Value Problems in Fluid Dynamics." Rep. STAN-CS-76-578. Computer Science Dept., School of Humanities and Sciences, Stanford University, Stanford, California.

Onishi, G. (1968). Numerical study on atmospheric boundary layer flow over inhomogeneous terrain. *J. Meteorol. Soc. Jpn.* **46**, 280–286.

Ookochi, Y. (1978). Preliminary test of typhoon forecast with a moving multi-nested grid (MNG). *J. Meteorol. Soc. Jpn.* **56**, 571–582.

Ookouchi, Y., Uryu, M., and Sawada, R. (1978). A numerical study of the effects of a mountain on the land and sea breezes. *J. Meteorol. Soc. Jpn.* **56**, 368–385.

Ooyama, K. (1971). A theory of parameterization of cumulus convection. *J. Meteorol. Soc. Jpn.* **49**, 744–756.

Orlanski, I. (1975). A rational subdivision of scales for atmospheric process. *Bull. Am. Meteorol. Soc.* **56**, 527–530.

Orlanski, I. (1976). A simple boundary condition for unbounded hyperbolic flows. *J. Comput. Phys.* **21**, 251-269.
Orlanski, I., Ross, B., and Polinsky, L. (1974). Diurnal variation of the planetary boundary layer in a mesoscale model. *J. Atmos. Sci.* **31**, 965-989.
Orszag, S. A. (1971). Numerical simulation of incompressible flows within simple boundaries: Accuracy. *J. Fluid Mech.* **49**, 76-112.
Orville, H. D. (1965). A numerical study of the initiation of cumulus clouds over mountainous terrain. *J. Atmos. Sci.* **22**, 684-699.
Orville, H. D. (1978). A review of hailstone-hailstorm numerical simulations. *Am. Meteorol. Soc. Monogr.* **38**, 49-61.
Orville, H. D., and Sloan, L. J. (1970). A numerical simulation of the life history of a rainstorm. *J. Atmos. Sci.* **27**, 1148-1159.
Otterman, J. (1974). Baring high albedo soils by desertification—a hypothesized desertification mechanism. *Science* **184**, 531-533.
Otterman, J. (1975). "Possible Rainfall Reduction through Reduced Surface Temperature due to Overgrazing," NASA Rep., Tech. Inf. Div., Code 250. Goddard Space Flight Center, Greenbelt, Maryland.
Paegle, J., Zdunkowski, W. G., and Welch, R. M. (1976). Implicit differencing of predictive equations of the boundary layer. *Mon. Weather Rev.* **104**, 1321-1324.
Pandolfo, J. P. (1966). Wind and temperature profiles for constant-flux boundary layers in lapse conditions with a variable eddy conductivity to eddy viscosity ratio. *J. Atmos. Sci.* **23**, 495-502.
Pandolfo, J. P., and Jacobs, C. A. (1973). "Tests of an Urban Meteorological Pollutant Model using CO Validation Data in the Los Angeles Metropolitan Area." Vol. I. The Center for the Environment and Man, Hartford, Connecticut.
Pandolfo, J. P., Jacobs, C. A., Ball, R. J., Atwater, M. A., and Sekorski, J. A. (1976). Refinement and validation of an urban meteorological-pollutant model. *U.S. Environ. Prot. Agency, Off. Res. Dev.* [*Rep.*] *EPA-600/4-76-037*, 1-21.
Parmenter, F. C. (1974). Observing and forecasting local effects from satellite data. *Prepr. Vol. Am. Meteorol. Soc. Conf. Weather Forecast. Anal., 5th 1974* pp. 46-49.
Parmenter, F. C., and Anderson, R. K. (1974). Mesoscale details in synoptic scale systems. *Prepr. Vol., Am. Meteorol. Soc. Conf. Weather Forecasting Anal., 5th, 1974* pp. 50-53.
Pasquill, F. (1961). The estimation of the dispersion of windborne material. *Meteorol. Mag.* **90**, 33-49.
Pearson, R. A. (1973). Properties of the sea breeze front as shown by a numerical model. *J. Atmos. Sci.* **30**, 1050-1060.
Pearson, R. A. (1974). Consistent boundary conditions for numerical models of systems that admit dispersive waves. *J. Atmos. Sci.* **31**, 1481-1489.
Peltier, W. R., and Clark, T. L. (1979). The evolution and stability of finite-amplitude mountain waves. Part II. Surface wave drag and severe downslope windstorms. *J. Atmos. Sci.* **36**, 1498-1529.
Perkey, D. J. (1976). A description and preliminary results from a fine-mesh model for forecasting quantitative precipitation. *Mon. Weather Rev.* **104**, 1513-1526.
Perkey, D. J., and Kreitzberg, C. W. (1976). A time-dependent lateral boundary scheme for limited-area primitive equation models. *Mon. Weather Rev.* **104**, 744-755.
Peterson, E. W. (1969). Modification of mean flow and turbulent energy by a change in surface roughness under conditions of neutral stability. *Q. J. R. Meteorol. Soc.* **95**, 561-575.
Peterson, J. T. (1970). Distribution of sulfur dioxide over metropolitan St. Louis, as

described by empirical eigenvectors, and its relation to meteorological parameters. *Atmos. Environ.* **4**, 501–518.

Philip, J. (1957). Evaporation and moisture and heat fields in the soil. *J. Meteorol.* **14**, 354–366.

Phillips, N. A. (1957). A coordinate system having some special advantages for numerical forecasting. *J. Meteorol.* **14**, 184–185.

Physik, W. (1976). A numerical model of the sea-breeze phenomenon over a lake or gulf. *J. Atmos. Sci.* **33**, 2107–2135.

Pickett, R. L., and Dossett, D. A. (1979). Mirex and the circulation of Lake Ontario. *J. Phys. Oceanogr.* **9**, 441–445.

Pielke, R. A. (1974a). A three-dimensional numerical model of the sea breezes over south Florida. *Mon. Weather Rev.* **102**, 115–139.

Pielke, R. A. (1974b). A comparison of three-dimensional and two-dimensional numerical predictions of sea breeze. *J. Atmos. Sci.* **31**, 1577–1585; Corrigendum to this paper published in *J. Atmos. Sci.* **33**, 1380 (1976).

Pielke, R. A. (1978). The role of man and machine in the Weather Service of the future. *Prepr. Vol., Am. Meteorol. Soc. Conf. Weather Forecast. Anal., 5th, 1978*, pp. 271–272.

Pielke, R. A. (1981). An overview of our current understanding of the physical interactions between the sea- and land-breeze and the coastal waters. *Ocean Manage* **6**, 87–100.

Pielke, R. A. (1982). "Mesoscale Numerical Modeling—An Introductory Survey." Academic Press, New York (to be submitted for publication).

Pielke, R. A., and Kennedy, E. (1980). "Mesoscale Terrain Features," Rep. No. UVA-ENV SCI-MESO-1980-1. Available from R. A. Pielke, Dept. Environ. Sci., Clark Hall, University of Virginia, Charlottesville.

Pielke, R. A., and Cotton, W. R. (1977). A mesoscale analysis over south Florida for a high rainfall event. *Mon. Weather Rev.* **105**, 343–362.

Pielke, R. A., and Mahrer, Y. (1975). Technique to represent the heated-planetary boundary layer in mesoscale models with coarse vertical resolution. *J. Atmos. Sci.* **32**, 2288–2308.

Pielke, R. A., and Mahrer, Y. (1978). Verification analysis of the University of Virginia three-dimensional mesoscale model prediction over south Florida for July 1, 1973. *Mon. Weather Rev.* **106**, 1568–1589.

Pielke, R. A., and Martin, C. L. (1981). The derivation of a terrain-following coordinate system for use in a hydrostatic model. *J. Atmos. Sci.* (in press).

Pilié, R. J., Mark, E. J., Rogers, C. W., Katz, U., and Kocmond, W. C. (1979). The formation of marine fog and the development of fog-stratus systems along the California coast. *J. Appl. Meteorol.* **18**, 1275–1286.

Plank, V. G. (1966). Wind conditions in situations of patternform and non-patternform cumulus convection. *Tellus* **18**, 1–12.

Price, J. C. (1979). Assessment of the urban heat island effect through the use of satellite data. *Mon. Weather Rev.* **107**, 1554–1557.

Project METROMEX (1976). METROMEX Update. *Bull. Am. Meteorol. Soc.* **57**, 304–308.

Purdom, J. F. W. (1976). Some uses of high-resolution GOES imagery in the mesoscale forecasting of convection and its behavior. *Mon. Weather Rev.* **104**, 1474–1483.

Purnell, D. K. (1976). Solution of the advective equation by upstream interpolation with a cubic spline. *Mon. Weather Rev.* **104**, 42–48.

Queney, P. (1947). "Theory of Perturbations in Stratified Currents with Applications to Air Flow over Mountain Barriers," Publication of the Dept. of Meteorology, University of Chicago, Misc. Rep. No. 23. Univ. of Chicago Press, Chicago, Illinois.

Queney, P. (1948). The problem of air flow over mountains: A summary of theoretical studies. *Bull. Am. Meteorol. Soc.* **29**, 16–26.

Raddatz, R. L., and Khandekar, M. L. (1979). Upslope enhanced extreme rainfall events over the Canadian western plains: a mesoscale numerical simulation. *Mon. Weather Rev.* **107**, 650–661.

Randerson, D., and Thompson, A. H. (1976). "Investigation of a Tiros III Photograph of the Florida Peninsula Taken on 14 July 1961," Sci. Rep. No. 6 prepared under Contract AF 19(604)-8450 for Air Force Cambridge Research Labs, Office of Aerospace Research, Bedford, Massachusetts by Texas A&M.

Rasmussen, E. (1979). The polar low as an extratropical CISK disturbance. *Q. J. R. Meteorol. Soc.* **105**, 531–549.

Raymond, D. J. (1975). A model for predicting the movement of continuously propagating convective storms. *J. Atmos. Sci.* **32**, 1308–1317.

Raymond, W. H., and Garder, A. (1976). Selective damping in a Galerkin method for solving wave problems with variable grids. *Mon. Weather Rev.* **104**, 1583–1590.

Raynor, G. S. (1971). Wind and temperature structure in a coniferous forest and a contiguous field. *For. Sci.* **17**, 351–363.

Reed, R. J. (1979). Cyclogenesis in polar air streams. *Mon. Weather Rev.* **107**, 38–52.

Reid, J. D., Grant, L. O., Pielke, R. A., and Mahrer, Y. (1976). Observations and numerical modeling of seeding agent delivery from ground based generators to orographic cloud base. *Proc. Int. Weather Modif. Conf., 1976* pp. 521–527.

Rhea, O. J. (1977). Orographic precipitation model for hydrometeorological use. Ph.D. Dissertation, Dept. of Atmos. Sci., Colorado State Univ., Fort Collins, Colorado.

Richtmyer, R. D., and Morton, K. W. (1967). "Difference Methods for Initial-Value Problems." Wiley (Interscience), New York.

Richwien, B. A. (1978). The damming effect of the Southern Appalachians. *Am. Meteorol. Soc. Conf. Proc. Weather Forcast. Anal.; Aviat. Meteorol., 1978* pp. 94–101.

Roach, W. T., and Slingo, A. (1979). A high resolution infrared radiative transfer scheme to study the interaction of radiation with cloud. *Q. J. R. Meteorol. Soc.* **105**, 603–614.

Robinson, G. D., ed. (1977). "Inadvertent Weather Modification Workshop," Final rep. to NSF under Grant No. ENV-77-10186. Prepared at The Center for the Environment and Man, Inc., Hartford, Connecticut.

Rodgers, E., Gentry, R. C., Shenk, W., and Oliver, V. (1979). The benefits of using short-interval satellite images to derive winds for tropical cyclones. *Mon. Weather Rev.* **107**, 575–584.

Rosenberg, N. (1974). "Microclimate: The Biological Environment." Wiley, New York.

Rosenthal, S. L. (1970). A circularly symmetric primitive equation model of tropical cyclone development containing an explicit water vapor cycle. *Mon. Weather Rev.* **98**, 643–663.

Rosenthal, S. L. (1971). The response of a tropical cyclone model to variations in boundary layer parameters, initial conditions, lateral boundary conditions and domain size. *Mon. Weather Rev.* **99**, 767–777.

Rosenthal, S. L. (1978). Numerical simulation of tropical cyclone development with latent heat release by the resolvable scales. I. Model description and preliminary results. *J. Atmos. Sci.* **35**, 258–271.

Rosenthal, S. L. (1979a). Cumulus effects in hurricane models—to parameterize or not to parameterize. (Unpublished manuscript.)

Rosenthal, S. L. (1979b). The sensitivity of simulated hurricane development to cumulus parameterization details. *Mon. Weather Rev.* **107**, 193–197.

Ross, B. B., and Orlanski, I. (1978). The circulation associated with a cold front. Part II. Moist case. *J. Atmos. Sci.* **35**, 445–465.

Sangster, W. E. (1977). An updated objective forecast technique for Colorado downslope winds. *NOAA Tech. Memo. NWS CR-61,* 1-24.

Sasaki, Y. (1970a). Some basic formalisms in numerical variational analysis. *Mon. Weather Rev.* **98,** 875-883.

Sasaki, Y. (1970b). Numerical variational analysis formulated under the constraints as determined by longwave equations and low-pass filter. *Mon. Weather Rev.* **98,** 884-898.

Sasaki, Y. (1970c). Numerical variational analysis with weak constraint and application to surface analysis of a severe gust front. *Mon. Weather Rev.* **98,** 899-910.

Sasaki, Y. (1971). VI. Computational problems, low-pass and band-pass filters in numerical variational optimization. *J. Meteorol. Soc. Jpn.* **49,** 766-774.

Sasaki, Y., and Lewis, J. M. (1970). Numerical variational objective analysis of the planetary boundary layer in conjunction with squall line formation. *J. Meteorol. Soc. Jpn.* **48,** 381-393.

Sasamori, T. (1970). A numerical study of atmospheric and soil boundary layers. *J. Atmos. Sci.* **27,** 1122-1137.

Sasamori, T. (1972). A linear harmonic analysis of atmospheric motion with radiative dissipation. *J. Meteorol. Soc. Jpn.* **50,** 505-517.

Sawai, T. (1978). Formation of the urban air mass and the associated local circulations. *J. Meteorol. Soc. Jpn.* **56,** 159-173.

Schaeffer, J. T. (1974). A simulative model of dryline motion. *J. Atmos. Sci.* **31,** 956-964.

Schere, K. L., and Demerjian, K. L. (1978). A photochemical box model for urban air quality simulation. *Jt. Conf. Sens. Environ. Pollut.* [*Conf. Proc.*], *4th, 1977* pp. 427-433 (reprint).

Schlesinger, R. E. (1973). A numerical model of deep moist convection. I. Comparative experiments for variable ambient moisture and wind shear. *J. Atmos. Sci.* **30,** 835-856.

Schlesinger, R. E. (1980). A three-dimensional numerical model of an isolated thunderstorm. II. Dynamics of updraft splitting and mesovortex couplet evolution. *J. Atmos. Sci.* **37,** 395-420.

Schmidt, F. H. (1947). An elementary theory of the land- and sea-breeze circulation. *J. Meteorol.* **4,** 9-15.

Schwartz, B. E. and L. F. Bosart, (1979). The diurnal variability of Florida rainfall. *Mon. Weather Rev.* **107,** 1535-1545.

Science News (1976). Dioxin toxicity data sent to aid Italy. *Sci. News (Washington, D.C.)* **110,** 359.

Scofield, R. A., and Weiss, C. E. (1977). A report on the Chesapeake Bay Region Nowcasting Experiment, *NOAA Tech. Memo. NESS 94.*

Scorer, R. S. (1949). Theory of waves in the lee of mountains. *Q. J. R. Meteorol. Soc.* **75,** 41-56.

Segal, M., and Pielke, R. A. (1980). Numerical model simulation of biometeorological heat load condition—summer day case study for the Chesapeake Bay area, *J. Appl. Meteorol.* (in press).

Segal, M., Mahrer, Y., and Pielke, R. A. (1980). A numerical model study of the meteorological patterns induced by a lake confined by mountains—the Dead Sea case. (In preparation.)

SethuRaman, S., and Cermak, J. E. (1973). "Stratified Shear Flows over a Simulated Three-dimensional Urban Heat Island," Proj. Themis Tech. Rep. No. 23. Fluid Dynamics and Diffusion Lab, Colorado State University, Fort Collins.

Shapiro, R. (1970). Smoothing, filtering and boundary effects. *Rev. Geophys. Space Phys.* **8,** 359-387.

Sheih, C. M. (1977). Mathematical modeling of particulate thermal coagulation and transport downstream of an urban source. *Atmos. Environ.* **11**, 1185–1190.
Sheih, C. M. (1978). A puff-on-cell model for computing pollutant transport and diffusion. *J. Appl. Meteorol.* **17**, 140–147.
Sheng, P. Y., Lick, W., Gedney, R. T., and Molls, F. B. (1978). Numerical computations of three-dimensional circulations in Lake Erie: A comparison of a free-surface model and a rigid-lid model. *J. Phys. Oceanogr.* **8**, 713–727.
Sherman, C. E. (1978). A mass-consistent model for wind fields over complex terrain. *J. Appl. Meteorol.* **17**, 312–319.
Shimanuki, A. (1969). Formulation of vertical distributions of wind velocity and eddy diffusivity near the ground. *J. Meteorol. Soc. Jpn.* **47**, 292–298.
Shreffler, J. H. (1979). Heat island convergence in St. Louis during calm periods. *J. Appl. Meteorol.* **18**, 1512–1520.
Simpson, J. (1976). Precipitation augmentation from cumulus clouds and systems: Scientific and technological foundation, 1975. *Adv. Geophys.* **19**, 1–72.
Simpson, J., Westcott, N. E., Clerman, R. J., and Pielke, R. A. (1980). On cumulus mergers. *Arch. Meteorol., Geophys. Bioklimatol., Ser. A* **29**, 1–40.
Simpson, R. H. (1978). On the computation of equivalent potential temperature. *Mon. Weather Rev.* **106**, 124–130.
Simpson, R. H., and Pielke, R. A. (1976). Hurricane development and movement. *Appl. Mech. Rev.* **29**, 601–609.
Sisterson, D. L., and Dirks, R. A. (1978). Structure of the daytime urban moisture field. *Atmos. Environ.* **12**, 1943–1949.
Skibin, D., and Hod, A. (1979). Subjective analysis of mesoscale flow patterns in northern Israel. *J. Appl. Meteorol.* **18**, 329–337.
Smith, P. J., and Lin, C. P. (1978). A comparison of synoptic-scale vertical motions computed by the kinematic method and two forms of the omega equation. *Mon. Weather Rev.* **106**, 1687–1694.
Smith, R. C. (1955). Theory of air flow over a heated land mass. *Q. J. R. Meteorol. Soc.* **81**, 382–395.
Smith, R. C. (1957). Air motion over a heated land mass: II. *Q. J. R. Meteorol. Soc.* **83**, 248–256.
Stephens, G. L., and Webster, P. J. (1981). Clouds and climate: Sensitivity of simple systems. *J. Atmos. Sci.* **38**, 235–247.
Stern, M. E., and Malkus, J. S. (1953). The flow of a stable atmosphere over a heated island. Part II. *J. Meteorol.* **10**, 105–120.
Sun, W.-Y. (1980). A forward-backward time integration scheme to treat internal gravity waves. *Mon. Weather Rev.* **108**, 402–407.
Sun, W.-Y., and Ogura, Y. (1979). Boundary-layer forcing as a possible trigger to a squall line formation. *J. Atmos. Sci.* **36**, 235–254.
Svendsen, H., and Thompson, R. O. R. Y. (1978). Wind-driven circulation in fjord. *J. Phys. Oceanogr.* **8**, 703–712.
Tag, P. M., Murray, F. W., and Kvenig, L. R. (1979). A comparison of several forms of eddy viscosity parameterization in a two-dimensional cloud model. *J. Appl. Meteorol.* **18**, 1429–1441.
Tapp, M. C., and White, P. W. (1976). A nonhydrostatic mesoscale model. *Q. J. R. Meteorol. Soc.* **102**, 277–296.
Taylor, P. A. (1974). Urban meteorological modelling—some relevant studies. *Adv. Geophys.* **18B**, 173–185.

Taylor, P. A. (1977b). Numerical studies of neutrally stratified planetary boundary-layer flow above gentle topography. I. Two-dimensional cases. *Bound. Layer Meteorol.* **12**, 37–60.
Taylor, P. A. (1977a). Some numerical studies of surface boundary-layer flow above gentle topography. *Bound. Layer Meteorol.* **11**, 439–465.
Taylor, P. A., and Gent, P. R. (1981). Modification of the boundary layer by orography. *In* "Orographic Effects in Planetary Flows" (R. Hide and P. White, eds.), pp. 143–165. WMO Publication.
Temperton, C. (1973). Some experiments in dynamic initialization for a simple primitive equation model. *Q. J. R. Meteorol. Soc.* **99**, 303–319.
Tennekes, H. (1974). The atmospheric boundary layer. *Phys. Today* pp. 52–63.
Thompson, W. T. (1979). "Effects of New York City on the Horizontal and Vertical Structure of Sea Breeze Fronts," Vol. III. Effects of Frictionally Retarded Sea Breeze and Synoptic Frontal Passages on Sulfur Dioxide Concentrations. R. D. Bornstein (P. I.). Report repared by the Dept. of Meteorology, San Jose State University, San Jose, California.
Threlkeld, J. L., and Jordan, R. C. (1958). Direct solar radiation available on clear days. *ASHRAE Trans.* **64**, 45–68.
Tuleya, R. E., and Kurihara, Y. (1978). A numerical simulation of the landfall of tropical cyclones. *J. Atmos. Sci.* **35**, 242–257.
Turner, B. D. (1969). Workbook of atmospheric dispersion estimates. *U.S. Public Health Serv. Publ. 999-AP-26*, 1–84.
Uccellini, L. W., Johnson, D. R., and Schlesinger, R. E. (1979). An isentropic and sigma coordinate hybrid numerical model: Model development and some initial tests. *J. Atmos. Sci.* **36**, 390–414.
Ulanski, S. L., and Garstang, M. (1978). The role of surface divergence and vorticity in the life cycle of convective rainfall. Part I. Observations and analysis. *J. Atmos. Sci.* **35**, 1047–1062.
Venkatram, A., and Viskanta, R. (1976). Radiative effects of pollutants on the planetary boundary layer, *U.S. Environ. Prot. Agency, EPA Off. Res. Dev. [Rep.] EPA-600/4-76-039*, 1–244.
Viskanta, R., and Daniel, R. A. (1980). Radiative effects of elevated pollutant layers on temperature structure and dispersion in an urban atmosphere. *J. Appl. Meteorol.* **19**, 53–70.
Viskanta, R., and Weirich, T. L. (1978). Feedback between radiatively interacting pollutants and their dispersion in the urban boundary layer. *WMO Symp. Boundary Layer Phys. Appl. Specific Probl. Air Pollut.*, WMO-No. 510, pp. 31–38.
Viskanta, R., Bergstrom, R. W., Jr., and Johnson, R. O. (1976). Modeling of the effects of pollutants and dispersion in urban atmosphere. *U.S. Environ. Prot. Agency, Off. Res. Dev. [Rep.] EPA* **EPA-600/4-76-002**, 1–107.
Viskanta, R., Bergstrom, R. W., Johnson, R. O. (1977). Effects of air pollution on thermal structure and dispersion in an urban planetary boundary layer. *Contrib. Atmos. Phys.* **50**, 419–440.
von Storch, H. (1978). Construction of optimal numerical filters fitted for noise damping in numerical simulation models. *Contrib. Atmos. Phys.* **51**, 189–197.
Vugts, H. F. (1980). A study of terrain inhomogeneity. *Bull. Am. Meteorol. Soc.* **61**, 568–569.
Vukovich, F. M., Dunn, J. W., III, and Crissman, B. W. (1976). A theoretical study of the St. Louis heat island: The wind and temperature distribution. *J. Appl. Meteorol.* **15**, 417–440.

Vukovich, F. M., King, W. J., Dunn, J. W., III, and Worth, J. J. B. (1979). Observations and simulations of the diurnal variation of the urban heat island circulation and associated variations of the ozone distribution: A case study. *J. Appl. Meteorol.* **18,** 836–854.

Wallace, J. M., and Hobbs, P. V. (1977). "Atmospheric Science: An Introductory Survey." Academic Press, New York.

Wang, D. P. (1979). Wind-driven circulation in the Chesapeake Bay, Winter 1975. *J. Phys. Oceanogr.* **9,** 564–572.

Wang, H.-H., Halpern, P., Douglas, J., Jr., and Dupont, T. (1972). Numerical solutions of the one-dimensional primitive equations using Galerkin approximations with localized basis functions. *Mon. Weather Rev.* **100,** 738–746.

Warner, C., Simpson, J., Martin, D. W., Suchman, D., Mosher, F. R., and Reinking, R. F. (1979). Shallow convection on Day 261 of GATE: mesoscale arcs. *Mon. Weather Rev.* **107,** 1617–1635.

Warner, T. J., Anthes, R. A., and McNab, A. L. (1978). Numerical simulations with a three-dimensional mesoscale model. *Mon. Weather Rev.* **106,** 1079–1099.

Weisberg, R. H. (1976). The nontidal flow in the Providence River of Narragansett Bay: A stochastic approach to estuarine circulation. *J. Phys. Oceanogr.* **6,** 721–734.

Weiss, C. E., and Purdom, J. F. W. (1974). The effect of early-morning cloudiness on squall-line activity. *Mon. Weather Rev.* **102,** 400–402.

Welch, R. M., Paegle, J., and Zdunkowski, W. G. (1978). Two-dimensional numerical simulation of the effects of air pollution upon the urban-rural complex. *Tellus* **30,** 136–150.

Wesely, M. L., and Lipschutz, R. C. (1976). An experimental study of the effects of aerosols on diffuse and direct solar radiation received during the summer near Chicago. *Atmos. Environ.* **10,** 981–987.

Wilson, T. A., and Houghton, D. D. (1979). Mesoscale wind fields for a severe storm situation determined from SMS cloud observations. *Mon. Weather Rev.* **107,** 1198–1209.

Wyngaard, J. C., and Coté, O. R. (1974). The evolution of a convective planetary boundary layer—a higher-order-closure model study. *Bound. Layer Meteorol.* **7,** 289–308.

Yamada, T. (1977). A numerical experiment on pollutant dispersion in a horizontally-homogeneous atmospheric boundary layer. *Atmos. Environ.* **11,** 1015–1024.

Yamada, T. (1978). A three-dimensional numerical study of complex atmospheric circulations produced by terrain. *Proc. Am. Meteorol. Soc. Conf. Sierra Nevada Meteorol. 1978* pp. 61–67.

Yamada, T. (1979). Prediction of the nocturnal surface inversion height. *J. Appl. Meteorol.* **18,** 526–531.

Yamada, T., and Mellor, G. (1975). A simulation of the Wangara atmospheric boundary layer data. *J. Atmos. Sci.* **32,** 2309–2329.

Yamada, T., and Meroney, R. N. (1971). "Numerical and Wind Tunnel Simulation of Response of Stratified Shear Layers to Nonhomogeneous Surface Features," Proj. THEMIS Tech. Rep. No. 9. Prepared by the Fluid Dynamics and Diffusion Laboratory, Colorado State University, Fort Collins.

Yamamoto, G. (1959). Theory of turbulent transfer in non-neutral conditions. *J. Meteorol. Soc. Jpn.* **37,** 60–69.

Yamamoto, G., and Shimanuki, A. (1969). Turbulent transfer in diabatic conditions. *J. Meteorol. Soc. Jpn.* **44,** 301–307.

Yamasaki, M. (1977). A preliminary experiment of the tropical cyclone without parameterizing the effects of cumulus convection. *J. Meteorol. Soc. Jpn.* **55,** 11–31.

Yap, D., and Oke, T. R. (1974). Sensible heat fluxes over an urban area—Vancouver, B.C. *J. Appl. Meteorol.* **13,** 880–890.

Yenai, M. (1975). Tropical meteorology. *Rev. Geophys. Space Phys.* **13**, 685–808.

Yu, T. (1977). A comparative study on parameterization of vertical turbulent exchange processes. *Mon. Weather Rev.* **105**, 57–66.

Yu, T. (1978). Determining height of the nocturnal boundary layer. *J. Appl. Meteorol.* **17**, 28–33.

Zdunkowski, W. G., Welch, R. M., and Paegle, J. (1976). One-dimensional numerical simulation of the effects of air pollution on the planetary boundary layer. *J. Atmos. Sci.* **33**, 2399–2414.

Zeman, O. (1979). Parameterization of the dynamics of stable boundary layers and nocturnal jets. *J. Atmos. Sci.* **36**, 792–804.

Zhou Xiao Ping (J. H. Golden (Ed.)) (1980). Severe storms research in China. *Bull. Am. Meteorol. Soc.* **61**, 12–21.

Zilitinkevich, S. S. (1970). "Dynamics of the Atmospheric Boundary Layer." Hydrometeorol. Publ. House, Leningrad.

Zipser, E. J. (1971). Internal structure of cloud clusters. "GATE Experimental Design Proposal," Interim Scientific Management Group, Vol. 2, Annex VII. WMO-ICSU, Geneva.

Zipser, E. J. (1977). Mesoscale and convective-scale downdrafts as distinct components of a squall line structure. *Mon. Weather Rev.* **105**, 1568–1589.

Zipser, E. J., and Gautier, C. (1978). Mesoscale events within a GATE tropical depression. *Mon. Weather Rev.* **106**, 789–805.

THE PREDICTABILITY PROBLEM: EFFECTS OF STOCHASTIC PERTURBATIONS IN MULTIEQUILIBRIUM SYSTEMS

RICHARD E. MORITZ[*]

Department of Geology and Geophysics
Yale University

AND

ALFONSO SUTERA

The Center for the Environment and Man, Inc.
Hartford, Connecticut

1. Introduction 345
2. The Predictability Problem 347
3. Prototype Model and Mathematical Concepts 349
4. Computational Techniques 359
5. Numerical Results 360
6. Concluding Remarks 374
   Appendix 376
   References 381

1. INTRODUCTION

The unceasing fluctuation of the Earth's atmosphere produces an endless sequence of flow patterns. Precise repetitions of previous states are conspicuously absent from the sequence, at least within our limited ability to discern them from data. Nonetheless, it is sometimes possible to identify flow regimes in a statistical sense, wherein some portion of the atmosphere adopts a configuration, persistent in time, that ultimately changes to a different configuration, distinguishable from its predecessor. An example is the irregular alternation between strong midtropospheric westerlies and blocking situations, in what was once known as the index cycle (see Charney *et al.*, 1981).

This observation poses basic questions about the predictability of the atmospheric system: What are the theoretical counterparts of persistent atmospheric patterns? Can the onset or duration of such configurations be predicted by theoretical means? What is the nature of mechanisms for in-

---

[*] *Present address:* Polar Science Center, University of Washington, Seattle, Washington 98105.

ternal energy transformation that lead to the termination of any given persistent regime?

The existence of multiple equilibrium solutions in dynamic–thermodynamic models of atmospheric physics is a topic of current interest, and seems to be a typical phenomenon when forcing, dissipation, and nonlinearity are retained in the governing equations (Vickroy and Dutton, 1979; Charney and Devore, 1979; Shirer and Dutton, 1979). Moreover, other investigators have found multiple solutions of periodic, aperiodic, or equilibrium type, coexisting with one another in the phase spaces of such models, over a range of the control parameters (e.g., Charney and Straus, 1980; Lorenz, 1980). In these cases it follows that a single deterministic solution, emanating from a particular initial condition, will realize only one of these possible regimes. To identify the persistence of some atmospheric patterns with oscillations near these equilibria is not an unreasonable hypothesis, but how are the transitions accomplished? At first sight, it appears that the study of possible mechanisms responsible for such transitions might lead to interesting insights into the predictability problem.

In this article we analyze a specific mechanism arising from the effects of stochastic perturbations on a deterministic flow that possesses more than one possible asymptotic solution regime, in the limit $t \to \infty$. The perturbations are introduced to represent the effects, on the model variables, of particular realizations of those scales of motion and physical processes not modeled deterministically, but which occur, nonetheless, in the atmosphere. We focus on the fundamental mechanism of transition between different configurations, using a highly idealized convection model as a prototype. Therefore, we choose parameter values and a random forcing process that facilitate clarity of demonstration, rather than quantitative comparison with atmospheric data. In more realistic models, designed to simulate closely some particular aspect of atmospheric motion, it may be possible to identify the major contributions to the random perturbations with a set of underlying instabilities (e.g., baroclinic instability, shear turbulence, convection) whose individual realizations are not predictable on the basis of the variables retained in the model. The statistical properties of the perturbations will vary, therefore, with the choice of deterministic variables and the point at which their frequency and wave number spectra are truncated in order to obtain solutions.

In a previous work one of us (Sutera, 1980, hereafter referred to as A.S.) noted the appearance of a new characteristic time, namely the "exit time" from the domain of attraction (see below) of a stable equilibrium flow, when small random perturbations were applied to a convection model with multiple equilibria. In this study we extend the analysis to the statistical properties of sample trajectories. In particular, we investigate the predictability of the system and its relation to the average exit time. Emphasis is placed on the distinction between uncertainty arising from probabilistic initial conditions and uncertainty in the trajectory of the system under stochastic forcing. These two sources of indeterminacy can affect predictability in fundamentally different ways, when the deterministic model possesses multiple domains of attraction. In the case of stochastic forcing, the sample trajectories produce "almost intransitive" statistics (Lorenz, 1967), and the sequence of transitions between different persistent regimes comprises a Markov chain.

In Section 2 we shall state our predictability problem. The prototype convection model and associated nomenclature are discussed in Section 3, along with the introduction of stochastic forcing. Section 4 comprises a description of our computational procedures, and is followed by the numerical results in Section 5. In Section 6 we close with a discussion of the implications of our results for modeling and predicting atmospheric behavior.

2. The Predictability Problem

The concept of atmospheric predictability eludes general definition. This is so because of the quite different information required of predictions in different circumstances, and the lack of universal agreement on a "zero skill" standard for comparison. It is generally recognized that the accuracy of atmospheric predictions is limited by two sources of uncertainty: lack of correspondence between physical processes occurring in the atmosphere and those represented in prediction models, and differences between the initial state of the atmosphere and the data used for model initialization. These concepts are discussed in detail by Pitcher (1977) in the context of probabilistic weather forecasting. We shall refer to these sources of uncertainty as "model uncertainty" and "initial uncertainty," respectively. Our study is concerned with the predictability limitations arising from the former source, and we are particularly interested in the longer time scales.

Most theoretical studies of atmospheric predictability tend to focus on the initial uncertainty and its propagation forward in time through the integration of an otherwise deterministic flow model (e.g., Thompson, 1957; Lorenz, 1963, 1969; Epstein, 1969; Gleeson, 1970; Fleming, 1971; Leith and Kraichnan, 1972). The philosophical justification for this view-point is perhaps most clearly understood in light of results obtained by Lorenz

(1969). Using a barotropic vorticity equation and assumption about the kinetic energy spectrum of atmospheric motion, Lorenz was able to show that the resulting solutions were observationally indistinguishable from a random process after a finite time interval, given any uncertainty in the initial data. One might therefore be tempted to attach only minor importance to model uncertainty, when even the "bare essentials" of the physics indicate the growth and dominance of initial uncertainty. These facts make it easier to appreciate the reasons for an apparent dichotomy in predictability research: probabilistic methods are applied routinely to study initial uncertainty, whereas progress in the area of model dynamics proceeds almost exclusively along deterministic lines.

Nevertheless, there appear to be limits on the predictability of deterministic, physical models, even in the absence of initial uncertainty. For instance, Robinson (1967) argues that the physical process of momentum diffusion (viscosity) limits the time interval on which justifiable predictions can be made by forward integration of Newton's second principle. After all, the principle asserts a proportionality relation between the acceleration and applied forces suffered by a material fluid parcel, and Robinson reasons that a parcel with characteristic dimension L, at the initial moment, does not exist as a physical entity after a time $T = L^2/8K$. Here K is the momentum diffusion coefficient appropriate to motions of scale L. Beyond this time the parcel has mixed, irreversibly, with surrounding fluid, and to speak of its acceleration or the applied forces is a meaningless exercise. For motions on the scale $L = 5000$ km, Robinson estimates $T \simeq 5$ days, in rough agreement with practical experience in large-scale prediction using dynamical methods.

The studies cited previously share the notion that predictability can be measured in terms of a time interval, beyond which the forecasts are no better than those obtained with a zero skill method (e.g., climatology). In the present work, we shall adopt as a predictability measure the decay time constant for autocorrelation functions of the predictands. Typical time series of atmospheric data exhibit exponential decay of autocorrelation (Leith, 1975) and our study is concerned with time-dependent models whose statistics share this property with the atmosphere. We assume, therefore, that some fraction of the total variance associated with a long realization of a predictand is composed of the covariance with the same variable at an earlier instant. When the covariance has dropped essentially to zero, all predictability associated with the earlier data has vanished. We stress that other definitions are equally valid, this particular choice being associated with the concept of "memory loss" from information theory. Obviously, we are identifying predictability with persistence, in a statistical sense, and in accordance with our earlier remarks about the importance of multiple, persistent regimes. This point of view departs from conventional meteorological definitions that are based on the rate at which pairs of solutions diverge when the initial conditions are slightly different. The emphasis here is on time scales that are, in some sense, long compared to at least one other characteristic time constant in the system considered. Once again the meteorological example that comes readily to mind is the persistence of large-scale blocking patterns (longer time scale) compared with the formation, movement, and decay of smaller scale baroclinic storms (shorter time scales). This point is discussed again in the Appendix. It is interesting to consider that most deterministic, numerical forecasting models show little or no skill beyond about 5 days, but Lorenz (1973) has shown the existence of some statistical predictability in large-scale height patterns at 12-day lags, based on observational data.

3. Prototype Model and Mathematical Concepts

Our choice of a prototype model is guided by the desire to represent the following atmospheric processes in simple form:

(1) input of energy by externally imposed thermal forcing;
(2) conversion of potential to kinetic energy;
(3) transfer of energy between different modes by nonlinear processes; and
(4) dissipation of energy by diffusive processes.

We adopt the approach used by Lorenz (1960) and study a particular case of the so-called minimum meteorological equations. Such systems possess the four properties cited above in their most basic manifestation, yet already they exhibit the mathematical phenomena of bifurcations to multiple equilibria and strange attractors (see below) with relatively few degrees of freedom. The minimum equations can be written

(3.1) $$\dot{x}_i = a_{ijk} x_j x_k + b_{ij} x_j + c_i$$

where we follow the Einstein convention of summation over repeated subscripts. Here $x_i(t)$ is an n-dimensional, time-dependent state vector, each component of which corresponds to a degree of freedom in the flow. The overdot signifies a time derivative and the coefficients a_{ijk}, b_{ij}, and c_i are taken to be time-independent control parameters. Typically the latter quantities represent the magnitude of external forcing, dissipation coefficients, topographic amplitudes, and the like. Ordinary differential equations like Eq. (3.1) are typical of numerical prediction models, where $x_i(t)$

represents grid point variables or spectral coefficients in a Galerkin expansion (e.g., Haltiner and Williams, 1980). We note, however, that time-dependent coefficients and nonquadratic nonlinearities may be present in more general models, including, say, seasonal variations of radiation and parameterization of radiative heating. In any case, the quadratic term arises when the generating partial differential equations contain advective terms, and in many cases of interest it satisfies

$$(3.2) \qquad a_{ijk}x_ix_jx_k = 0$$

indicating that the nonlinear interactions do not alter the total energy $E = \Sigma_i x_i^2/2$ of the system. For a viscous, conducting fluid, the dissipative processes are represented by $b_{ij}x_j$ and their contribution $b_{ij}x_ix_j$ to the change in E will be negative. External forcing is typically represented by the inhomogeneous term c_i after, at most, a linear transformation of variables.

We shall be concerned with the asymptotic behavior of the deterministic solutions of Eq. (3.1) because it is vitally connected to the predictability when stochastic perturbations are applied. Therefore, it will be convenient to introduce a few elementary terms from the theory of dynamical systems to facilitate the discussion. Application of these concepts to meteorological predictability was introduced by Lorenz (1963). Let us rewrite Eq. (3.1) in the vector form

$$(3.3) \qquad \dot{\mathbf{x}} = \mathbf{b}(\mathbf{x})$$

with the identifications $\mathbf{x}(t) = (x_1, x_2, \ldots, x_n)$, $\mathbf{b} = (a_{1jk}x_jx_k + b_{1j}x_j + c_1, \ldots, a_{njk}x_jx_k + b_{nj}x_j + c_n)$. Equation (3.3) is a classical dynamical system, wherein the state vector $\mathbf{x}$ corresponds to a point in the n-dimensional space $\mathbf{R}^n$. A unique solution $\mathbf{x}(t, \mathbf{x}_0)$ emanates from each initial condition $\mathbf{x}_0$, tracing a curve (trajectory or orbit) through the phase space. Moreover, the vectorial field $\mathbf{b}(\mathbf{x})$ defines through Eq. (3.3) a flow in the phase space, everywhere tangent to the trajectories. Clearly any collection of states of this system may be regarded as a point set $S = \{x^1, x^2, \ldots, x^m, \ldots\}$ in the topological space M defined by $\mathbf{R}^n$ and its open subsets (superscripts here denote members of the set). In this same spirit we observe that the flow $\mathbf{b}(\mathbf{x})$ defines a one-parameter group of transformations, say $T_t: M \to M$ that maps the phase space onto itself. Simplification, understanding, and, most importantly, physical insight can be enhanced if it is possible to partition the phase space into a limited number of distinct regions, inside each of which the trajectories exhibit similar behavior with respect to the properties that we hope to predict. The natural starting point from which to build such a partition consists in the identification of attractor sets in the phase space. It is well known in the mathematical theory of mechanics that, for uniformly bounded trajectories, there exist certain minimal, positively invariant limit sets, denoted $\omega^+(x)$, such that

(3.4) $$\lim_{t \to \infty} T_t x \to \omega^+(x)$$

(e.g., Abraham, 1967), where x is any point in M. Intuitively, the trajectory starting from x passes repeatedly through successively smaller neighborhoods of each and every point in $\omega^+(x)$ as time progresses. Let us fix the set ω^+ for the moment and consider the set of all initial conditions $\{x\} = S$ for which Eq. (3.4) is true. If S occupies a nonzero volume in the phase space $\mathbf{R}^n$, then some suitably chosen volume containing any point in S is also within S. In this case $\omega^+(S)$ is a physically realizable (observable) limit set, given the inevitable initial uncertainty due to measurement errors. Following Lorenz' (1980) definition, we call such limit sets attractors, and S becomes the domain of attraction.

The simplest example of an attractor is the asymptotically stable steady state, say $\hat{\mathbf{x}}$, satisfying

(3.5) $$\mathbf{b}(\hat{\mathbf{x}}) = \mathbf{0}$$

and the usual condition that all eigenvalues λ_i, $i = 1, 2, \ldots, m$, of the matrix $\mathbf{A}$, obtained from the linearization

(3.6) $$\begin{aligned} \mathbf{x}' &= \mathbf{x}(t) - \hat{\mathbf{x}} \\ \dot{\mathbf{x}}' &= \mathbf{A}\mathbf{x}' + O(\mathbf{x}'^2) \end{aligned}$$

lie in the left half of the complex plane.

Our interest here is in natural partitions that may exist if more than one domain of attraction is present in regions of the phase space corresponding to initial conditions one might encounter in practice. Therefore, let us suppose that there exists a finite volume in $\mathbf{R}^n$, bounded by a surface Σ on which the normal component of $\mathbf{b}$ is inwardly directed at all points. Furthermore, we assume that Σ encloses all initial conditions of physical interest. These conditions are typically encountered in forced dissipative systems, where the dissipation dominates the time derivatives whenever the total energy exceeds some threshold (Lorenz, 1963, 1980). In particular, it is true for Eq. (3.1) if Eq. (3.2) holds. Given the monotonic increase of viscous dissipation with kinetic energy and of infrared radiation with temperature, it seems likely that some such constraints should also apply to the atmosphere itself. In this case we may identify points inside Σ with the set M, and it is evident that no trajectories can leave M. Therefore, all attractors whose domains intersect M must be inside Σ, and every point of M must be in some such domain, or in a set with zero phase volume, whose limit sets cannot be attractors. If there is only one attractor inside Σ, then the partition defined by its domain of attraction S_1 provides no predictability information beyond that provided by the location of Σ. With probability 1, all initial conditions of interest approach the single attractor, and predictability must be studied in terms of trajectories on the attractor itself. In the case of multiple equilibria, or, more generally, multiple attractors in Σ it follows from the minimality of limit sets and the uniqueness of trajectories that the domains of attraction are disjoint, i.e., $S_1 \cap S_2 = \phi$ (null set). If the physical model, represented by the choice of x and the processes modeled by b(x), is well-founded, then two equally relevant initial conditions in Σ can lead to different asymptotic regimes, each of which represents some basic physical balance. Moreover, the combinations of components x_i that occur in ω_1^+ and ω_2^+ are mutually exclusive, rendering predictions on x and statistics from long integrations of the model critically dependent on the location of the initial value x_0, unless the two attractors are extremely close to one another throughout the phase space. Obviously the existence of multiple attractors and the locations of $S_1, S_2, \ldots$ are fundamental predictability information. Furthermore, the detailed physical processes involved in the approach toward some attractor may be exceedingly complicated, yet the important physical balance represented by the attractor itself might be more easily comprehended. For instance, Charney and Devore (1979) identified multiple, stable fixed points in the phase space of a barotropic, quasi-geostrophic model of atmospheric flow over topography. One equilibrium corresponds to a balance between externally imposed momentum forcing and the pressure gradient across the topography, while the second one balances the forcing against boundary-layer friction. In the prototype primitive equation model studied by Lorenz (1980, see also the Appendix) the limit set consists of time-dependent, aperiodic trajectories on an infinite complex of surfaces. Lorenz was able to show that these trajectories are part of an invariant set that is free of gravity waves, and that the state variables here satisfy a form of nonlinear balance equation that defines the attractor set.

It is evident that the choice of a model is of the utmost importance with regard to the ultimate behavior and predictability of the solutions. Returning to the minimum equations (3.1), we note that the vector $x_i(t)$ must be truncated in any particular problem due to certain physical and computational constraints. Of course, the molecular limit imposes the ultimate truncation point on the Navier–Stokes system, and in practice the statistical effects of molecular motions are modeled as diffusive terms with empirical coefficients (viscosity and diffusivity).

In large-scale meteorology, the emphasis is on predicting scales of motion many orders of magnitude greater than the smallest turbulent fluctuations, and it is usually presumed that the molecular limit is irrelevant.

Truncation is therefore effected at some macroscopic length scale, typically on the order of 250 km or more in the horizontal dimension. By analogy with the molecular problem, the statistical effects of all neglected scales of motion and physical processes are represented by empirical parameterization formulas, closed in the set of variables retained for prediction. Considering now the parameterization relationship between, say, a turbulent "stress" and the gradient of a large-scale velocity component, we note that universal functions and constants are identified, using similarity hypotheses, and their parameters are estimated by empirical–statistical summarization of ensemble observational data. Each member of the ensemble, called a realization, associates a particular stress with a particular velocity gradient. Ensembles comprising all realizations with fixed velocity gradients (i.e., all similar flows) are inevitably characterized by variability in the measured stresses. The best that a parameterization can achieve, then, is to estimate the ensemble expectation for the stress and its moments, over all such similar flows. The atmosphere, on the other hand, is a particular realization, associated with a particular stress at any time. The particular stress will, of course, deviate from its expectation in any realization, by an amount that can only be described by a probability distribution (Monin and Yaglom, 1971). It is well known that a deterministic system, subject to random fluctuations with expectation 0, will not remain close to the deterministic trajectory indefinitely. Therefore, we shall add a fluctuating term $f_i(t)$ to Eq. (3.1) as the simplest representation of deviations from the mathematical expectation for the trajectories. The most familiar process of this type, and the one chosen here solely for clarity of demonstration, is the Gaussian random process, acting with equal variance on all frequencies (i.e., possessing a "white noise" spectrum). Therefore, we assume that the best possible parameterizations of neglected scales and processes appear in the terms of Eq. (3.1) and that the fluctuations about these expectation values are modeled by $f_i(t)$.

It may be worthwhile to point out that the actual coefficients used to parameterize small-scale effects in large-scale models are estimated on the basis of data obtained under rather idealized circumstances, for example, horizontally homogeneous and steady large-scale flow. Testimony to the rarity with which such conditions are encountered in the atmosphere can be found in standard books on micrometeorology (e.g., Haugen, 1973), where one notes the great care that must be taken in order to satisfy just these two requirements in an experimental study. If scales smaller than 250 km are neglected, unresolved phenomena such as super-cell thunderstorms (hardly steady and homogeneous flow) may have considerable influence on the larger-scale variables. Thus it seems intuitively clear that one should consider the role of terms such as $f_i(t)$, and their representation as a stochastic process appears to be the most tractable course for studying model uncertainty.

Summarizing the previous discussion, Eq. (3.1) is rewritten

(3.7) $$\dot{x}_i = a_{ijk}x_j x_k + b_{ij}x_j + c_i + f_i(t)$$

where $f_i(t)$ is the Gaussian random process discussed above. Equation (3.7) is a Langevin equation and it is extremely versatile in the case in which the object of our calculations is to compute fluctuations of its linearization around a steady state (if any exist), i.e., near solutions of the algebraic equations

(3.8) $$a_{ijk}x_j x_k + b_{ij}x_j + c_i = 0$$

However, in the nonlinear case the use of Eq. (3.7) leads to apparent mathematical contradictions [due essentially to the nowhere-differentiability of $f_i(t)$] and necessitates introduction of the Ito-calculus (Schuss, 1980). The differences between the two approaches are subtle, and their mathematical and physical explanation requires more discussion than we need for our purposes, though we feel that the key of our argumentation might be explained along this line. We shall return to this point in another work. The Ito-calculus requires that Eq. (3.7) be written in vectorial form

(3.9) $$d\mathbf{x} = \mathbf{b}(\mathbf{x})\,dt + \boldsymbol{\epsilon}^{1/2}(\mathbf{x})\,dW$$

where dW is the differential of the Wiener process $W(t)$ (Doob, 1952). The increments of the Wiener process, defined by

(3.10) $$dW = W(t + dt) - W(t)$$

are mutually independent, Gaussian-distributed quantities with expectation 0 and variance dt, so that the variance of the process $\boldsymbol{\epsilon}^{1/2}(\mathbf{x})\,dW$ becomes $|\boldsymbol{\epsilon}(\mathbf{x})|\,dt$. Hereafter we assume $\boldsymbol{\epsilon}^{1/2}(\mathbf{x}) = \epsilon^{1/2}$, a constant scalar parameter, representing a scale factor for the strength of typical stochastic fluctuations relative to the deterministic flow $\mathbf{b}(\mathbf{x})$. If ϵ is small, Eq. (3.9) can be studied as a stochastic perturbation of the deterministic process $\mathbf{b}(\mathbf{x})$ (Ven'Tsel and Friedlin, 1970). Hence the local velocity of the trajectories through phase space is given by $\mathbf{b}$, representing the expectation for the physics and parameterizations embodied in the model. The random perturbations represent deviations of any particular realization due to processes not governed by $\mathbf{x}$.

Following A.S., we focus on the system of equations that have come to be known as the "Lorenz model." The system is a subset of equations derived by Saltzman (1962), who analyzed finite-amplitude Bénard convection in a fluid layer by expanding the Oberbeck–Boussinesq vorticity and thermodynamic equations as double Fourier series in the (x, z) plane. Lorenz (1963) truncated Saltzman's system of ordinary differential equations for the trigonometric mode amplitudes, retaining only the following nondimensional set, with $\epsilon = 0$:

(3.11)
$$dX = (-\sigma X + \sigma Y)\, dt + \epsilon^{1/2}\, dW$$
$$dY = (rX - XZ - Y)\, dt + \epsilon^{1/2}\, dW$$
$$dZ = (XY - bZ)\, dt + \epsilon^{1/2}\, dW$$

where $X(t)$ represents the amplitude of the convective motion for a single, two-dimensional roll, $Y(t)$ measures the temperature difference between ascending and descending branches of the roll, and $Z(t)$ is proportional to the departure of horizontally averaged, vertical temperature profile from the linear stratification imposed by specifying a Rayleigh number

$$R = g\alpha H^3\, \Delta T/\kappa\nu$$

Here g is the acceleration of gravity, α is the coefficient of thermal expansion, ΔT is the imposed temperature difference between the upper and lower boundaries of a fluid layer of depth H, and κ and ν are the thermal diffusivity and molecular viscosity, respectively. Energy is supplied to the system through the parameter r, representing the Rayleigh number in the dimensionless form $r = R/R_c$. R_c is the critical value for onset of convection. σ is the Prandtl number and $b = 4/(1 + a^2)$ varies with the aspect ratio a of the convective rolls. Following Saltzman and Lorenz, we take $b = \frac{8}{3}$ as fixed, in which case X and Y correspond to the most unstable convective modes found by Lord Rayleigh (1916) in his analysis of the stability of viscous fluids heated uniformly from below. The dot signifies a derivative with respect to the nondimensional time $t = (\pi^2(1 + a^2)\kappa/H^2)t'$, where t' is the dimensional time.

The derivation of Eqs. (3.11) can be found in Saltzman (1962) and Lorenz (1963) along with discussion of the validity of the various approximations. These equations probably have some relevance for laboratory Bénard convection when r is only slightly greater than 1 and surface tension effects are minimized (Busse, 1972). For our purposes it is sufficient to note that these are a special case of the minimum equations, and we study them as a prototype in the spirit proposed by Lorenz (1960). Furthermore, in the 17 years since Lorenz' analysis of the nonperiodic solutions of Eqs. (3.11) (with $\epsilon = 0$), these equations have been the object of study by numerous mathematicians and physicists (see Rabinovich, 1978, for an illuminating review) because they produce an unpredictable, seemingly nonperiodic signal, in the absence of stochastic perturbation. We note with interest that Lorenz (1980) finds special parameter ranges wherein equations appropriate to the large-scale quasi-horizontal motions of the atmosphere reduce to systems like (3.11) (see Appendix).

The first task in any program of analysis is to find the steady states of the deterministic system. The equilibria and corresponding eigenvalues of Eqs. (3.11) were analyzed thoroughly by Lorenz (1963). Briefly, three steady solutions are possible, depending on r, and are given in the format (X, Y, Z):

$$
\begin{aligned}
& \mathbf{C}^0 = (0, 0, 0) \quad \text{for} \quad r \geq 0 \\
& \mathbf{C}^+ = ((b(r-1))^{1/2}, (b(r-1))^{1/2}, r-1) \quad \text{for} \quad r \geq 1 \\
& \mathbf{C}^- = (-(b(r-1))^{1/2}, -(b(r-1))^{1/2}, r-1) \quad \text{for} \quad r \geq 1
\end{aligned}
\tag{3.12}
$$

Bifurcation occurs at $r = 1$, where three steady states coalesce, two of them becoming imaginary (hence unphysical here). Since b is fixed, we can portray the bifurcation sequence in terms of a single control parameter r (Fig. 1).

The next obvious calculation is perturbative, i.e., to find the eigenvalues and stability in the neighborhoods of the steady solutions. Again citing Lorenz, $\mathbf{C}^0$ is asymptotically stable with three real eigenvalues λ_ℓ, $\ell = 1, 2, 3$, when $r < 1$, but one eigenvalue changes to positive at $r = 1$, giving Rayleigh's condition for onset of convection. At $\mathbf{C}^+$ and $\mathbf{C}^-$ identical eigenvalues are found, now dependent on both r and σ. For the larger values of r and σ considered here, one eigenvalue is real and the others are complex conjugates. For cases with r and σ near 1, all three are real and negative. When $\sigma < b + 1$ the real parts are negative for all r, giving stable, steady convection. For $\sigma > b + 1$ there exists a critical value

$$r_c = \sigma(\sigma + b + 3)/(\sigma - b - 1) \tag{3.13}$$

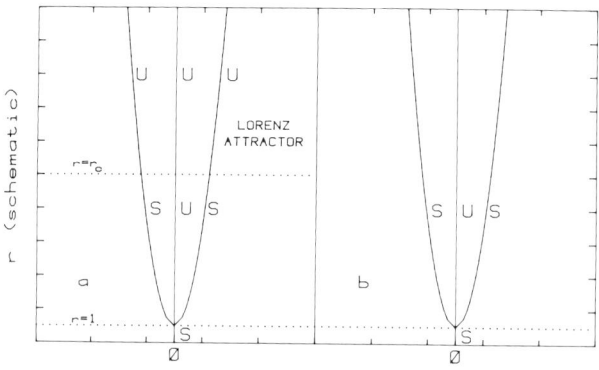

FIG. 1. Bifurcation diagrams for equilibrium X and control parameter r. Stable (S) and unstable (U) equilibria are indicated for (a) $\sigma > b + 1$ and (b) $\sigma < b + 1$.

at which the eigenvalues become pure imaginary. For $r > r_c$ all three equilibria are unstable. Marsden and McCracken (1976) proved that a subcritical Hopf bifurcation occurs at r_c, so that unstable periodic orbits exist in some neighborhood of $r = r_c$, and $\mathbf{C}^+$ and $\mathbf{C}^-$ become unstable for $r > r_c$. Two further properties, both noted by Lorenz, are of interest. First, Eqs. (3.11) can be manipulated to show that all trajectories are eventually confined inside a spherical phase volume with maximum radius

(3.14) $$\rho \leq b(r + \sigma)/K$$

where $K = \min(2\sigma, 2, 2b)$. This was to be expected in view of the discussion of the energy threshold, beyond which dissipation dominates the time derivatives. Second, any phase volume, defined by a set of initial conditions, decays exponentially to zero following the trajectories, so that all of the attractor sets have dimension less than three. These facts are also discussed by Rabinovich (1978), for example.

Much more difficult questions now arise about the structure of the attractor sets when all three equilibria are unstable, yet the solution remains bounded. This fundamentally nonlinear problem was studied deeply by Lorenz, and has engaged many authors since then (see Manneville and Pomeau, 1980, for a fairly complete list of recent articles). For our purposes, it is sufficient to say that when $\sigma > b + 1$, an attractor set appears in the phase space, on which the solutions are apparently nonperiodic, when r is near the critical value r_c. Topologically, this attractor set gives the appearance of an infinite complex of surfaces in $\mathbf{R}^n$, whose intersection with an arbitrary curve is a Cantor set. Such structures have been dubbed "strange attractors" (see Lorenz, 1980). In parts of this parameter range there may be two other attractors ($\mathbf{C}^+$ and $\mathbf{C}^-$), so that different initial conditions produce different asymptotic solutions. At $r = r_c$ all initial conditions apparently find the nonperiodic attractor, hereafter called the Lorenz attractor. It is worth noting that most of the results regarding the detailed bifurcations and statistical properties of the attractors have been derived from numerical integration of Eqs. (3.11), illustrating our earlier statement about the analytical difficulties attending even the most primitive representations of nonlinear interaction.

The properties discussed above are summarized, schematically, in Fig. 1. The Lorenz model provides the possibility to study single or multiple equilibria, stable or unstable, and in the presence or absence of the Lorenz attractor. All of these properties are realizable within the framework of the minimum equations (3.1), representing several basic atmospheric processes in prototype form.

A crucial point of our analysis is the concept of almost-intransitivity. It is convenient to give a formal definition because it will help to clarify the rationale followed in the choice of parameters (namely r and σ) in our numerical experiments.

The minimal limit sets and, therefore, all of the attractors are invariant with respect to the flow Eqs. (3.11) when $\epsilon = 0$. If $\mathbf{C}^+$ and $\mathbf{C}^-$ are the only attractors present, then we expect that their domains of attraction separate the phase space into equal, disjoint volumes, given the symmetry of Eqs. (3.11) under $X \to -X$, $Y \to -Y$, $Z \to Z$. If we imagine an invariant probability measure $P(\mathbf{x})$ corresponding to the asymptotic occurrence of state $\mathbf{x}$ from randomly chosen initial conditions, then it will comprise two Dirac delta functions $P(\mathbf{C}^+) = \frac{1}{2}\delta(\mathbf{C}^+ - \mathbf{x})$ and $P(\mathbf{C}^-) = \frac{1}{2}\delta(\mathbf{C}^- - \mathbf{x})$ on the steady states, and $P(\mathbf{x}) = 0$ elsewhere. Now, an arbitrary measure-preserving transformation T_t satisfies the Birkhoff individual ergodic theorem (Halmos, 1956) if T_t is metrically transitive with respect to all such invariant sets. Aside from the mathematical technicalities, the theorem implies that if T_t is metrically transitive, it is impossible to decompose the space $\mathbf{R}^n$ into two or more disjoint sets, each with positive measure. In our example a set has measure proportional to the volume of initial conditions that produce trajectories that become trapped in the set as $t \to \infty$. Small stochastic perturbations, whose physical relevance was discussed previously, play the mathematical role of transfiguring the intransitive system ($\epsilon = 0$) into the almost intransitive system ($\epsilon > 0$). Because Eqs. (3.11) are a diffusion process (Doob, 1952), the trajectories ultimately visit all of phase space, eliminating any decomposition into disjoint sets of positive measure. Nonetheless, a manifest nonuniformity survives in the form of $P(\mathbf{x})$, representing the crucial physical balances inherent in the deterministic part of Eqs. (3.11). Qualitatively, almost-intransitivity implies the existence of "leaks" in the disjoint sets* of the intransitive system. The leaks may result from stochastic perturbations or, for instance, from the appearance of a homoclinic trajectory in the bifurcation sequence of a deterministic dynamical system. In any event, the characteristic signature of almost-intransitivity is the existence of grossly nonuniform invariant probability distributions, with the implication that long residence in some locality of phase space is followed by transition to another locality and long residence there.

The relevance of studying almost intransitive behavior is, in our view, twofold: first, we seek *a priori* knowledge of the regions occupied by the system with high probability and, equally important, we search for the characteristic times for transitions between the various regions. To demonstrate the nature and relevance of the transitions, we have analyzed the system (3.11) for different parameter values, studying statistical quantities usually considered in analyses of any physical signal.

---

* The authors acknowledge J. Dutton for this graphic description.

4. Computational Techniques

Our basic data sets comprise discrete time series of X, Y, and Z, generated by numerical integration of the system (3.11). Our integrator is a fourth-order Runge–Kutta (RK) algorithm for systems of ordinary differential equations (Ince, 1956, p. 546). All computations were performed in FORTRAN double precision, with 16-decimal-place accuracy.

The RK procedure is applied to current values X_t, Y_t, Z_t to obtain the deterministic iterates $\tilde{X}_{t+\Delta t}$, $\tilde{Y}_{t+\Delta t}$, $\tilde{Z}_{t+\Delta t}$. To each of these latter quantities we add the products of $\epsilon^{1/2}$ and Gaussian-distributed random variables with mean 0 and variance 1, representing dW, to obtain the new current values $X_{t+\Delta t}$, $Y_{t+\Delta t}$, $Z_{t+\Delta t}$. Random variables are generated by a standard numerical routine that combines a multiplicative congruential generator and a shift-register generator to improve randomness. We measure the stochastic forcing in terms of its variance per unit time γ. Noting that our representation of dW has variance 1 rather than Δt, this is the appropriate normalization. Taking expectation values $E(\ )$, we obtain

$$(4.1) \qquad \gamma \Delta t = E[(\epsilon^{1/2} dW)^2] = \epsilon$$

using the fact that ϵ is a constant parameter and $E[(dW)^2] = 1$. Therefore

$$(4.2) \qquad \gamma = \epsilon/\Delta t$$

is the desired parameter, used to compare stochastic forcing strength between cases with different Δt and ϵ.

Our study of almost-intransitivity requires that we keep track of the "exit time," that is, the time interval between transitions from one domain of attraction to another. When $\mathbf{C}^+$ and $\mathbf{C}^-$ are stable equilibria and the Lorenz attractor is not present ($\sigma < b + 1$), this time is identical to the "first passage time" in the theory of stochastic differential equations (Gihman and Skorohod, 1972). However, delimitation of the boundaries of these domains is, for Eqs. (3.11), a very difficult task. Therefore, we adopted the following rule: the exit time is recorded as the time interval between successive crossings [denoted by $(X \to X)$] of the planes $X(t) = X^+$ and $X(t) = X^-$, satisfying either $(X^+ \to X^-)$ or $(X^- \to X^+)$, and where X^+ and X^- correspond to the X coordinates of $\mathbf{C}^+$ and $\mathbf{C}^-$, respectively [Eqs. (3.12)]. In the deterministic cases with $\sigma > b + 1$ and $r > r_c$ (the Lorenz attractor) our numerical results (Fig. 12) indicate that $X = 0$ plays a fundamental role in separating temporarily the oscillations around $\mathbf{C}^+$ from those around $\mathbf{C}^-$, even though neither equilibrium point has a domain of attraction with nonzero phase volume. Thus the "exit time" can be recorded by the same rule in both cases, corresponding to a temporary localization time when the Lorenz attractor is present.

We shall distinguish sample parameters from asymptotic (infinite time) statistics. The sample mean of any quantity ξ is denoted by

(4.3) $$\bar{\xi}(t_0, T) = \frac{1}{N} \sum_{i=1}^{N} \xi_i$$

where $t = t_0 + i \, \Delta t$ and we show explicitly the dependence on initial value $\xi(t_0) = \xi^*$ and averaging time interval T. The sampling time interval for the series is $\Delta t = T/N$, for N discrete points. The time average is

(4.4) $$\langle \xi \rangle = \lim_{T \to \infty} \bar{\xi}(t_0, T)$$

Our estimator of the sample autocovariance function is

(4.5) $$\hat{C}_{\xi\xi}(t_0, T, \tau) = \frac{1}{N} \left[\sum_{i=1}^{N-K} \xi_i \xi_{i+k} - \bar{\xi} \sum_{i=1}^{N-K} (\xi_i + \xi_{i+k}) + (N - K)\bar{\xi}^2 \right]$$

(Jenkins and Watts, 1968), so the sample variance is just $\hat{C}_{\xi\xi}(t_0, T, 0) \equiv \hat{\sigma}_\xi^2$ and $\tau = k \, \Delta t$ is the lag time. The sample autocorrelation function is

(4.6) $$\hat{r}_{\xi\xi}(t_0, T, \tau) = \frac{\hat{C}_{\xi\xi}(t_0, T, \tau)}{\hat{\sigma}_\xi^2}$$

Asymptotically as $T \to \infty$ a transitive system has the function $r_{\xi\xi}(\tau)$, independent of the initial conditions. Finally, we estimate the power spectra

(4.7) $$\hat{P}(t_0, T, \omega) = T|a_\omega|^2$$

where the complex Fourier coefficients

(4.8) $$a_\omega = \frac{1}{N} \sum_{j=1}^{N} \xi_j \exp(-2\pi i (j-1)(k-1)/N)$$

are computed by fast Fourier transform methods. The frequencies are given by $\omega(k) = (k - 1)/N \, \Delta t$.

5. Numerical Results

The numerical experiments discussed below are summarized in Table I. Primary comparisons are made between the single equilibrium case S-1 and the multiple equilibrium case S-2 (corresponding to points below and above the dotted line in Fig. 1b, respectively).

The deterministic curves $X(t)$ (Figs. 2 and 3) derive from cases D-1 and D-2, respectively. We chose initial conditions away from the equilibria to make clear the asymptotic approach to a steady state in the absence of stochastic forcing. In Fig. 2 the equilibrium coordinate $X^0 = 0$ is approached. Two solutions are plotted in Fig. 3, demonstrating the exis-

TABLE I. PARAMETER VALUES FOR NUMERICAL EXPERIMENTS

Code	r	σ	$\epsilon^{1/2}$	Δt	γ	T^*	$\bar{\tau}_e$
			Stochastic cases				
S-1	0.9	1.0	0.03	0.05	0.018	1000	—
S-2	1.1	1.0	0.03	0.05	0.018	7500	197.3
S-3	1.1	1.0	0.02	0.005	0.08	2500	39.0
S-4	1.5	1.0	0.15	0.02	1.125	500	12.6
			Deterministic cases				
D-1	0.9	1.0		0.05		25	—
D-2	1.1	1.0		0.05		55	∞
D-3	24.75	10.0		0.005		1000	2.07
D-4	30.00	10.0		0.005		250	1.71

tence of two domains of attraction. The only nontrivial predictability problem arises when the initial values $\mathbf{X}^*$ are near the boundary between domains of attraction of $\mathbf{C}^-$ and $\mathbf{C}^+$, introducing uncertainty about the ultimate solution. Even here, however, one need only wait until the relevant domain is evident to obtain complete predictability, because the phase points always move away from the boundary.

The situation is more interesting when stochastic forcing is applied. In Fig. 4 we present $X(t)$ for case S-1. The initial condition matched that of D-1. During the first 10 time units the curve moves toward X^0 on a path somewhat similar to that in Fig. 2, but is subsequently driven back towards $X = +0.5$ by the noise. Fluctuations of similar magnitude occur irregularly throughout the graph. We note the persistence in the vicinity of $X = 0$ during $200 \leq t \leq 280$. A similar plot was generated for case S-2,

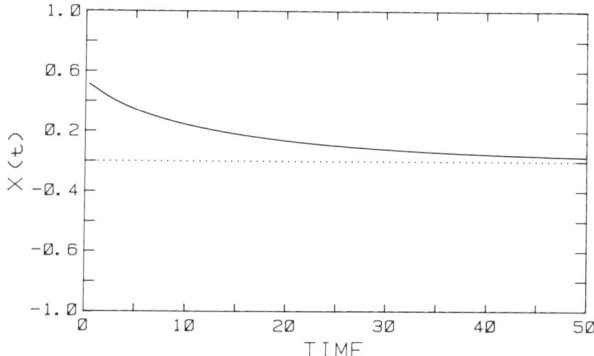

FIG. 2. Numerical solution $X(t)$ for case D-1, with a single equilibrium X^0 (dotted line). $r = 0.9$, $\sigma = 1$.

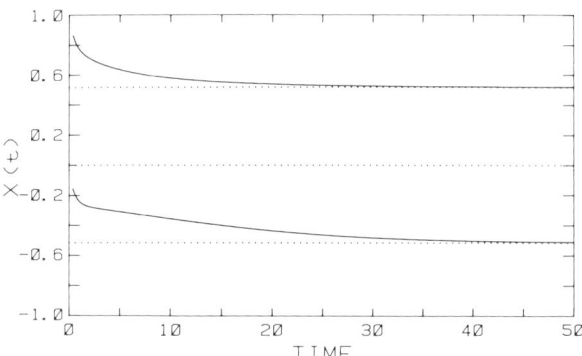

FIG. 3. Numerical solutions $X(t)$ for case D-2, showing asymptotic approach to X^+ and X^- from initial values in respective domains of attraction. $r = 1.1$, $\sigma = 1$.

where two equilibria are present (Fig. 5). This solution has the same initial value and noise realization as the previous case. We see immediately that the deterministic flow exerts a profound influence on the solution, even under significant stochastic perturbations. For example, during the initial 10 time units $X(t)$ stays close to $X^+ = 0.516$, in marked contrast to Fig. 4. Moreover, the interval $200 \leq t \leq 280$ is now characterized by fluctuations near X^+ rather than X^0. A qualitative property of the curve is its tendency to persist near X^+ or X^-, followed by much shorter transitional intervals wherein X^0 is crossed. These phenomena are all the more pronounced when we utilize information from all three variables X, Y, and Z. The distances

(5.1)
$$d0(t) = |\mathbf{X} - \mathbf{C}^0|$$
$$d1(t) = |\mathbf{X} - \mathbf{C}^+|$$
$$d2(t) = |\mathbf{X} - \mathbf{C}^-|$$

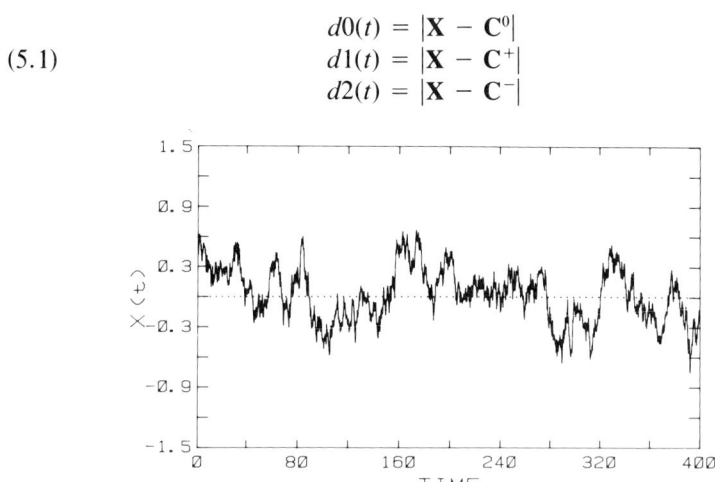

FIG. 4. Numerical simulation of process $X(t)$ for case S-1, showing fluctuations around the single equilibrium X^0. $r = 0.9$, $\sigma = 1$.

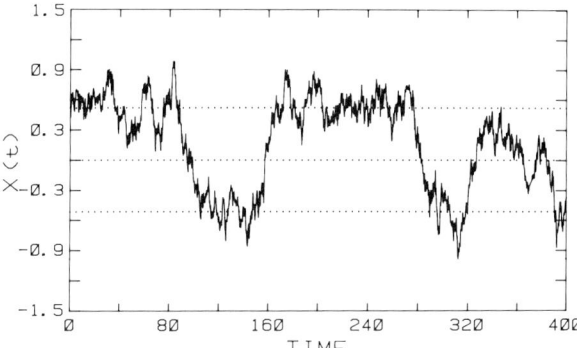

FIG. 5. Numerical simulation of process $X(t)$ for case S-2, showing persistence near X^+ and X^-, and rapid transitions. $r = 1.1$, $\sigma = 1$.

between the phase point and each steady state are plotted against time in Figs. 6–8. Figure 6 shows the initial 800 time units for case S-1. There is no obvious tendency for persistence except near $d0 = 0.15$, representing fluctuations near C^0. The corresponding plot for case S-2 is presented in Fig. 7. Here the system is usually far from C^0, because of the instability of the latter point, although infrequent approaches to zero occur, generally followed by quick departures to larger distance. Figure 8 illustrates clearly the persistence near C^+ and C^-, and the sharp transitions between the two. The upper and lower curves vary in opposition, one usually remaining near zero while the other is close to maximum. Our counting procedure (Section 5) recorded exits $t = 105, 166, 290, 339, 390, 522,$ and 586. These times are marked El–E7 in Figs. 7 and 8. The system is almost intransitive because, although the Ito equations (3.11) satisfy rigorously the Birkhoff theorem, a manifest nonuniformity is evident in the phase space. Clearly the probability weightings of regions near C^+ and C^- are

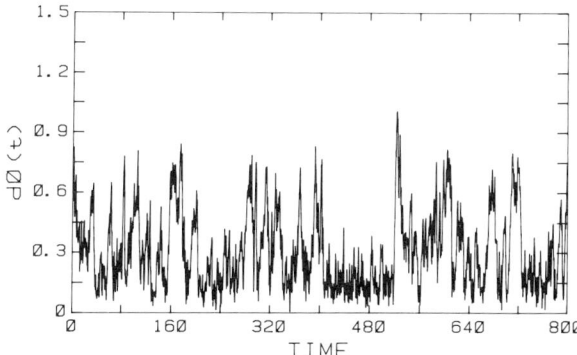

FIG. 6. Distance $d0(t)$ of the phase point from the stable equilibrium C^0, case S-1, showing persistence at small values. $r = 0.9$, $\sigma = 1$.

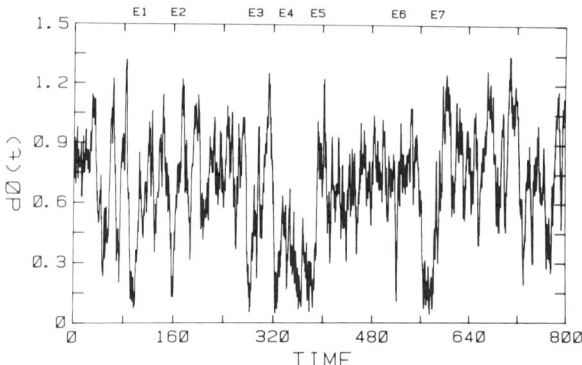

FIG. 7. Same as Fig. 6 only for case S-2, showing persistence at large distance. Exits are denoted by E1–E7. $r = 1.1$, $\sigma = 1$.

large compared to intermediate regions. Furthermore, the exit times appear to provide a useful index of the almost-intransitivity, in that they quantify the persistence time due to nonuniformities near $\mathbf{C}^+$ and $\mathbf{C}^-$. The mean exit time, computed from integration over 7500 time units and 38 exits, is $\bar{\tau}_e = 197.3$ (Table I) for S-2. The corresponding standard deviation is of the same order, namely $\hat{\sigma}_{\tau_e} = 179.1$, a fact that can be appreciated by noting the variability on Fig. 8. For instance, only 50 time units elapsed between E3 and E4, while 428.3 units were required to achieve transition from E7 to E8.

The exit mechanism can be understood qualitatively by comparing Figs. 7 and 8 at the time of exit. In each case, the transition occurs when $d0$ is close to a minimum (near zero), indicating that the phase point approaches the unstable state $\mathbf{C}^0$. The eigenvalues of this point (Table II) are

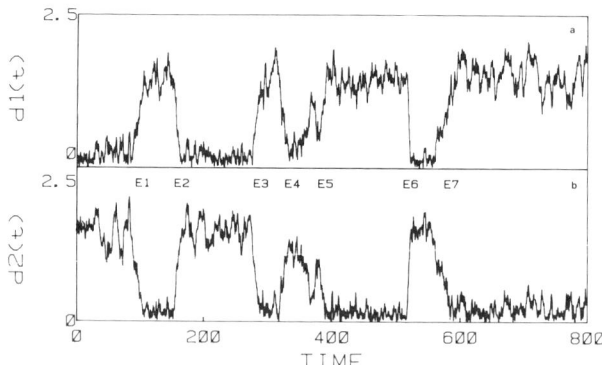

FIG. 8. Distances $d1(t)$ (a) and $d2(t)$ (b) of the phase point from $\mathbf{C}^+$ and $\mathbf{C}^-$, respectively, for case S-2. Persistence near the stable equilibria and sharp transitions are evident. $r = 1.1$, $\sigma = 1$.

TABLE II. EIGENVALUES OF THE LINEARIZED EQUATIONS NEAR THE STEADY STATES

Code	C^0			C^+, C^-		
	λ_1	λ_2	λ_3	λ_1	λ_2	λ_3
S-1	−2.67	−0.051	−1.95	—	—	—
S-2	−2.67	0.049	−2.05	−0.104	−2.56	−2.00
S-3	−2.67	0.049	−2.05	−0.104	−2.56	−2.00
S-4	−2.67	0.225	−2.23	−0.667	−2.00	−2.00
D-1	−2.67	−0.051	−1.95	—	—	—
D-2	−2.67	0.049	−2.05	−0.104	−2.56	−2.00
D-3	−2.67	10.90	−21.90	−13.70	$\begin{bmatrix} 4 \cdot 10^{-4} \\ 9.63i \end{bmatrix}$	$\begin{bmatrix} 4 \cdot 10^{-4} \\ -9.63i \end{bmatrix}$
D-4	−2.67	12.40	−23.40	−14.00	$\begin{bmatrix} 0.147 \\ 10.5i \end{bmatrix}$	$\begin{bmatrix} 0.147 \\ -10.5i \end{bmatrix}$

real, two having negative signs and one positive. The corresponding eigenvectors of the linearized problem near C^0 define two directions of stability (approaching C^0) and one direction of instability (leaving C^0). Actually the unstable direction is parallel to one of the stable directions, both of which lie on the radial vector from C^0 to $X(t)$, projected on the plane $Z = 0$. The remaining (stable) direction extends along the line $(0, 0, Z)$, whereon the trajectories approach C^0 as $Z_0 e^{-b(t-t_0)}$. The most probable close approach to C^0 in the stochastic cases is along this stable direction, because the deterministic flow aids the approach (Ven'Tsel and Freidlin, 1970). However, as the phase point draws near to C^0, stable and unstable directions converge to a point, so the trajectories can depart from the neighborhood of C^0 on quite different paths. When noise is present, small perturbations often find paths leading into the opposite basin of attraction. The close correspondence between exit times and minima in $d0$ provides qualitative support that our argument is correct. A similar effect was found by A.S. with $r = 23$ and small stochastic forcing, and is also a property of the deterministic Lorenz attractor.

We turn now to the predictability of systems S-1 and S-2. For this purpose we computed the autocorrelations $\hat{r}_{xx}(0, T, \tau)$ using different lengths of record T. The respective plots are presented in Figs. 9 and 10 for $0 \leq \tau \leq 400$. Different values of T were chosen to demonstrate the effects of the exits on the predictability time

$$\tau_p \equiv \left[\frac{1}{\hat{r}_{xx}} \frac{\partial \hat{r}_{xx}}{\partial \tau} \right]^{-1} \tag{5.2}$$

that describes the decay time constant for autocorrelation. We calculate τ_p at the first lag. The leftmost curve ($T = 100$) in Fig. 9 decays rapidly

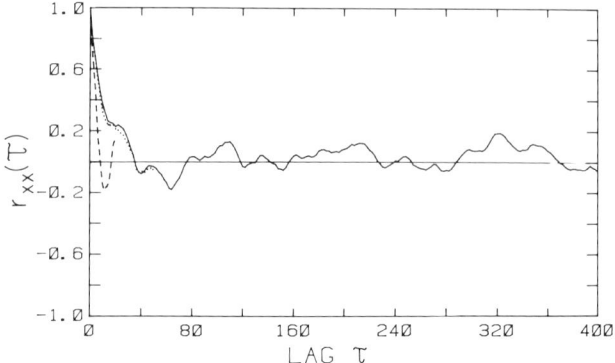

FIG. 9. Sample autocorrelations for $X(t)$ for case S-1, showing rapid loss of "memory" of initial values. $T = 100$ (----); 200 (·····); 1000 (——). $r = 0.9$, $\sigma = 1$.

($\tau_p = 5.2$) and shows a tendency to oscillate around zero at the larger lags. The latter effect is clearly associated with the initial 100 units of our noise realization (Fig. 4) and disappears when T is increased to 200. The curves for $T = 200$ and 1000 are very close over the initial 20 lag times, indicating that our calculated values τ_p (here centered on $\tau = 1$) should be insensitive to further increases in T. The falloff of the autocorrelation implies loss of predictability, and for S-1 this corresponds to rapid approach toward $\mathbf{C}^0$, after which time random fluctuations dominate the signal. The nonlinear system (3.11) is, in this case, not unlike the Langevin equation obtained by linearization around $\mathbf{C}^0$. It has been pointed out (Hasselman, 1976; North and Cahalan, 1981) that the best prediction for the latter type of equation is the trivial one, namely that any initial condition produces

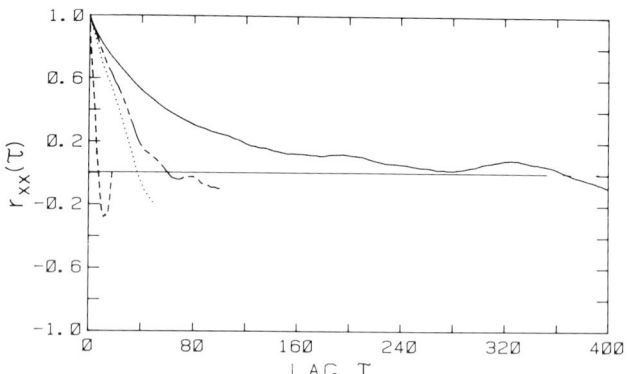

FIG. 10. Same as Fig. 9 only for case S-2. "Memory" decays less rapidly. $T = 100$ (----); 200 (·····); 500 (—·—·—); 7250 (——). $r = 1.1$, $\sigma = 1$.

trajectories approaching C^0 asymptotically, with persistence limited by the exponential decay time of the deterministic solution. When the solution approaches C^0, where the deterministic flow vanishes the noise is free to determine the subsequent states of the system, independent of the initial values. Predictive information attending the initial conditions is then lost.

The results for S-2 stand in sharp contrast to those shown in Fig. 9, for, although the autocorrelation is similar in the two cases with $T = 100$, it changes markedly as T increases to 7250. Our long realization S-2 has predictability times $\tau_p(0, 100) = 4.7$ and $\tau_p(0, 7250) = 41.3$, illustrating the crucial role played by the change of parameter from $r = 0.9$ to $r = 1.1$, while holding constant the noise realization, σ, and the initial values. Evidently the manifestation of multiple attractors (C^+ and C^-) changes fundamentally the predictability of the system, when the first exit occurs around $t = 105$ for S-2. Table III puts the role of the exit time in evidence, showing sample statistics calculated from different lengths of record. For $T = 100$, S-1 and S-2 possess quite similar variance and autocovariance. The sample means differ, of course, because the coordinates of the relevant attractors C^0 and C^+ differ ($X^0 = 0$; $X^+ = 0.516$). Trajectories tend to approach these stable equilibria, hence the sample means temporarily approximate to the values X^0 and X^+ in the respective cases. The departures of X from these sample means often change sign within very short time intervals, because the deterministic flow is zero on the equilibria and positive perturbations have the same probability as negative ones in each increment $dW(t)$. The fluctuations therefore reduce the autocovariance after only a few lag times. When the first exit occurs in S-2, a dramatic change in statistics occurs because the sample mean $\bar{X}$ shifts toward X^0 and the phase point subsequently persists in the domain of attraction of C^-. Now the covariance increases (Table III) because departures of $X(t)$ from X^0 have similar signs for periods on the order of the mean exit time.

TABLE III. STATISTICAL PARAMETERS DERIVED FROM NUMERICAL SOLUTIONS $X(t)$

T	$\bar{X}(0, T)$	$\hat{\sigma}_x^2(0, T)$	$\hat{C}_{xx}(0, T, 2.5)$	$\hat{r}_{xx}(0, T, 2.5)$	τ_p
		Stochastic case S-1			
100	0.1292	0.0583	0.0386	0.6624	5.22
200	0.0857	0.0748	0.0581	0.7773	7.94
1000	−0.0510	0.0655	0.0547	0.7744	9.25
		Stochastic case S-2			
100	0.4763	0.0536	0.0342	0.6394	4.71
200	0.2155	0.2265	0.2082	0.9190	23.68
500	0.0363	0.2415	0.2246	0.9302	26.92
7250	−0.0211	0.2385	0.2244	0.9413	41.30

Correlations between data separated by relatively large time intervals are systematically greater than for S-1 at similar lags. At lag times long relative to the mean exit time, the autocorrelation ultimately approaches zero, as it must in view of the fact that τ_e is itself a random variable whose variance is the same order of magnitude as its mean (Table I). The spectrum of $X(t)$ for case S-2 is shown in Fig. 11. Here the shape reminds one of the familiar "red noise" spectrum, wherein persistence concentrates the power at the lowest frequencies. We note the absence of a spectral peak at the frequency $\omega_e = 2\pi/\bar{\tau}_e$, corresponding to the mean exit time, which is to be expected when the variance $\hat{\sigma}_{\tau_e}^2$ is large (Benzi et al., 1981).

In case S-2 with two stable equilibria, initial values $\mathbf{X}^*$ convey two useful pieces of predictive information. First, the domain of attraction is determined by $\mathbf{X}^*$, knowledge of which is sufficient to make a prediction about the statistics over time intervals commensurate with $\bar{\tau}_e$. During such time intervals, we expect that the sample mean, for example, will remain close to the relevant attractor, in the sense that variations will be small compared to the variance of the total process, reaching all the attractors. Furthermore, we can predict an expectation time for transition to the other domain of attraction. These predictions should be useful for time intervals that scale with $\bar{\tau}_e$, but we must remember that the large value of $\hat{\sigma}_{\tau_e}^2$ limits their utility in specific realizations. Secondly, the position of $\mathbf{X}^*$ relative to the nearest attractor provides information about the future states $\mathbf{X}(t)$, rather than $\bar{X}$, but this information decays much more rapidly than τ_p because of the noise.

The numerical results presented above indicate clearly the role of multiple, stable equilibria in producing almost-intransitivity when stochastic forcing is applied to the system. Moreover, the predictive information conveyed by initial conditions persists, in a statistical sense, well beyond the predictability time appropriate for departures from a *single* equilib-

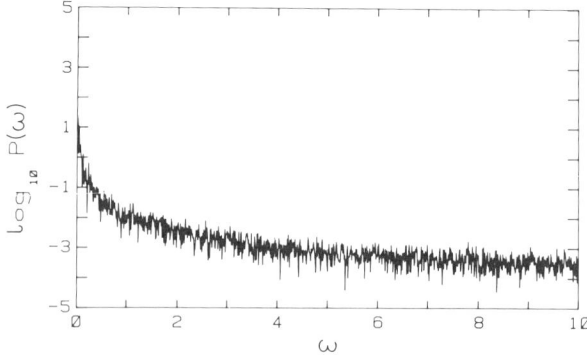

FIG. 11. Power spectrum of $X(t)$ for case S-2. $r = 1.1$, $\sigma = 1$, $0 < t < 2100$.

rium. It appears that the mean exit time $\bar{\tau}_e$ plays a major role in determining the asymptotic ($T \to \infty$) predictability, because it characterizes the time interval on which sample statistics are likely to describe conditions in the domain of attraction containing $\mathbf{X}^*$. Recall that our goals are to determine the convective states occupied with maximum likelihood and the characteristic time required for transition between states. Our results indicate that this fundamental predictability information is determined by the structure of the attractor sets (and the noise) without specific reference to particular initial conditions or the sensitivity of the solutions thereto.

So far we have considered cases corresponding to the bifurcation sequence of Fig. 1b ($\sigma < b + 1$). An alternative scheme is depicted in Fig. 1a, where the Lorenz attractor is realized for $r > r_c$. Trajectories in this regime have been thoroughly analyzed (numerically) by others, and here we confine our discussion to a rough analogy between case D-3 (Table I) and the stochastic case S-2.

Figure 12 shows the signal $X(t)$ on $200 \leq t \leq 400$. The details bear scant resemblance to, say, Fig. 5 (S-2), except that the solution is temporarily "confined" on a given side of the plane $X = 0$, followed by a sharp transition to the other side, and so on. During this confinement, the variables undergo oscillations around $\mathbf{C}^+$ or $\mathbf{C}^-$, with periods corresponding to the imaginary part of the eigenvalues on these equilibria (Table II). That these oscillations enclose the steady states is easily seen from the (X, Y) projection of a typical trajectory (Fig. 13).

We may describe the signal $X(t)$ as almost intransitive if we consider, for example, that sample statistics calculated from segment A (Fig. 12) differ markedly from those on segment B, due to the nonuniformity of the

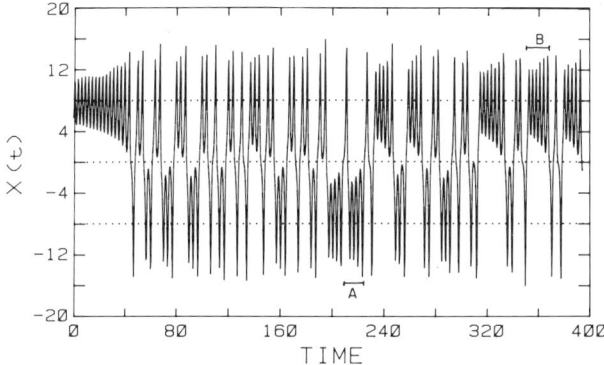

FIG. 12. Numerical solution $X(t)$ for case D-3. Unstable steady states denoted by dotted lines. A and B indicate temporary confinement on opposite sides of $X = 0$. $r = 24.75$, $\sigma = 10$.

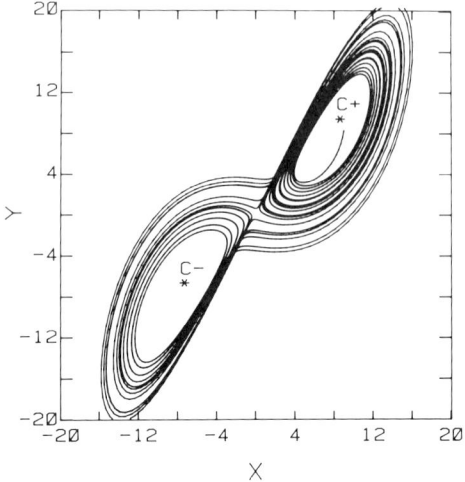

FIG. 13. Numerical solution in $X-Y$ phase plane for case D-3. Oscillations around C^+ and C^- are evident. $r = 24.75$, $\sigma = 10$.

attractor set in phase space. The analog of exit from a domain of attraction is the "flip" across the plane $X = 0$, representing reversal of the convective flow in the fluid layer.

Our solution was initialized near C^+ and executed oscillations about this point until the first crossing of $X = 0$ at $t = 214$. The power spectrum of $X(t)$ on this time interval (Fig. 14) exhibits a dominant spectral peak at the eigenvalue of C^+ (and C^-), and harmonics thereof. The peaks disappear when the interval $214 \leq t \leq 614$ is considered, despite the obvious

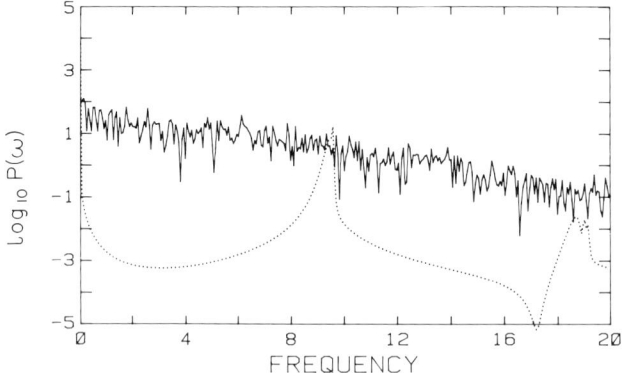

FIG. 14. Power spectra for $X(t)$ for case D-3. Dotted curve ($0 < t < 205$) shows oscillations prior to first "flip;" solid curve ($214 < t < 614$) shows lack of spectral peaks when flipping occurs. $r = 24.75$, $\sigma = 10$.

importance of the basic frequency over short time intervals (Fig. 12). $X(t)$ resembles to some degree a sequence of regular oscillations imposed on a "square wave." However, the "period" of the square wave corresponds to the flipping time across $X = 0$, and the latter quantity occurs in a nonperiodic sequence. This last assertion follows from Lorenz' (1963) conclusion that the solutions in this regime are nonperiodic. Therefore, the disappearance of the spectral peaks is understandable, as well as the absence of peaks at the flipping frequency $\omega_f = 2\pi/\bar{\tau}_e$. The unpredictability of a nonperiodic system is confirmed by inspection of the autocorrelation function (Fig. 15). Here the correlation decays very steeply with increasing lag time, giving the short predictability time $\tau_p = 0.448$.

S-2 and D-3 are analogous in that both exhibit a tendency to be confined in limited regions of the phase space until an exit (or flip) occurs, i.e., both processes are almost intransitive. Furthermore, the decay times for predictability seem to scale with the exit time in both cases. Figure 16 shows this result, where we plot τ_p versus $\bar{\tau}_e$ for the cases S-2, S-3, S-4, D-3, and D-4.

Lorenz (1963) described an empirical technique that might appear, at first sight, to allow prediction of the sequence of flipping times in case D-3. One plots each local maximum M_n of $Z(t)$ against the succeeding value M_{n+1} to obtain a smooth, apparently single-valued function (Fig. 17). As Lorenz notes, in principle one could begin with an initial value M_0 and iterate the mapping to obtain the future maxima in sequence. There is no reference to time in this, only to the number of iterations. However, we recall that $Z(t)$ oscillates around $\mathbf{C}^+$ and $\mathbf{C}^-$ (between flips) with a frequency closely approximated by the imaginary part of the eigenvalues at these points. Denoting the corresponding period by T_0, we can calculate

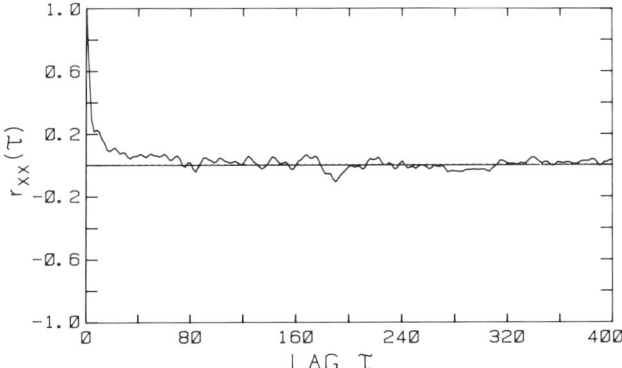

FIG. 15. Sample autocorrelation for $X(t)$ for case D-3 in "flipping" regime. Rapid loss of "memory" is indicated. $r = 24.75$, $\sigma = 10$, $214 < t < 614$.

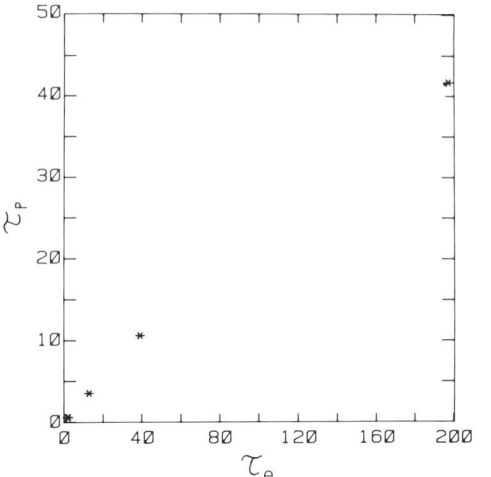

FIG. 16. Numerical values of exit time versus predictability time. D-4 and D-3 plot over one another at lower left.

the time interval for the Nth maximum following M_0 as $t = NT_0$. Furthermore, Lorenz (1980) points out that each time the mapping point traverses beneath the cusp in Fig. 17, $X(t)$ flips from near $\mathbf{C}^+$ to near $\mathbf{C}^-$, or *vice versa*. Therefore the initial value M_0, a curve fitted to the points in Fig. 17, and the eigenvalues of the steady states are, in principle, sufficient information to formulate a prediction scheme, giving the complete sequence of

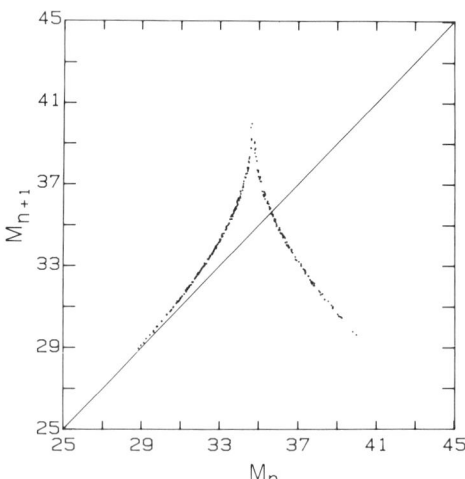

FIG. 17. Numerically determined values of successive maxima in $Z(t)$ for case D-3. $r = 24.75$, $\sigma = 10$.

flipping times. However, Lorenz (1963) used an idealized representation of Fig. 17, comprising two straight lines, intersecting at the cusp and having slopes of 2 and -2, respectively, to show that all periodic mapping sequences are unstable. Furthermore, his results indicated that any uncertainty in M_0 would grow in proportion to the product of the slopes of the mapping curve at successive iterations. It is also possible to illustrate the unpredictability and the instability of periodic sequences by simply constructing iterations of the mapping. For this purpose, we shall use the empirical curve $M_{n+1} = \Lambda(M_n)$ fitted by Yorke and Yorke (1979) to the aperiodic set of trajectories with $r = 24$, $\sigma = 10$, $b = \frac{8}{3}$. The function Λ is given by

$$(5.3) \quad \Lambda(M_n) = d_1 - d_2|M_n - M^*|^{d_3}[1 + d_4|M_n - M^*| - d_5|M_n - M^*|^2]$$

where $M^* = 33.795$, $d_1 = 40.724$, $d_2 = 5.936$, $d_3 = 0.285$, $d_4 = 0.0571$, and $d_5 = 0.00106$ are the fitted parameters. Equation (5.3) was used to generate the first, second, and fifth iterations of the mapping, shown in Figs. 18 and 19. The intersection of the mapping with the diagonal shows periodic solutions, where $M_N = M_0$. In this case the "period" is N. The first iteration has one periodic point, while the second iteration possesses two. Each successive application of the mapping produces two cusps from one and their associated intersections with the diagonal, indicating that 2^{N-1} periodic points are present with period N. Notice that the function begins to approach a set of near-vertical, closely spaced spikes, whose maxima and minima have almost identical abcissae (M_0). This fact

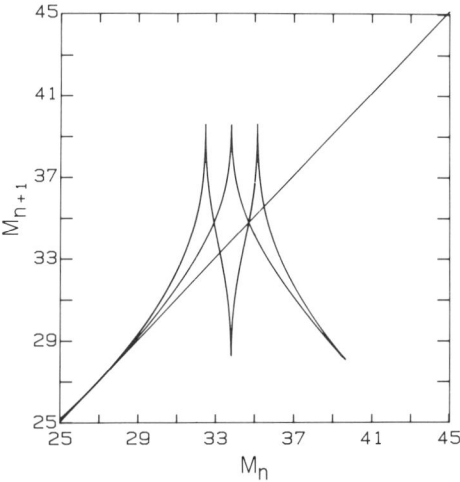

FIG. 18. First and second iterations of Eq. (5.3), showing doubling of cusps and periodic points.

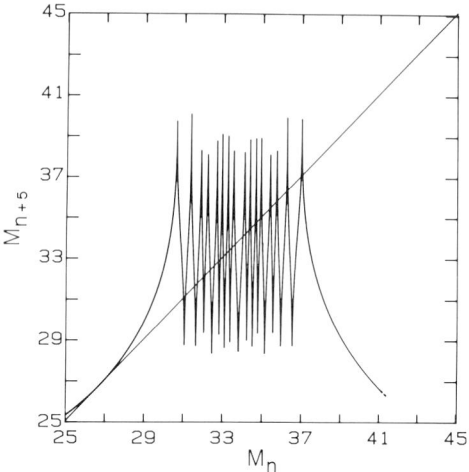

FIG. 19. Same as Fig. 18, except for the fifth iteration.

illustrates the instability of the mapping, in that two initial conditions, no matter how close initially, eventually diverge from one another to a distance equivalent to the entire range of the diagram. Furthermore, two points with very different abcissae will eventually have nearly identical ordinates. Thus we have a graphical illustration of the nature of a limit set. Clearly, it would be of some advantage to obtain predictability information that is, in some way, independent of the initial values. Let us remark at this point that the mapping shown in Fig. 17 represents the indicator of the Lorenz attractor, in that it provides the evidence that periodic solutions are everywhere unstable. This point is discussed again in the Appendix.

6. Concluding Remarks

In this work we have presented numerical evidence that the exit time is an important parameter for describing the almost-intransitivity of a system with multiple stable equilibria, driven by stochastic perturbations. Also, the flipping time seems to play a similar role with respect to the autocorrelation of the Lorenz attractor. In terms of predictability, knowledge of these parameters would enhance the ability to predict statistics of the system over relatively long time intervals. It remains to investigate the possibility of *a priori* calculation of this parameter. Beginning with the stochastic case, two equations are available:

$$\tfrac{1}{2}\epsilon\nabla^2\bar{\tau}_e + \mathbf{b}\cdot\boldsymbol{\nabla}(\bar{\tau}_e) = -1 \qquad (6.1)$$

(6.2) $$\bar{\tau}_e(\partial \mathbf{A}) = 0$$

where $\mathbf{A}$ is the boundary of any neighborhood of the initial data. Similarly, the standard deviation of τ_e follows as the solution of

(6.3) $$\tfrac{1}{2}\epsilon \nabla^2 M_2 + \mathbf{b} \cdot \boldsymbol{\nabla}(M_2) = -2\bar{\tau}_e$$

where M_2 is the second moment of τ_e, subject to the same boundary condition as the mean exit time (Gihman and Skorohod, 1972; Schuss, 1980). Here the gradient operators apply to the state variable $\mathbf{X}$, i.e., $\boldsymbol{\nabla} = \mathbf{i}\, \partial/\partial X_1 + \mathbf{j}\, \partial/\partial X_2 + \mathbf{k}\, \partial/\partial X_3$ and $\mathbf{b}(\mathbf{X})$ is, of course, a known coefficient. These equations do not seem unreasonable, even for the Lorenz model, and a solution may perhaps be found. To date, no corresponding techniques exist for studying the statistics of deterministic trajectories on the strange attractor that determine the flipping time. It is suggested, then, that some practical advantages of computation may attend the probabilistic approach.

Let us summarize our arguments, in light of the results above. The atmosphere does not settle down to a steady state, nor does it appear to be periodic. Our inability to deal with infinite-dimensional, nonlinear equations requires that we model the fluctuating atmosphere with truncated equations, wherein certain scales of motion and physical processes are neglected. Typically, the deterministic components of such models possess multiple domains of attraction, at least in cases where the structure of the phase space can be investigated. Important physical balances are therefore represented by the attractors, whose stability typically depends on certain model parameters. Stochastic perturbations are surely present if we identify any of these models with the real atmosphere, but they are not necessarily relevant for simulating fluctuations if all the equilibria are unstable (Fig. 1a). Alternatively, large fluctuations and persistent regimes may occur if the system is subject to such perturbations, in the presence of multiple attractors (Fig. 1b). A certain degree of similarity can exist between the two cases, as we noted above, illustrating the difficulties encountered in trying to decide the question from atmospheric data alone. A crucial objective would seem to be the determination of major bifurcation sequences in more realistic models. It seems well established that certain particular steady solutions are baroclinically unstable, but might not these be analogous to $\mathbf{C}^0$ in the Lorenz model, with stable solutions appearing above the bifurcation? If it is found that such bifurcations are relevant to atmospheric prediction, then it will be of great advantage to theoretical meteorology, because there will be a possibility for *a priori* calculations of analytical parameters such as the exit time, in terms of control parameters and noise levels that should be estimable from atmospheric data. We close with the remark that further investigations, using stochastic techniques applied to more realistic models of atmospheric flow should be illuminating.

Appendix

In this appendix we shall discuss another example of the effect of stochastic perturbations on deterministic models of atmospheric flow. This topic, closely related to that of the main text, is motivated by the latest article of Lorenz (1980). Let us digress briefly and attempt to summarize the main points.

Lorenz studied the structure of attractor sets for a truncation of the shallow-water equations [and therefore a prototype of primitive equation (PE) models on which general circulation models are based]. This system possessed nine components, namely: three spatial modes for each of the variables representing horizontal divergence, vorticity, and geopotential height. At the lower boundary a fixed, single-mode topography was specified, and zonally uniform thermal forcing was simulated by a mass source, proportional to the parameter F_1. We shall refer to this model, which we do not reproduce here, as C, and the reader may find details in Lorenz' article.

In order to investigate the nature of quasi-geostrophic (QG) equilibrium, Lorenz derived a QG approximation to C that turns out to be mathematically identical to the model described in Eqs. (3.11) of the present work. Lorenz refers to this model as D, in recognition of its close resemblance to the simple, three-mode model of Benard convection, and we will follow this convention. The dynamical system (3.11) of the main text can be identified with D, provided that the following parameters are chosen

$$\sigma = 3$$
$$b = 1$$

The Rayleigh number r is linearly related to F_1, so the latter quantity plays the role of bifurcation parameter in C and D.

Lorenz chose F_1 such that the general solution to C was aperiodic and the attractor was strange. The limit set for these PE trajectories was found to consist mainly of states near geostrophic balance, and was more accurately characterized as being completely free of gravity waves. Subsequent comparisons were made between the attractor sets of C and D, based on the similarity of shape between the curves relating successive maxima in $Z(t)$ (see Section 5, main text). After all, this mapping is the only indicator of the aperiodic nature of the trajectories and, hence, the strangeness of the attractor. A difficulty arose in the comparison between C and D, related to the parameter F_1. In D, the critical value of r [Eq. (3.13)] corresponds to $F_1 = 0.10785$ in Lorenz' article. Above this value all of the equilibria are unstable and numerical results indicated the presence of a strange attractor. By contrast, the strange attractor in C appears in the range $0.056 < F_1 \leq 0.10$, larger values producing stable periodic solutions. Of course, it is impossible to be certain, without exhausting all possible initial conditions, that values of $F_1 > 0.10$ exclude strange attractors, because their domains of attraction would necessarily be disjoint from that of the periodic solutions, found numerically by Lorenz. For our purpose, it is sufficient to assume that there is no strange attractor present when $F_1 > 0.10$ in model C. Lorenz argues that it is not unreasonable to adjust F_1 in order to make the comparison between C and D. It is our intention here to offer a different, but suggestive, approach, wherein the adjustment of F_1 may be unnecessary.

For this purpose, we performed a numerical simulation, as described in Section 4, with the parameters $\sigma = 10$, $b = \frac{8}{3}$, $r = 23.5$, $\epsilon^{1/2} = 0.04$, $\gamma = 0.32$. The deterministic behavior here is qualitatively identical to model D in the subcritical range, i.e. $F_1 = 0.10$. The total variance of the resulting stochastic processes $X(t)$ and $Y(t)$ exceeded 50 dimensionless units squared, indicating that our choice for γ is a modest perturbation. We initialized our sample trajectory in the domain of attraction of $\mathbf{C}^+$ and integrated forward in time through 1000 dimensionless units.

The time series $X(t)$ is presented in Fig. 20. The main character of the strange attractor described by Lorenz is established after the stochastic perturbation has acted for a sufficiently long time (roughly 160 units). After leaving the domain of attraction of $\mathbf{C}^+$, the sample trajectory encounters a strange set on which the characteristic flipping motion is sustained over time intervals comparable to those in which the solution oscillates near one or another stable equilibrium. Remarkably, the flipping time is very much shorter than the exit time. It also appears that there coexist at least two statistical regimes, dominated by the stochastic perturbations and the strange set, respectively. This point is more clearly illustrated by a projection of a segment of the trajectory on some plane $Z = $ constant (Fig. 21). The strange attractor behavior appears to be qualitatively similar to that found by Lorenz. On the other hand, the two prominent "holes" appearing around $\mathbf{C}^+$ and $\mathbf{C}^-$ in Lorenz' graphs are filled by our sample orbit, indicating the presence of a second statistical regime. Finally, we show in Fig. 22 the plot of successive maxima M_{n+1} versus M_n. The characteristic shape, indicating instability of periodic sequences, is preserved qualitatively, but now has a distribution about the smooth

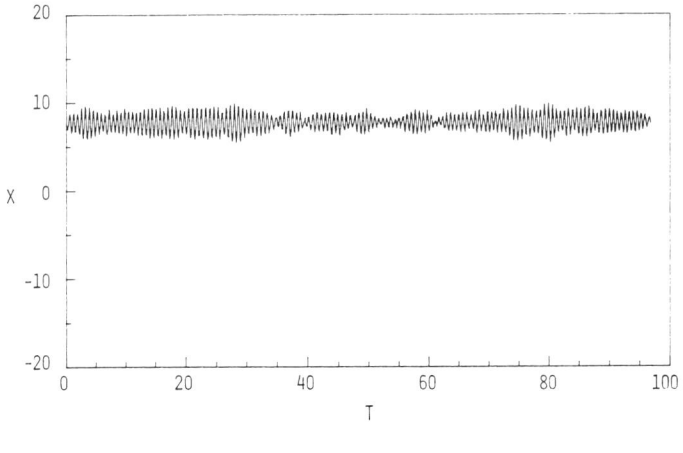

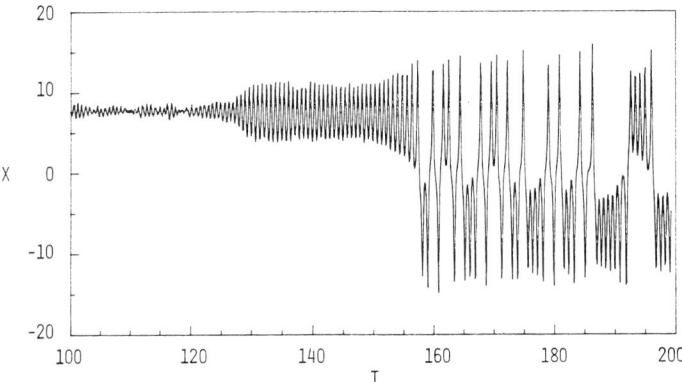

FIG. 20. Numerical simulation of process $X(t)$ for $r = 23.5$, $\sigma = 10$, $b = \frac{8}{3}$, and $\gamma = 0.32$.

curve due to the noise. Clearly the stochastic forcing enlarges the set of initial conditions whose trajectories encounter the strange set, when Eqs. (3.11) are perturbed to form a diffusion process.

The phenomena observed in our simulation and in D can be explained, topologically speaking, with reference to certain known properties of D. It is not our intention here to produce an essay on model D, but only to describe the gross properties of bifurcations on r (F_1). The reader can satisfy his desire for rigor elsewhere (e.g., Afraimovich *et al.*, 1977).

In the main text (Section 3) we considered that the domain of attraction of a stable, steady state, say x, is a well-defined entity. Furthermore, it is always possible to choose a spherical phase volume surrounding x

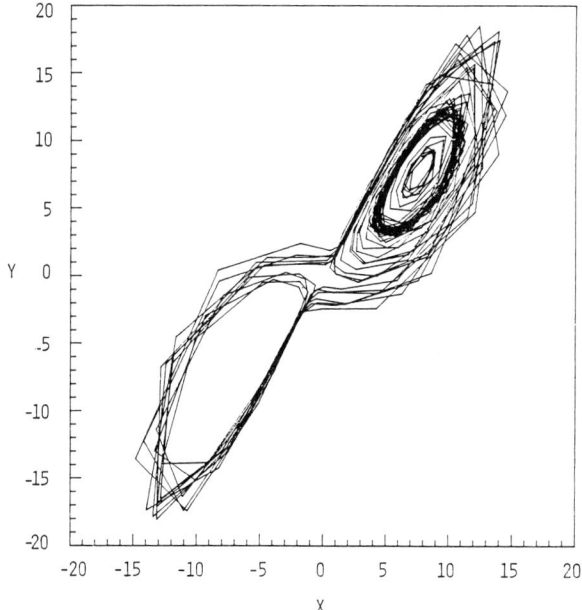

FIG. 21. X–Y phase plane projection; same parameters as in Fig. 20.

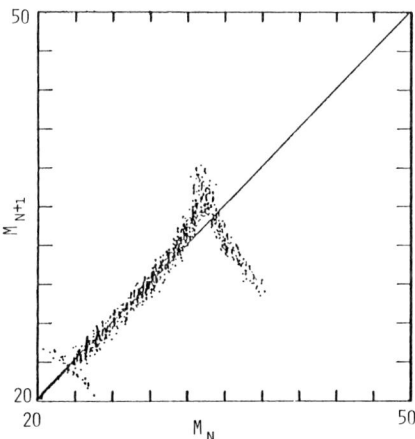

FIG. 22. Numerically determined values of successive maxima in $Z(t)$; same parameters as in Fig. 20.

wherein the trajectories are trapped, approaching x as $t \to \infty$. On the other hand, the definition of domain of attraction does not impose any constraint on the geometric complexity of the set, stating only that the set must occupy phase volume. In other words, the domain of attraction can possess any structure it likes, regardless of our desire for simplicity. In particular, it can be the intersection of a smooth manifold (see Leith, 1980) of nonzero phase volume with a Cantor set (see Lorenz, 1980). If we consider model D with subcritical r, recent numerical studies indicate that deterministic trajectories can exhibit the chaotic flipping motion for a finite time, after which the final decay toward $\mathbf{C}^+$ or $\mathbf{C}^-$ occurs (Yorke and Yorke, 1979). This behavior has been interpreted as the extension below the subcritical Hopf bifurcation, at $r = r_c$, of the regime where in the aperiodic mapping M_{n+1} versus M_n exists. Trajectories of this type are obtained if initial conditions are selected from a particular set, whose phase volume decreases with r, eventually collapsing to a neighborhood of $\mathbf{C}^0$ when $r \simeq 13.926$ (with $\sigma = 10$, $b = \frac{8}{3}$). From these considerations, it appears that initial conditions outside this set lead directly to the stable equilibria, while within this set one sees transient excursions near the strange set, followed by eventual approach to equilibrium. These, then, are the solutions found in D with $F_1 = 0.10$ (Lorenz, 1980, Fig. 10, thin curve). Further, it seems likely that, at the very least, the attraction domains of $\mathbf{C}^+$ and $\mathbf{C}^-$ are, in the subcritical regime, folded around one another near the strange set itself. If this is true then, precisely on the strange set, the phase point cannot decide whether to approach $\mathbf{C}^+$ or $\mathbf{C}^-$, so it does neither, and stays on the strange set. In practice, it would leave this set, due to round-off error in the integration. However, our results indicate that it leaves quite slowly, and, with stochastic perturbations, this unstable manifold exhibits a kind of attractiveness, leading to a third probability maximum in the phase space. Therefore, the union of the attraction domains in the deterministic case would seem to be the entire phase space minus an infinite set of surfaces with Cantor structure, although we cannot claim to have proven this. Application of the noise removes this complication, as demonstrated by our simulation results, leaving only a single attractor with a very rich, probabilistic, internal structure. Furthermore, the noise removes the discontinuity at the critical value of $r(F_1)$ in the sense that either of models C or D should exhibit the aperiodic flipping motion at a single value of forcing, given small stochastic perturbations.

The previous results lead us to speculate on the nature of trajectories in more realistic models of atmospheric flow. The enlargment of the parameter range for which strange and stochastic behavior mix the initial conditions renders the system unpredictable in the usual, deterministic sense.

This is not merely a computational exercise, but also leads us to consider carefully the nature of the perturbations on which the predictability depends. Loss of memory by the system of its initial conditions does not impose the ultimate limit to atmospheric predictability, but lack of knowledge of the internal physics of the atmospheric system (both deterministic and stochastic components) can comprise such a constraint. This poses a question about the acceptability of the autocorrelation decay time as a predictability measure. It may be more productive to better investigate the probability distributions of variables in the phase space, identifying those regions where basic physical balances produce maxima. Then the distributions of temporal parameters, such as flipping time or exit time, provide the basis for probabilistic prediction, closely tied to our knowledge of the physical system.

In closing, we should like to quote from the final section of Eady's famous article on cyclone waves, because it advocates the probabilistic approach with a clarity and style that we are unable to better.

> "Although we cannot, with complete certainty, say anything about long-term developments . . . it does not follow that all possible developments are equally probable. On the contrary, we may infer that probabilities are very unequally distributed (and therefore that information of this kind may be, from a practical point of view, almost as good as 'guaranteed' information).
>
> . . . the rather formidable task facing theoretical meteorology (is) that of discovering the nature of and determining quantitatively all the forecastable regularities of a 'permanently unstable' (i.e., permanently turbulent) system. We can be certain that these regularities are necessarily statistical and to this extent our technique must resemble statistical mechanics" (Eady, 1949, p. 52).

Acknowledgments

We wish to thank B. Saltzman, G. D. Robinson, J. Pandolfo, and J. Dutton for their help in improving the manuscript. We are grateful for useful discussions with R. Benzi, E. N. Lorenz, and M. Ghil. This research was supported by the Division of Atmospheric Sciences, National Science Foundation under grant ATM-7925013 at Yale University, and grant ATM-7918875 at the Center for the Environment and Man, Inc. The first author is grateful for support from the Polar Science Center, during final preparation of the manuscript.

References

Abraham, R. (1967). "Foundations of Mechanics." Benjamin, New York.
Afraimovich, V. S., Bykov, V. V., and Shil'nikov, L. P. (1977). *Sov. Phys.—Dokl. (Engl. Transl.)* **22**, 253.
Benzi, R., Parisi, G., Sutera, A., and Vulpiani, A. (1981). Stochastic resonance in climatic change. *Tellus* (in press).
Busse, F. H. (1972). The oscillatory instability of convection rolls in a low prandtl number fluid. *J. Fluid Mech.* **52**, 97–112.

Charney, J. G., and Devore, J. G. (1979). Multiple flow equilibria in the atmosphere and blocking. *J. Atmos. Sci.* **36**, 1205–1216.

Charney, J. G., and Straus, D. M. (1980). Form-drag instability, multiple equilibria and propagating planetary waves in baroclinic, orographically forced, planetary wave systems. *J. Atmos. Sci.* **37**, 1157–1176.

Charney, J. G., Shukla, J., and Mo, K. C. (1981). Comparison of a barotropic blocking theory with observation. *J. Atmos. Sci.* **38**, 762–779.

Doob, J. L. (1952). "Stochastic Processes." Wiley, New York.

Eady, E. T. (1949). Long waves and cyclone waves. *Tellus* **1**, 33–52.

Epstein, E. S. (1969). Stochastic dynamic prediction. *Tellus* **21**, 739–759.

Fleming, R. J. (1971). On stochastic dynamic prediction. I. The engergetics of uncertainty and the question of closure. *Mon. Weather Rev.* **99**, 851–872.

Gihman, I. I., and Skorohod, A. V. (1972). "Stochastic Differential Equations." Springer-Verlag, Berlin and New York.

Gleeson, T. A. (1970). Statistical-dynamical predictions. *J. Appl. Meteorol.* **9**, 333–344.

Halmos, P. R. (1956). "Lectures on Ergodic Theory." Chelsea Publ. Co., New York.

Haltiner, G. J. and Williams, R. T. (1980). "Numerical Prediction and Dynamic Meteorology." Wiley, New York.

Hasselman, K. (1976). Stochastic climate models. Part 1. The theory. *Tellus* **18**, 473–484.

Haugen, D. A. (1973). "Workshop on Micrometeorology." Am. Meteorol. Soc., Boston, Massachusetts.

Ince, E. L. (1956). "Ordinary Differential Equations." Dover, New York.

Jenkins, G. M., and Watts, D. G. (1968). "Spectral Analysis and its Applications." Holden-Day, San Francisco, California.

Leith, C. E. (1975). Climate response and fluctuation dissipation. *J. Atmos. Sci.* **32**, 2022–2026.

Leith, C. E. (1980). Nonlinear normal mode initialization and quasi-geostrophic theory. *J. Atmos. Sci.* **37**, 958–968.

Leith, C. E., and Kraichnan, R. H. (1972). Predictability of turbulent flows. *J. Atmos. Sci.* **29**, 1041–1058.

Lorenz, E. N. (1960). Maximum simplification of the dynamic equations. *Tellus* **12**, 243–254.

Lorenz, E. N. (1963). Deterministic, nonperiodic flow. *J. Atmos. Sci.* **20**, 130–141.

Lorenz, E. N. (1967). Climatic determinism. *Meteorol. Monogr.* **8**, 1–3.

Lorenz, E. N. (1969). The predictability of a flow which possesses many scales of motion. *Tellus* **21**, 289–307.

Lorenz, E. N. (1973). On the existence of extended range predictability. *J. Appl. Meteorol.* **12**, 543–546.

Lorenz, E. N. (1980). Attractor sets and quasi-geostrophic equilibrium. *J. Atmos. Sci.* **37**, 1685–1699.

Manneville, P., and Pomeau, Y. (1980). Different ways to turbulence in dissipative dynamical systems. *Physica D (Amsterdam)* **1**, 219–226.

Marsden, J. E., and McCracken, M. (1976). "The Hopf Bifurcation and its Applications." Springer-Verlag, Berlin and New York.

Monin, A. S., and Yaglom, A. M. (1971). "Statistical Fluid Mechanics." MIT Press, Cambridge, Massachusetts.

North, G. R., and Cahalan, R. F. (1981). Predictability in a solvable climate model. *J. Atmos. Sci.* **38**, 504–513.

Pitcher, E. J. (1977). Application of stochastic dynamic prediction to real data. *J. Atmos. Sci.* **34**, 3–21.

Rabinovich, M. I. (1978). Stochastic self oscillations and turbulence. *Sov. Phys.—Usp. (Engl. Transl.)* **21,** 443–468.

Rayleigh, Lord (1916). On convective currents in a horizontal layer of fluid when the higher temperature is on the under side. *Philos. Mag.* [6] **32,** 529–546.

Robinson, G. D. (1967). Some current projects for global meteorological observation and experiment. *Q. J. R. Meteorol. Soc.* **93,** 409–418.

Saltzman, B. (1962). Finite amplitude free convection as an initial value problem-I. *J. Atmos. Sci.* **19,** 329–341.

Schuss, Z. (1980). Singular perturbation method in stochastic differential equations of mathematical physics. *SIAM Rev.* **22,** 119–155.

Shirer, H. N. and Dutton, J. A. (1979). The branching hierarchy of multiple solutions in a model of moist convection. *J. Atmos. Sci.* **36,** 1705–1721.

Sutera, A. (1980). Stochastic perturbation of a pure convective motion. *J. Atmos. Sci.* **37,** 245–249.

Thompson, P. D. (1957). Uncertainty of initial state as a factor in the predictability of large scale atmospheric flow patterns. *Tellus* **9,** 275–295.

Ven'Tsel, G. A., and Freidlin, M. I. (1970). On small random perturbation of dynamic system. *Russ. Math. Surv. (Engl. Transl.)* **25,** 1–55.

Vickroy, J. G., and Dutton, J. A. (1979). Bifurcation and catastrophe in a simple, forced, dissipative quasi-geostrophic flow. *J. Atmos. Sci.* **36,** 42–52.

Yorke, J. A., and Yorke, A. D. (1979). Metastable chaos: The transition to sustained chaotic behavior in the Lorenz model. *J. Stat. Phys.* **21,** 263–277.

INDEX

A

Aerosol materials, conservation of, in mesoscale atmospheric models, 187–189, 191
Air pollution, 307, 310–312
 deposition, 229
 and radiation, 244, 247
Albedo, 242–245
Almost-intransitivity, 357–359, 364, 368, 374
Alternating-field demagnetization, 48–53, 58–60, 62, 66–67, 70
Anelastic approximation to conservation of mass relation, 190
Anhysteretic remanent magnetization (ARM), 50, 63–68
Archean times, global spreading, 5–6, 11–14, 19
[40]Argon degassing calculations, in construction of Earth's thermal history, 4–6, 15, 19
ARM, *see* Anhysteretic remanent magnetization
Atlantic-type continental shelf
 steady currents, 139
 storm currents over, 166–170
Atmosphere
 mesoscale models, 186, 278–280
  conservation relations, 187–204
  moist mesoscale systems, 248–253
  planetary boundary layer, 226–239
  synoptically-induced systems, 290, 313–317
  terrain-induced systems, 290–313
 predictability of atmospheric system, 345–383
  computational techniques, 359–360
  numerical results, 360–374
  prototype model and mathematical concepts, 349–358
 radiation, 239–248
Attractor, in atmospheric predictability problem, 351, 357–358, 368–369, 376–377

B

Ballistic magnetometer, 41–43
Basalt, lunar, 65
Bathystrophic tide, 167
Birkhoff individual ergodic theorem, 358
Blackbody, 239
Blocking of airstream, 301
Blocking temperature, 84–85
Boundary conditions, in mesoscale modeling, 272, 276–285, 302–303; *see also* Planetary boundary layer
Boundary-layer problem, continental shelf circulation as, 149–155
Box model, 307
Bulk aerodynamic formulation, 232

C

Camille, Hurricane, 316–317
Chemical demagnetization, 54–55
Chemical remanent magnetization (CRM), 28–29, 64–66, 94
Chinook effect, 303
Clouds, mesoscale modeling, 313–314
 in moist mesoscale system, 248–253
 and radiation, 240, 244, 247
Coastal jet, 120–127
Coastal ocean, circulation in, 101–183
 steady currents, 139–176
  frictional adjustment, 141–143
  parallel flow over straight shelf, 147–149
  shelf circulation as boundary-layer problem, 149
  stratified fluid, 159–162
  thermohaline circulation, 155–159
  vorticity tendencies, 143–147

385

surface fronts, 130–133
trapped waves and propagating fronts, 126–139
upwelling, downwelling, and coastal jets, 119–126
wind-driven transient currents, 103–118
Complete results magnetometer, 39–40
Computer, on-line signal processing in rock magnetism studies, 95
Conservation relations, for mesoscale atmospheric models, 187–204
Continental drift, rate of, 5
Continental shelf, 102; *see also* Oregon shelf
circulation as boundary-layer problem, 149–155
Pacific type, longshore wind impulses, 115–118
steady parallel flow over, 147–149
storm currents over, 166–170
Convection of heat, in the Earth, 4–5
kinematic convective models, 5–6, 19
parameterized convection, 6–11, 13, 19
Core of the Earth, formation of, 4, 14–19
mechanism of, 15–18
timing of, 15
Coriolis effect, in airflow over rough terrain, 302
Coriolis force, in coastal ocean circulation, 104–107, 110, 116, 121, 156
CRM, *see* Chemical remanent magnetization
Crust of the Earth
Archean, 14
formation of, 2
Cryogenic magnetometer, 33–36, 93–95
Curie temperature, 28, 84–85, 90
Cyclone, mesoscale modeling, 316–317
Cyclonic mean flow, in coastal oceans, 172–173

D

Deep continuity equation, 190
Deposition velocity, of planetary boundary layer, 229
Depositional remanent magnetization (DRM), 28–29, 64–66
Diagnostic model, 273, 304–305
Diastrophism, in mesoscale modeling, 276
Dioxin, 307

Direct-field demagnetization, 55
Discrete-level representation in mesoscale modeling, 269
Downwelling, in coastal ocean, 119–125
Drag coefficient, in planetary boundary layer model, 232
DRM, *see* Depositional remanent magnetization

E

Earth, *see also* Core of the Earth; Crust of the Earth; Mantle of the Earth; Rotation of the Earth
degassing, 15, 17–19
thermal evolution, 1–23
core formation, 14–19
global heat flow estimates, 2
models, 3–14
and viscosity, 4
Earthquake, and seismomagnetic effect, 89–90
Ekman layer (transition layer)
of atmospheric boundary layer, 231–239
in coastal ocean, 106–107
Erie, Lake
sea level gradients, 108
winds, 313
Eulerian average current, 172
Eulerian studies, coastal circulation, 102
Exchange coefficients, in planetary boundary layer model, 228–229, 231–233, 236–237
Exit time, in atmospheric predictability problem, 346–347, 359, 364, 367–369, 371–372, 374–375, 377
Exner function, 201
Extratropical storm, 166–167, 170

F

First-order closure representation, of planetary boundary layer, 231, 237
Flipping time, in atmospheric predictability problem, 371–374, 377
Flux-gate magnetometer, 34
Föhn effect, 303
Friction, and coastal ocean circulation, 141–143
Friction temperature, of planetary boundary layer, 228

Friction velocity, of planetary boundary layer, 228

G

Gaseous materials, conservation of, in mesoscale atmospheric models, 187–189, 191
Geostrophic adjustment, of coastal ocean currents, 104–105
Gray body, 239
Great Lakes, *see also* names of specific Lakes
 climatology of coastal currents, 139
 snowstorms, 312–313
 thermohaline circulation, 155
 upwelling, 124
 wind-driven coastal currents, 110–115
 wind-driven transient currents, 103
Grid and domain structure, in mesoscale modeling, 269–272
Grid-volume averaging, 199

H

Haruna dacite, 70
Heat, conservation of, in mesoscale atmospheric models, 187–189, 191
Heat, convection of, *see* Convection of heat
Heat island effect, 307–309
Hematite, 71, 80, 87
High-field susceptibility measurements, 76–77, 79–80
Hopkinson effect, in low-field susceptibility, 80, 84
Horizontally homogeneous steady-state boundary-layer theory, 227, 236
Huron, Lake, mean flow patterns, 172
Hurricane, 166–170, 316–317
Hydrostatic equation, 197

I

IFYGL, *see* International Field Year on the Great Lakes
Ilmenite, 65
Incompressible form of conservation of mass relation, 190
Inertial wave, 215
Initialization, in mesoscale modeling, 272–276
Internal boundary layer, 236

Internal Kelvin wave, 129, 131, 133–134, 137
International Field Year on the Great Lakes (IFYGL), 110
Inversion, and planetary boundary layer, 234
IRM (isothermal remanent magnetization), *see* Saturation remanence
Iron
 in core of the Earth, 15–16
 in mantle of the Earth, 18–19
Isentropic representation, 222
Isobaric representation, 222
Isothermal remanent magnetization, *see* Saturation remanence

K

Katabatic wind, *see* Nocturnal drainage flow
Kinematic model, *see* Diagnostic model
Kinetic energy, in mesoscale model, 286–288

L

Lagrangian studies, coastal circulation, 102
Lake Ontario, 113
Lake effect mesoscale model, 312–313
Laminar sublayer, of planetary boundary layer, 226–227
Land breeze, mesoscale model, 291–292, 294, 299–300
Land surface, mesoscale modeling, 281
Layer-domain-averaged variable, 201
Layered representation, in mesoscale modeling, 269
Lead isotopes, and formation of Earth's core, 15
Lithosphere
 oceanic, 2
 in parameterized convection calculations, 7
Long cores, continuous measurement of, 94–95
Long-wave radiation, mesoscale modeling, 240, 245–248
Lorenz model, in atmospheric predictability, 354–355, 357
Low-field susceptibility measurements, 76–80

INDEX

Low-temperature demagnetization, 55
Lunar rocks, 63, 65

M

Maghemite, 64–65
Magnetic domain, 91–92
Magnetic hysteresis, 27, 49, 81–83, 94
Magnetic susceptibility, 76–80
Magnetite, 65, 80, 87–89
Magnetocrystalline anisotropy, 87–90
Magnetocrystalline energy, 87
Magnetometer
  ballistic and resonance, 41–43
  for rotating rock samples, 36–41
  for static rock samples, 31–36
Magnetostriction, 87–90
Mantle of the Earth
  current heat flow, 10–11
  ferric iron and water, 18–19
  thermal evolution, 1–6, 16–19
  viscosity, 7, 19
Mass, conservation of, in mesoscale atmospheric models, 187–191
Meissner effect, in superconductivity, 34
Mesoscale numerical modeling, 185–344
  basic set of equations, 187–198
  boundary and initial conditions, 269–285
  coordinate representation, 221–226
  examples of mesoscale models, 290–317
  linear models, 209–221
  methods of solution, 253–269
  model evaluation, 285–290
  moist thermodynamics, 248–253
  physical models, 204–208
  planetary boundary-layer parameterization, 226–239
  radiation parameterization, 239–248
Metabreccia, lunar, 65
METROMEX program, 308
Michigan, Lake, propagation of warm front, 137–138
Mid-Atlantic Bight, 103
  circulation, 137, 143, 162–166
  extratropical storms, 170
  winter salinity distribution, 155
Mineral, saturation magnetization, 27
Minimum meteorological equations, 349, 352, 355
Morin transition, 80
Motion, conservation of, in mesoscale atmospheric models, 187–189, 191–198
Mountain-valley wind, mesoscale model, 295–300

N

Natural remanent magnetization (NRM), 28–56
  ballistic and resonance magnetometers, 41–43
  magnetometer for static samples, 31–36
  selective demagnetization, 43–56
Nested grid, in mesoscale modeling, 271
Newton-Raphson procedure, 285
Nocturnal drainage flow (katabatic wind), 295–296, 298
No-slip conditions, in mesoscale modeling, 281
NRM, see Natural remanent magnetization
Nudging coefficient, in mesoscale modeling, 274

O

Ocean, coastal, see Coastal ocean, circulation in
Ontario, Lake, 103
  air–lake interactions, simulations of, 280
  current reversals, 126
  mean circulation, 170–174
  sea level gradients, 108–109
  surface water temperature, 132, 140
  upwelling, downwelling, and coastal jets, 125–127
  wave propagation, 137–139
  wind-driven coastal currents, 110–114
Oregon shelf, 103
  circulation, 134–137
  mean summer circulation, 174–176
  upwelling, 124–125
  wind-driven transient currents, 115–118
Overspecified boundary conditions, in mesoscale modeling, 276–277

P

Pacific-type continental shelf, 115–118
Paleodeclination, 30
Paleomagnetism, see Rock magnetism and paleomagnetism
Paramagnetism, 84
Parameterization, 191

in moist mesoscale systems, 250–253
planetary boundary layer, 226–239
radiation, 239–248
Parameter remanence coercivity, 28
Parasitic grid representation, in mesoscale modeling, 271
Partial thermoremanent magnetization (PTRM), 29, 69, 71, 94
PDRM, see Postdepositional remanent magnetization
Piezomagnetism, 89–90
Planetary boundary layer, in mesoscale models, 226–239
Plate tectonics, and parameterized convection calculations, 6–7
Polar-wandering curve, 5
Postdepositional remanent magnetization (PDRM), 64
^{40}Potassium, in construction of Earth's thermal history, 6
Potential temperature, in mesoscale modeling, 281, 302–303
Precambrian paleomagnetism, 92–93
Precipitation
  acidity, 307
  lake effects, 312–313
  in mountainous regions, 303–305
Primary Air Quality Standards, 310–311
Primitive equations from conservation laws, 201
PTRM, see Partial thermoremanent magnetization
Pycnocline
  surface intersection, 130, 133
  and upwelling and downwelling of coastal waters, 119, 121–125
Pyrrhotite, 87

Q

Quasi-geostrophic equilibrium, 376

R

Radiation parameterization, in mesoscale numerical modeling, 239–248
Radiative boundary condition, in mesoscale modeling, 277
Radioactivity, and heat content of the Earth, 3–4, 11

Radiosonde measurements, in mesoscale modeling, 274–275
Remagnetization circle, 53
Remanence ratio, 82
Resonance magnetometer, 41–43
Reynold's assumption, 200
Rock generator, 37; see also Spinner magnetometer
Rock magnetism and paleomagnetism, 25–99
  artificial remanent magnetization, 56–75
  future trends in studies of, 92–95
  magnetometer for rotating samples, 36–41
  natural remanent magnetization, 29–56
  parameters measured in presence of a field, 76–92
  terminology, 26–29
Rossby number, 192, 194
Rotation of the Earth
  and currents in shallow seas, 102, 119
  and trapped waves, 129
Roughness height, of planetary boundary layer, 227

S

Satellite imagery, in mesoscale modeling, 273, 308–309
Saturation coercivity, 27–28
Saturation magnetization, 27–28, 81
Saturation remanence (isothermal remanent magnetization), 27–29, 60–63, 66–68, 82
Scale analysis, 190
Sea, shallow, see Coastal ocean, circulation in
Sea breeze, mesoscale model, 291–295, 299–300
Seafloor spreading, 5–6
Sea level
  coastal, 134, 167
  and storms, 166–169
  and wind-driven coastal currents, 108, 116–117
  wind stress and, in coastal area, 128
Second-order closure representation, of planetary boundary layer, 231, 237
Seismomagnetic effect, 89–90
Set theory, in evaluation of mesoscale model, 289

Shallow continuity equation, 190
Shallow sea, see Coastal ocean, circulation in
Shaw's method of paleointensity determination, 75
Short-wave radiation, mesoscale modeling, 240–245
Sigma representation, 222
Silicate, in Earth's core and mantle, 16–19
Slope flow, mountain-valley wind, 295–296
Snowfall
  eastern North America, 314
  lake effects, 312–313
Soil, mesoscale modeling, 281–285
Solar radiation, 240–245
Soundproof approximation to conservation of mass relation, 190
Spin glasses, magnetic properties, 26
Spinner magnetometer, 37–40
Sponge boundary condition, in mesoscale modeling, 277
Squall line, mesoscale modeling, 315–316
SQUID, see Superconducting quantum interference device
Staggered grid, in mesoscale modeling, 272
Storage test, see Zero-field demagnetization
Storm surge, 166–170
Strange attractor, in atmosphere predictability problem, 357, 377
Stratified fluid, mean circulation of, 159–162
Stratified water column, response to wind, 119–121
Stratosphere, mesoscale modeling, 278
Stretch-grid formulation, in mesoscale modeling, 270–271
Subgrid scale correlation terms, 200
Subgrid scale perturbation, 199
Superconducting quantum interference device (SQUID), in cryogenic magnetometer, 33–36
Surface layer, of planetary boundary layer, 227–231
Synoptically-induced mesoscale models, 290, 313–317

T

TCRM, see Thermochemical remanent magnetization
Terrain-following coordinate, 222
Terrain-induced mesoscale models, 290–313
Thelliers' method of paleointensity determination, 71–74
Thermal demagnetization, 46–48, 50–53, 58–62, 66, 70
Thermal evolution of the earth, see Earth, thermal evolution
Thermal fluctuation analysis, 85–87
Thermochemical remanent magnetization (TCRM), 69
Thermodynamics, in mesoscale numerical modeling, 248–253
Thermohaline circulation in shallow sea, 155–159
Thermomagnetic properties of geological specimens, 83–87
Thermoremanent magnetization (TRM), 28–29, 64–75
Titanomaghemite, 64–65
Titanomagnetite, 87
Transition layer, see Ekman layer
TRM, see Thermoremanent magnetization
Tropical cyclone, mesoscale modeling, 316–317
Tropopause, mesoscale modeling, 278
Troposphere
  mesoscale modeling, 278
  radiation, 240
Turbidity, atmospheric, 241
Turbulent Ekman layer, in coastal ocean, 106–107
Two-dimensional mesoscale model, 269–270

U

Upwelling, 138–139
  in coastal ocean, 119–125
Urban canyon, 283
Urban circulation, 306–312

V

Valley wind, 295–296
Vegetation, mesoscale modeling, 281–284
Viscous remanent magnetization (VRM), 29, 56–60, 94
Volcanism, liberation of Earth's heat, 16–17, 19

Vorticity, and coastal ocean circulation, 143–147
VRM, *see* Viscous remanent magnetization

W

Water, conservation of, in mesoscale atmospheric models, 187–189, 191
Water column, stratified, response to wind, 119–121
Water properties, and coastal circulation, 119
Water surface, mesoscale modeling, 279–281
Wave, trapped, propagation of, in coastal oceans, 126–139
Wave motion, in mesoscale model of atmosphere, 209–217
Well-posed boundary conditions, in mesoscale modeling, 276
Wet deposition, of planetary boundary layer, 229
Wien's displacement law, 240
Wind
  linearized model of sea and land breezes, 217–221
  in mesoscale modeling, 272–273
  forced airflow over rough terrain, 300–306
  lake effects, 312–313
  land and sea breezes over flat terrain, 291–295
  mountain–valley winds, 295–300
  urban circulation, 306–312
  water surfaces, 280–281
  response of stratified water column to, 119–121
Wind-driven coastal ocean currents, 103–118
  coastal constraint, 105–106
  Great Lakes, 110–115
  longshore pressure gradients, 108–109
  Pacific-type continental shelves, 115–118
  quasi-geostrophic model, 104–105
  velocity distribution, 106–108
Wind profile, in surface layer of planetary boundary layer, 229–230

Z

Zenith angle, 241
Zero-field demagnetization (storage test), 44–45
Zero-plane displacement of surface layer of planetary boundary layer, 230

RAYMOND H. FOGLER LIBRARY
DATE DUE